Guide to
Science and Technology
in the USSR

Guide to Science and Technology in the USSR

A reference guide to science and technology in the Soviet Union

Editor: Sarah White, PhD

Q127
R9
G8
1971

FRANCIS HODGSON

Scientific and Technical Publishers since 1884

Guide to Science and Technology in the USSR
published by Francis Hodgson Limited
PO Box 74, Guernsey, British Isles

First edition December 1971

International Standard Book Number: 0 85280 280 3
Library of Congress Catalogue Card No: 76–165289

© Copyright 1971 Francis Hodgson Limited

This publication is protected under international copyright law and may not be reproduced, in whole or in part, by any means including expressly photocopying, recording, or on any information storage system, without the previous written permission of the publishers.

MADE AND PRINTED BY OFFSET IN GREAT BRITAIN BY
WILLIAM CLOWES & SONS, LIMITED, LONDON, BECCLES AND COLCHESTER

Acknowledgements

THE editor and publishers wish to thank all those who have assisted in the preparation of this volume, in particular: Mr A. I. Akimov, science attaché at the Soviet Embassy in London for general advice; Mr George Kuznetsov of *Soviet News* and Mrs Iris Smith of Novosti Press Agency; Mr Dennis Baldrey of the Ministry of Public Building and Works for assistance with the chapter on construction; Sir Stanley Brown, Chairman of the Central Electricity Generating Board for information used in the chapter on power; Miss Angela Conning of the Confederation of British Industry for assistance in contacting the various working parties set up between the Soviet and British governments; Mr James Gregory of Furzedown College, London, and his publishers, George G. Harrap & Co Ltd, for permission to use maps from his book *Russian Land, Soviet People* (London 1968); Mr Wade B. Holland of the Rand Corporation, California, for permission to quote material published in *Soviet Cybernetics Review*; Mr H. R. Mathys, deputy chairman of Courtaulds for providing information used in the section on patents; Professor Grigory Tokaty of the Aeronautics Department, City University, London, for providing information and references for the chapters on space and aviation; Mr E. Toneri of the British Railways Board for providing information for the section on railways. Our thanks are due to Mr R. J. Fifield for reading the typescript before publication.

The editor and publishers would also like to thank the following Soviet organisations which replied to their requests for information: Abastumani Astrophysical Observatory; All-Union Research Institute of Mining, Geomechanics and Surveying; Flax Institute, Torzhok; Gorky Institute of Water Transport Engineers; Georgian State University; Georgian Academy of Sciences; Institute of Cybernetics, Kiev; Institute of Geology, Kiev; Institute of Mathematics, Siberian Branch of the Academy of Sciences; Institute of Mining Mechanics, Georgian Academy of Sciences; Institute of Russian Literature, Pushkin House; Kislovodsk Station of the Pulkovo Observatory; Leningrad Academy of Timber Technology; Lithuanian Academy of Sciences; Main Geophysical Observatory, Leningrad; Ministry of Construction Material Industry; Mr E. O. Paton of the Institute of Electric Welding; State Hydrological Institute, Leningrad; and Tartu Astronomical Observatory.

The Editor would also like to mention the following sources as having been of particular use:

Books and Articles
Artemiev, E: 'Letter from the USSR', *Industrial Property*, pp. 222–226, August 1969; *Directory of Selected Scientific Institutions in the USSR*, prepared by Battelle Memorial Institute for the National Science Foundation Charles Merrill, Columbus, Ohio 1963;

Gregory, James S: *Russian Land, Soviet People*, George Harrap, London 1968; Hemy, G. W.: 'The Soviet Chemical Industry' *Chemistry and Industry*, pp. 207–215, 17 February 1968; Mathys, H. R: *A Commentary on a visit to 'The Committee of Inventions and Discoveries under the USSR Council of Ministers'* in Moscow, February 1969; *Narodnoe Khozaistvo SSR v 1969 godu*, Moscow 1970; Nozhko, K., Monoszon, E., Zhamin, V., and Severtsev, V: *Educational Planning in the USSR*, UNESCO, Paris 1968; Petrov, V. and Ushakov, S: *Transport in the USSR*, Novosti, Moscow; Petrovsky, B: *Public Health in the USSR*, Novosti, Moscow; Prokofyev, M: *Public Education*, Novosti, Moscow; *Review of the Soviet Space Program*, Report of the Committee on Science and Astronautics, US House of Representatives, prepared by the Science Policy Research Division Legislative Reference Service Library of Congress, Washington 1967; *Science Policy and Organisation of Research in the USSR*, UNESCO, Paris, 1967; Tokaty, G. A: *Aeronautical Engineering Education and Research in the USSR*, Lawrence, Kansas 1960; *USSR Questions and Answers*, Novosti, Moscow 1967; Vvedensky, B. A. chief ed: *Soyuz Sovetskikh Sotsialisticheskikh Respublik 1917–1967*, Moscow 1967; *The World of Learning, 1969–1970*, Europa Publications, London 1970; Yelyutin, V: *Higher Education in the USSR*, Novosti, Moscow; Zaleski, E., Kozlowski, J. P., Wienert, H., Davies, R. W., Berry, M. J., and Amann, R: *Science Policy in the USSR*, OECD, Paris 1969; and Zvorykin, A. A: 'The Organisation of Scientific Work in the USSR', *Impact of Science on Society*, UNESCO, vol. XV (1965) no. 2.

Journals and Series
Bulletin of the Institute for the Study of the USSR, Munich, West Germany: *NLL Translations Bulletin*, National Lending Library for Science and Technology, Boston Spa, Yorkshire; *Novosti Booklets*, Moscow; *Soviet Cybernetics Review*, Rand Corporation, California.

Contents

1	Introduction	9
2	Organisation and Planning of Science and Technology	16
3	Structure of the Research Bodies	36
4	Education	58
	Appendix	71
5	The Natural Sciences	74
6	Space Research	85
7	Military Affairs	96
8	Computer Industry	100
9	Power Industry	109
10	Metallurgy and Engineering	128
11	Chemical Industry	137
12	Timber, Cellulose and Woodworking Industry	146
13	Food and Consumer Industries	148
14	Construction Industry	153
15	Agriculture	160
16	Medical Services	178
17	Transport	183
18	Communications Industry	202
19	Patents, Information, Libraries and Museums	206
20	The Regions	216
	I Russian Soviet Federal Socialist Republic	216
	II The Ukraine	220
	III Byelorussia	222

IV	Uzbekistan	224
V	Kazakhstan	225
VI	Georgia	227
VII	Azerbaijan	229
VIII	Lithuania	230
IX	Moldavia	231
X	Latvia	233
XI	Kirghizia	234
XII	Tadjikistan	236
XIII	Armenia	237
XIV	Turkmenia	239
XV	Estonia	240

Index of Establishments mentioned in the Text   243

English Cover to Cover Translations of Soviet Journals   261

Publishers of English Cover to Cover Translations of Soviet Journals   277

Academies of Sciences of the USSR   285

1 Introduction

(I) AN EMPHASIS ON SCIENCE AND TECHNOLOGY

When the Bolshevik Party took power in 1917, its members recognised that science was an integral part of the modern state. They were the first government to do so - a percipience derived from their Marxist philosophy.

What this meant in practice was that from the very beginning, even in the middle of the economic chaos caused by the civil war, wars of intervention and famine, attention was paid to the development of science and technology. The most famous example of this attitude was Lenin's statement at a Moscow Party Conference in 1920 that "Communism equals Soviet power plus electrification of the whole country". In other words, Soviet power needed massive supplies of electric power for industry, agriculture and transport, if the economy was to be developed to the high level of technology demanded by a communist society. At the time of the revolution there were some 35 power stations in the Soviet Union, the majority of them in the western part of the country. In 1920 the State Commission for the Electrification of Russia (GOELRO) was set up, and in the initial stages more than 100 specialists worked together to draw up the plans. These later became the model for the first five-year and succeeding plans.

The Soviet government also started organising scientific work throughout the country soon after the revolution. In April 1919 Lenin drew up his *Outline of a Plan for Scientific and Technical Work*. In this he defined the main directions and forms of work of the Academy of Sciences under the new conditions. The main task of its scientists was to be the study of the country's natural productive forces. Without knowledge of these the development of the economy would be impossible.. At the same time a number of new research institutes and laboratories were established. They were all designed to promote the development of key branches of the economy. For example, the Central Aerodynamic Institute, founded in 1918 on the initiative of N.E. Zhukovsky and S.A. Chaplygin, provided the basis for a national aviation industry. In the same year the State Optical Institute, the Central Chemical Laboratory, the Moscow Academy of Mining and the Physico-Technical Institute under Ioffe were all established. A year later came the Nizhegorod Radio Laboratory and in 1921 the All-Union Electrotechnical Institute and the All-Union Heat Engineering Institute. Each of these contributed to the development of its own particular branch of industry; many of them later gave birth to new institutes forming centres of scientific research and training throughout the country.

Government promotion and direction of science in this way was not entirely foreign to Russia, though the enthusiasm of the Soviet regime and the importance it attached to science were certainly of a new order. Previous to the revolution, most scientists had worked within the university-academy framework, dependent on the Tsarist government for funds and support. "Gentlemen" scientists with their own independent incomes were extremely rare.

The main centres of scientific activity were the Academy of Sciences, the universities and a number of higher technical institutes. The academy had been founded in 1725 on the initiative of Peter the Great. Lack of indigenous Russian scientists at that time meant that all its initial members were foreign scientists. The foreign bias of the academy continued throughout the nineteenth century, even after the development of a strong nucleus of good native born Russian scientists. Dmitri Mendeleev, for example, was nominated for election as an academician in 1880, but his nomination failed causing a tremendous scandal in Russian scientific circles over the role of the Imperial Academy. The scientific role of the academy, prior to the revolution, was essentially practical, reflecting the government's main interest in science.

The universities and institutes became the chief centres of original scientific research. There were eight universities in all. Moscow University had been founded in 1755, Dorpat in 1802, Vilna in 1803 (to be replaced by Kiev in 1830), Kazan and Kharkov in 1805, St. Petersburg in 1819, Odessa in 1865 and Tomsk in 1888. Their organisation was modelled on that of the German universities, though they never achieved the latter's autonomy. The university staff were always subject to the vagaries of the tsarist autocracy. Nicolas I banned the use of the word 'progress' for a time; students who sent a telegram and wreath to Darwin's funeral in 1882 were regarded by the government as dangerous revolutionaries.

The number of Russian scientists in the nineteenth century was small; however, the overall level of their work, in particular from the middle of the century onwards, was as good as any in Europe. Mendeleev in chemistry, Sechenov and Pavlov in physiology, Timiriazev in plant physiology, Alexander Kovalevsky in embryology, Mechnikov in parasitology, Lobachevsky in geometry, Tsiolkovsky in rocketry - these are all names well known in their fields, and they are not isolated cases. Some, such as Pavlov, Timiriazev and Tsiolkovsky bridged the gap between Russian and Soviet science.

As for technology, Russia presented no consistent picture at the time of the revolution. The country was basically agricultural, much of this still being organised along feudal lines, although attempts at reform had been started before the First World War. On the other hand, the country had a comparatively broad industrial base. Industry had only begun to develop rapidly after the emancipation of the serfs in 1861 had provided the necessary free labour for the factories in the cities. Much of it was foreign-owned and there were large factories employing thousands of workers and using modern industrial techniques. The Trans-Siberian railway was completed in 1917.

The First World War, Civil War and wars of intervention caused an almost complete breakdown of industry. The majority of the workers were only one or two generations removed from the peasantry and many returned to the country in order to survive. By 1926, however, the New Economic Policy of the Soviet government had restored the two main branches of the national economy - industry and agriculture - to their pre-war level. But the USSR was still a basically agricultural country; 18 per cent of the population worked in the towns, the remaining 82 per cent in the country. The political position of the government was such that trade and foreign investment could only play a very small part in building up the economy. So the capital outlay for heavy industrial and consumer development - the necessary prerequisites for the advanced technological society envisioned by the communists - the food for the city, and the small surplus needed to pay for necessary imports, all had to come from agricultural production.

Government belief in the importance of science for the development of a modern technologically-based industry meant that from the very beginning (despite the economic difficulties) a greater proportion of the national wealth was allocated to science than in any other country at that time. In return science was expected to contribute to the national economy, and this it did in many ways. Thus, during the Civil War, the Academy of Sciences had initiated a thorough study of the Kursk magnetic anomaly which led to the discovery of vast deposits of iron ore. Later surveys identified sources for many of the raw materials needed by industry.

The role of the research institutes was to produce a nucleus of good practising scientists. Teaching was secondary, and in fact fell more to the lot of the universities. Numbers were so small at first that brilliant individuals left a tremendous mark. Ioffe, Vavilov and Landau each ran more than one institute. The pressures on such individual scientists were enormous. For example, by 1929 to 1930 the staff of Ioffe's Physico-Technical Institute in Leningrad had grown from the original eight to 2,000, 700 of whom were physicists and 1,300 assistants. New institutes were split off, either as extensions of existing sections, such as the Electro-Physical Institute, or as totally new organisations, such as the institutes founded in Kharkov, Dniepropetrovsk, Tomsk and Samarkand. The directors of the various institutes formed a general co-ordinating committee of which Ioffe was chairman.

This must have been a very exciting period for Soviet science and technology. Government support was generous. Ioffe recorded that in 1930 it took him just ten minutes to obtain the funds needed for his institute to begin research into nuclear physics from the chairman of the Supreme Council of National Economy. There was free association with scientists of other countries, and many foreign scientists came to work in the Soviet Union for a year or two. There was tremendous scope and enthusiasm for research and development. In production, the base of heavy industry was being laid; and in agriculture collectivisation was in full swing.

The purges of 1937, the Second World War, and the post-war imprisonments must have had their toll of scientists and scientific activity, though little is clear about the period. The rise of Lysenko was accompanied by the dismissal, disappearance and death of Academician N.I. Vavilov, the Soviet Union's most distinguished geneticist. Other scientists found themselves doing their research from special prisons.

Kapitsa the physicist was placed under a kind of house arrest. Information on foreign scientific and technological development was available to scientists in their own particular fields, but there was not the same open exchange and movement as before.

Since the war the main Soviet scientific achievements and the most publicised have been in the fields of nuclear physics and space. In 1949 the USSR exploded her first nuclear weapon, and four years later a prototype hydrogen bomb - only nine months after the United States. On the more peaceful side, her first atomic power station went into operation in 1954; and in 1956 Academician Kurchatov was the first in any country to describe openly research on the extremely difficult and still unsolved problem of using controlled nuclear fusion as a power source. The Soviet Union has also built powerful accelerators for her nuclear physicists. The first electron accelerator, a 30-MeV synchrotron, was built in 1947. Ten years later they had a 10-GeV synchrophasotron, and in 1968 the most powerful accelerator in the world (70GeV) went into operation at Serpukhov near Moscow.

Soviet achievements in space are well known. Its first two Sputniks went up in 1957, the second being more than three hundred times the weight of the first American satellite and carrying the dog Laika. The monitoring of Laika's heart and respiration rates provided the first physiological data from space on the effects of weightlessness. In 1959 the Russians photographed the back of the Moon for the first time; two years later Yuri Gagarin was the first man in space.

Another important field of Soviet research is that of energetics. The vision of the State Commission for the Electrification of Russia (GOELRO) is still present in the overall aim of providing an electricity supply to the whole country and unifying it into a single system. There has been an extensive programme of hydro-electric development, in particular on the Volga and large Siberian rivers. In fact East Siberia will soon have the three largest hydro-electric power stations in the world with the completion of the installations at Krasnoyarsk (6,000 MW), Ust Ilim (4,500 MW) and Bratsk (4,500 MW). They have had to develop new cooling systems for steam and gas turbines, ways for the complex generation of steam turbines so that they provide both electricity and heat; and methods for high-voltage transmission of electricity to make long-distant transport both feasible and economic.

Other scientific and technological developments have included irrigation schemes in

Central Asia, research on floating ice caps in the Arctic, the discovery of enormous oil and gas resources in Western Siberia - in other words mapping, exploring and trying to conquer the more inaccessible and inhospitable regions of the country.

Two areas which do not have a very high record of achievement are the biological and chemical sciences. In the current five-year plans the importance of their development has been stressed. Biological research is necessary for agriculture, medicine and the food industries; on the chemical side the need is for fertilisers and synthetic materials.

Although progress in both science and technology is uneven, the Soviet Union has, during the past decade or so, clearly reached a new stage in her development. Whereas before she was working to catch up with the advanced industrial countries - either making good in areas devastated by war or just developing fields that had not existed before - she is now basically up among the front runners. For the whole of society, including science and technology, this demands a new approach, a new emphasis on innovation and the application of new ideas. It is no longer just a question of catching up in methods and quantity of production, they are now faced with the problem of being originators. The new emphasis demands new methods of organisation, and the country is now in the process of this readjustment. For example, the economic reforms introduced in 1965 are an attempt to encourage innovation in industry by giving individual enterprises more independence and responsibility. Whereas before, the plan specified the quantity of goods to be produced, the plant now has to show it can also sell its goods. There are various incentive schemes to encourage reducing manpower and increasing automation and mechanisation. Closer links between industry and research are being promoted.

In education the comprehensive schools which provide compulsory eight-year schooling and which had the task of educating a population, nearly three quarters of which had been illiterate 50 years ago, are now being supplemented by special schools for pupils gifted in one particular field. They include maths, physics, and biology special schools. Further encouragement of science and a means for detecting talented individuals are the childrens' "olympiads". These were first held in maths but have now been extended to other subjects. At the primary school level the first four-years' teaching syllabus has recently been compressed into three years. In addition algebra and geometry are being taught to encourage the use of such mathematical concepts from an early age.

The building of scientific cities such as Akademgorodok, near Novosibirsk in Siberia, has provided tremendous scope for the enthusiastic younger scientists, who tend to get bogged down in the scientific hierarchy of Moscow and Leningrad. Similar science centres are planned for Irkutsk and Vladivostock. And there is also to be an agricultural city near Akademgorodok. Apart from providing excellent research facilities, these centres of scientific excellence are also intended to reduce the distance between the laboratory and the factory. They have their own testing facilities where new ideas can be piloted and growing pains eliminated, as far as possible before the innovation reaches the factory floor.

All these measures illustrate the importance attached to science in the Soviet Union. It is one of the most significant factors in the building of the state - a factor that determines the rate of development of the national economy and which ensures further technological progress.

(II) POLITICAL, SOCIAL AND ECONOMIC BACKGROUND

(i) Area and Population

The Soviet Union is the largest country in the world, occupying 22,400,000 square kilometres. On January 15, 1970, the population was more than 241,000,000. This is an average density of 10.9 persons per square kilometre. Between 1957 and 1970, the population grew by about 40,000,000 - an average of more than 3,000,000 a year.

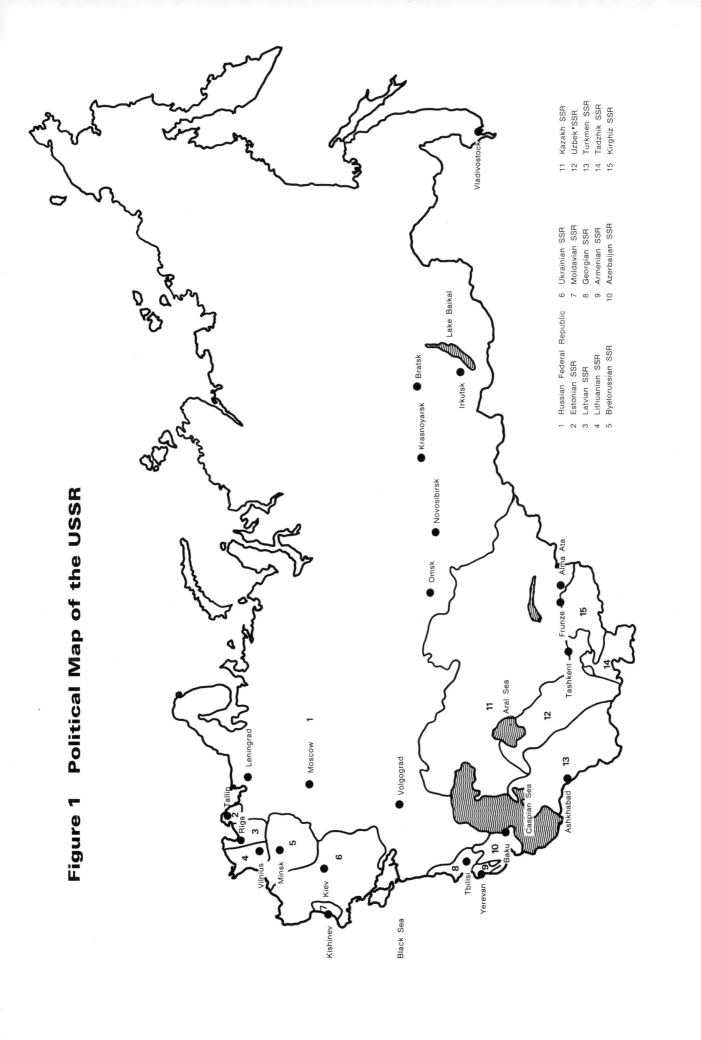

(ii) Political Structure

The Soviet state came into existence in 1917, following the Bolshevik Revolution of that year. In December 1922, the *Union of Soviet Socialist Republics (USSR)* was created by the peoples of the former Russian Empire. The USSR today consists of 15 *Union Republics*: the Russian Soviet Federal Socialist Republic (RSFSR), the Ukrainian Soviet Socialist Republic (UkSSR), the Byelorussian SSR, the Uzbek SSR, the Kazakh SSR, the Georgian SSR, the Azerbaijan SSR, the Lithuanian SSR, the Moldavian SSR, the Latvian SSR, the Kirgiz SSR, the Tadzhik SSR, the Armenian SSR, the Turkmen SSR, and the Estonian SSR (see also Fig. 1.)

The USSR is a multi-national state. Throughout the country, there are the same organs of state authority and administration, a single 'all-union' legal code, one national economy, one army, and a single 'union' citizenship.

The all-union bodies deal with international relations, the organisation of defence, planning of the national economy, trade, the management of banks and other commercial enterprises of an all-union importance, transport and communications, state security, and so on. The USSR can admit new republics to the union, control the observance of the constitution and ensure that the constitutions of the republics are in line with it.

The Union Republics are legally independent states, able to maintain their sovereign rights beyond the competence of the all-union bodies. They have their own constitution, and organs of government.

There are also *Autonomous Republics*. These enjoy political autonomy as part of a Union Republic. They usually represent some particular nationality making up a majority in that region.

The bodies of state power are the *Soviets (or Councils) of Workers' Deputies*. There are 50,000 Soviets in the country ranging from local Soviets to the Supreme Soviet of the USSR. They are all elected bodies.

The Supreme Soviet of the USSR is the highest body of state power in the country. It has the right to issue all-union laws, to alter the constitution of the USSR, admit new republics into the USSR, organise the defence of the USSR, approve the budget, direct the monetary and credit system, administer transport and communications, define basic principles of land tenure and use of mineral wealth and other resources.

The Supreme Soviet elects a *Presidium*, which performs its functions in the periods between sessions. It also appoints the *Council of Ministers of the USSR* which is the supreme executive and administrative body of state power in the country. It further elects the Supreme Court of the USSR and appoints the Procurator General of the USSR, who supervises the running of the legal system throughout the country.

The Supreme Soviet consists of two chambers - the Soviet of the Union and the Soviet of Nationalities. The deputies of the Soviet of the Union are elected throughout the country on the basis of one deputy for every 300,000 inhabitants. In the Soviet of Nationalities there are 32 deputies elected by each Union republic irrespective of size or population. 11 for each autonomous republic, five deputies for each autonomous region and one for each national district. All in all there are 1,500 deputies to the Supreme Soviet of the USSR elected for a term of four years.

The *Presidium* is the highest permanently functioning body of state power in the country when the Supreme Soviet is not in session. It has 37 members consisting of the chairman 15 vice-chairmen representing each Union republic, a secretary and 20 members.

The *USSR Council of Ministers* is the governing body of the country. It is the highest executive body of state power, responsible and accountable to the Supreme Soviet or to its Presidium between sessions. The Council of Ministers has around 90 members. These consist of the chairman, the first vice-chairman, the ordinary vice-chairman of the Republican Councils of Ministers, the head of the State Bank, and the Central Statistical Agency.

The *Communist Party of the Soviet Union (CPSU)* is a voluntary, militant union of communists, whose aim is the construction of a communist society. CPSU is the only political party in the Soviet Union. It is the directing nucleus of all organisations within the country - the soviets, trade unions, Young Communist League, and co-operative societies.

The 24th Congress of the CPSU was held at the end of March and beginning of April 1971. It discussed and approved the report of the CPSU Central Committee, including the directives for the next five-year plan (1971-75).

(iii) Economic Structure

The economic basis of the USSR is socialist, with state ownership of the implements and means of production. Land, minerals, forests, water, plants and factories, mines pits, state farms, banks, transport, means of communication, municipal enterprises, the majority of houses in towns and industrial areas all constitute State socialist property. Buildings and enterprises of an economic character, which are part of kolkhozes and co-operative organisations, their machinery and equipment, livestock, crops and farms all make up the co-operative or kolkhoz socialist property. National state property of the means of production, as distinct from co-operative, kolkhoz or personal property, makes up about 90 per cent of all means of production.

The state budget plays the leading role in the finance system of the USSR. It covers the all-union, republican and regional budgets. Income from the socialist sector of the economy provides the main sources of finance for the country. In 1965 the accumulation in the form of profits, turnover taxes, social insurance funds and other means was about 1,250,000,000,000 roubles. Revenue from the socialist sector of the economy provides about 90 per cent of all income in the state budget.

2 Organisation and Planning of Science and Technology*

Belief in the possibility of a rational centralised system is the principal feature of the structure of science and technology in the Soviet Union. All activities in these fields are planned and carried out according to the national economic programme.

The general direction of scientific and technological research throughout the country is the responsibility of the *Council of Ministers of the USSR*, which examines and sanctions the principal trends of development in science and technology, the development plan for science and technology, the financing of research in the fields of science and technology, and the principal measures for improving the management of the development of science and technology in the USSR.

The elaboration of directives for the development of science and technology by the Council of Ministers is based on the programme of the *Communist Party of the Soviet Union (CPSU)* and the directives for the development of the national economy in the USSR adopted at the *CPSU Congresses*, and directly supported by the *State Committee for Science and Technology*, the *State Planning Committee (Gosplan USSR)*, the *Academy of Sciences* and the *Central Statistical Board (CSU)*. The chairmen of the State Committee for Science and Technology, Gosplan and the CSU are automatically members of the Council of Ministers.

The State Committee for Science and Technology, together with the Academy of Sciences, draws up proposals for the development of the principal trends of science and technology and, in collaboration with other organisations, prepares the way for a practical utilisation of the achievements of scientific research of importance to the national economy. These proposals are examined by the Council of Ministers and, once sanctioned, become the guiding lines for working out the five-year and annual plans for both specialised sectors and the country as a whole.

In the course of this work proposals from the ministries and their departments, scientific establishments throughout the country and from individual scientists are put into effect. Proposals in the process of elaboration are examined by the scientific and technical councils attached to the State Committee for Science and Technology.

The majority of academicians and corresponding members of the Academy of Sciences and the republic academies, top specialists working in administration and a considerable number of workers from scientific establishments, higher educational institutions and directors of industrial enterprises and organisations, participate in the discussions

---

* Most of the information in this chapter is drawn from two important sources: Zaleski et al. *Science Policy in the USSR* (OECD, Paris, 1969); and *Science Policy and Organisation of Research in the USSR* (UNESCO, Paris, 1967).

Figure 2 Central organs interested in Planning of Research and Development in the Soviet Union

Administrative Subordination *(December 1966)*

Source: Zaleski et al.: *Science Policy in the USSR*, (OECD, Paris, 1969)

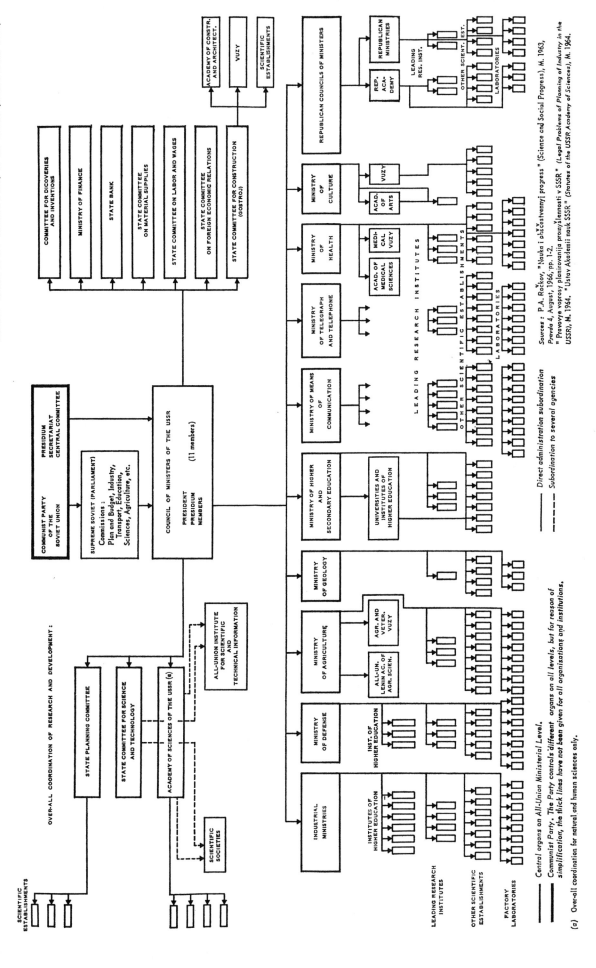

over the proposals in the science councils attached to the State Committee for Science and Technology. The committee, however, decides on which scientific and technical problems its efforts and those of other interested bodies should be concentrated. It organises the execution of the most important complex problems whose solution depends on the joint efforts of a number of different ministries.

(I) STRUCTURE AND FUNCTIONS OF BODIES CONCERNED WITH SCIENCE POLICY

2(I) COMMUNIST PARTY

Communist Party control over research and development planning and policy is exercised at all levels, as part of its overall control over the economic, social and cultural activities of the country. There is a hierarchy of party organisations from the Central Committee down to the party branch through which the control is exercised; it is also supplemented by other closely related organisations such as the Komsomol (Youth League of the Communist Party) or the Committee on Popular Control.

(i) Department for Science and Establishments of Higher Education

The Department for Science and Establishments of Higher Education is the most important specialised body for control of science and education. It is attached to the Secretariat of the Central Committee of the CPSU. It has its own Secretariat, subdivisions, and a series of 'instructors' permanently employed in supervising a given sector of scientific or educational activity.

(ii) State Committee on Popular Control

The State Committee on Popular Control is a Union-Republican body with branches in the republics and regions. It depends on the republican and regional party committees, as well as on administrative authorities such as the councils of ministers of the republics, or executive committees on the regions. The committee executes party and government directives on matters such as cost economy and the implementation of annual plans, and thus serves as a kind of watchdog to prevent defects and errors, or misuse of party directives.

(iii) Groups for Co-operation with Popular Control

Groups for Co-operation with Popular Control are organised in individual enterprises such as universities, collective farms, etc. They are composed of party and non-party members. Within a university, for example, their job would be to supervise the quality of teaching, to develop scientific research and facilitate economic utilisation of research results in production.

2(2) SUPREME SOVIET

The Supreme Soviet, the Soviet equivalent of a parliament, is officially the highest authority in the country. Its limited powers were slightly increased in August 1966 with the creation of its Permanent Commissions. These include those for Education, Science and Culture, the Plan and the Budget, Industry, Transport and Communications, and Agriculture, which will actively co-operate in examining the State Plan for Science and Technology. The head of the Central Party Committee's Department for Science and Establishments of Higher Education, Sergei P. Trapeznikov, is also president of the Permanent Commission for Education, Science and Culture.

2(3) COUNCIL OF MINISTERS OF USSR

Appointed by the Supreme Soviet, the Council of Ministers of USSR is the highest executive and administrative organ of state power. It is composed of ministers, the chairmen of the more important state committees and state organs, and the presidents of the Republican councils of ministers.

The all-union ministries exercise authority or control a particular branch of industry, at State Level, including union republics. Certain specialised ministries may exist both for the union and in individual republics.

The all-union ministries are the following: aviation industry, automobile industry, foreign trade, gas industry, civil aviation, machine building, machine building for the light food and household equipment industries, merchant marine, defence industry, general machine building, instrument making, automatic devices and control systems, railroad communications, radio industry, medium machine building, machine and hand-tools industry, communal, building and roadbuilding machinery, shipbuilding industry, tractor and agricultural machinery construction, transport construction, heavy power and transport machine building, chemical and oil equipment production, electronics industry, and electrical engineering industry.

The Republican ministers are responsible for the following areas: higher and secondary specialised education, geology, health, foreign affairs, culture, light industry, the timber and wood processing industry, melioration and water economy, special installation and construction projects, meat and milk industry, oil-extracting industry, oil-processing and petrochemical industry, defence, preservation of public order, food industry, building materials industry, education, fisheries, postal services and telecommunications, agriculture, trade, coal industry, finance, chemical industry, non-ferrous metallurgy, power and electrification, construction of heavy industry enterprises, industrial construction, agricultural construction, construction, and the medical industry.

In certain union republics, some of the foregoing ministries may not exist if there are no enterprises of that branch in the republic.

The State Committee and state organs whose chairmen sit on the Council are: the State Planning Committee (Gosplan), the State Committee for Building Affairs (Gosstroi), the State Committee for Material and Technical Supply, the People's Control Committee, the State Committee for Labour and Wages, the State Committee for Science and Technology, the State Committee for Vocational and Technical Education, the State Purchasing Committee, the State Forestry Committee, the State Committee for Foreign Economic Relations, the Committee for State Security, the All-Union Association for the Sale of Agricultural Machinery to State and Collective Farms (Selkhoztekhnika), the State Bank, and the Central Statistical Authority.

The Council of Ministers examines and sanctions the principal trends of development in science and technology, the development plan for science and technology, the financing of research in the fields of science and technology, and the principal measures for improving the management of the development of science and technology in the USSR. The proposals concerning policies in these fields come chiefly from the State Committee for Science and Technology and the Academy of Sciences. Once they have been examined and sanctioned by the Council of Ministers they become the guiding lines for drawing up the five-year and annual plans for scientific and technological development.

Ministries and departments which are responsible for one particular branch of industry, put into effect a single policy for the whole country in that branch of the national economy. They also co-ordinate the work of developing science and technology within this branch. They compile a list of the most important specialised scientific and technological problems for which scientific research, design and project work is envisaged in the national economy plan, and then report on this to the Republican council of ministers and other organisations concerned. To carry out this work, the ministry or department has at its disposal a network of scientific research establishments, design offices and enterprises.

2(4) REPUBLICAN COUNCILS OF MINISTERS

Each union republic has its own council of ministers. These direct the work of the scientific research establishments, design offices and similar bodies under the authority of the republic. When organising scientific and technological research within a republic, the council of ministers of that republic is guided by the academy of sciences and planning committee of that republic.

2(5) PRESIDIUM OF THE COUNCIL OF MINISTERS

The real power of decision within the Council of Ministers is concentrated in its Presidium, which is made up of the Chairman (A.N. Kosygin) and eleven Vice-Chairmen. Problems of science and technology are involved at this highest level of government. For example, the chairman of the State Committee for Science and Technology is also an *ex officio* Presidium member as vice-chairman of the Council of Ministers. Other vice-chairmen represent the interests of armaments, Gosplan, and material and equipment supplies.

2(6) STATE COMMITTEE FOR SCIENCE AND TECHNOLOGY

Throughout the 1930s and 1940s there was no single national body to overlook and co-ordinate research and development. Serious attempts to establish overall planning of science are a recent phenomenon. The State Committee for the Co-ordination of Research (now the State Committee for Science and Technology), set up in 1961, was the first central authority which had the potential to act as a powerful supervisor over all science. This committee together with the State Planning Committee (Gosplan) and the Academy of Sciences of the USSR are the main bodies co-ordinating the plans for scientific research and development at present.

The main functions of the former committee were to co-ordinate research and development, improve the planning and introduction of new techniques and end any projects which were no longer necessary. The new committee has to ensure "the cohesion of state policy in the field of scientific and technical progress".* One of its purposes is to determine basic tendencies in the development of science and technology, it organises the dissemination of scientific information; and it has an overall responsibility to increase the efficiency of scientific research and to introduce new technology into the economy with the maximum economic benefit and with minimum expenditure.

The State Committee for Science and Technology has at its disposal a reserve fund from the State budget, (representing 2 per cent of the general expenditure on science) and a general reserve of scientists and their corresponding wage funds, which can be used for unscheduled research work. These reserves can be increased during the year by cancelling grants for current non-useful research, since these grants no longer need to be returned to the Ministry of Finance.

In collaboration with Gosplan and the State Committee for Materials Supplies, the committee participates in the distribution of special equipment for the most important research projects. It also works with the Academy of Sciences, ministries and other organisations to scrutinise the principal projects completed by academy, university and higher educational institutions, in order to organise successive stages of research and development and to study the possibilities of introducing the results into the economy. The committee has, in fact, the decision and policy-making powers to draw up the list of the most important research projects and research institutes.

The General Assembly of the Committee and the College of the Committee - whose members are of high governmental, administrative and academic rank - are its chief policy-making bodies. The president or chairman of the general assembly is nominated by the Supreme

---

*(Zaleski et al: *Science Policy in the USSR*, P 58, OECD Paris 1969)

Soviet. He is an *ex officio* vice-chairman of the Council of Ministers. The vice-chairmen of the committee's assembly are nominated by the Council of Ministers. The rest of the membership is made up as follows: *ex officio* members - president of the Academy of Sciences, vice-president of Gosplan, minister of Higher and Secondary Specialised Education and president of the State Committee for Inventions and Discoveries; nominated members - leading personalities from industry and prominent scientists; probable members - heads of the committee's departments. The College of the committee is nominated by the Council of Ministers. Its specific functions are not clear.

Thirty three science councils are attached to the State Committee for Science and Technology. They have a consultative function and their members are mainly made up of prominent scientists and directors of leading research establishments.

The most important institutions attached to the committee are those concerned with information, the Committee for the Exhibition of National Economic Achievements, and the All-Union Council of Scientific and Technical Societies. The State Energy Publishing House and the Consortium of Publishing Houses of Scientific Material were attached tot he State Committee for Co-ordination of Scientific Research. It is not clear whether they remained attached to the State Committee for Science and Technology after the 1965 changes.

2(7) ACADEMY OF SCIENCES OF THE USSR

The structure and functioning of the Academy of Sciences of the USSR is described in detail in the following chapter. Here it is sufficient to mention its role in the co-ordination of science policy. The academy has overall control over research into the natural and social sciences and over the training of scientific workers. It also has the right to supervise the research activities of the republican academies and of higher educational institutes. It maintains contacts with research in other countries. Other functions include the creation of science councils, contracting for research in connection with important projects, and presenting information on research plans. The status of the academy appears to be similar to that of a union ministry or other organisation subordinated to the USSR Council of Ministers.

2(8) COUNCIL FOR THE CO-ORDINATION OF SCIENTIFIC ACTIVITIES OF THE REPUBLICAN ACADEMIES OF SCIENCES

The Council for the Co-ordination of Scientific Activities of the Republican Academies of Sciences is a non-executive body created in 1947. In 1955 it was decided that the council would meet at least once a year to co-ordinate the work of all the academies, (the USSR Academy, the Republican academies, branch academies, and affiliates), to decide on priority projects, and to help introduce the results of research into the economy. Later it was also asked to make recommendations on the correct balance between pure and applied research, to improve the method of planning and co-ordination, and to further the work of the science councils.

Since 1963, when its composition changed, the council has been composed of representatives of the USSR Academy and the Republican academies. It has no permanent presidium of its own. Between sessions its powers are taken over by the USSR Academy Presidium, to which it is subordinated in any case. The terms of reference of the council remained essentially the same as before.

2(9) STATE PLANNING COMMITTEE (GOSPLAN)

Research and development of science and technology - being an integral part of the economy - are generally co-ordinated along with other economic activities in the National Economic Plan. The Plan is the responsibility of the State Planning Committee (Gosplan).

From 1947 the government has made a number of attempts to create a separate co-ordinating agency for science and technology. But even after the State Committee for the Co-ordination of Scientific Research was set up in 1961, no clear cut separation was made between planning for the introduction of new technology and the activities of Gosplan. For a few years the State Committee had certain powers for working with Gosplan in the drawing up of plans for science and technology. But since 1966 the planning powers of the State Committee for Science and Technology have been limited to elaborating perspective plans for basic scientific and technical problems only; general prospective plans are now the responsibility of Gosplan alone. In fact, a special *Department for the Comprehensive Planning of the Introduction of New Technology into the National Economy* was created within Gosplan in 1966 - proof of its increased powers in planning for research and development.

Gosplan works together with the State Committee for Science and Technology to draw up the all-union plans for scientific research. The two organisations make sure that the proposed plans for the development of science and technology fit in with the overall state plans for the development of the national economy, and they make provisions for the necessary finance and supply of materials. In addition, Gosplan puts forward in its five-year and annual plans practical measures for testing new machinery, equipment, technological processes, and materials and for introducing them into production. It also plans for the development of mechanisation and automation. These measures are worked out in collaboration with the ministries and the Republican councils of ministers.

2(IO) SCIENCE COUNCILS

Science councils are a well-established feature of the Soviet science scene. In their early history they were usually formed as committees which met regularly to advise on the development of a particular branch of science. The number of science councils grew spontaneously until the mid-1960s when it was decided to re-organise them round particular complex problems of the pure and applied sciences. The system was further re-organised when republican councils were absorbed into the new 'complex' councils.

The main purpose of the majority of science councils is to co-ordinate the scientific activities being carried out in all research establishments taking part in the solution of one particular project. Another purpose is to provide solutions for those problems considered by the government to be the most important at that moment. Thus their function is directly influenced by government policy, and "in a very real sense they can be considered as vehicles for government policy".* They exist on different levels throughout the research and development structure, but it is the ones for complex and interdisciplinary problems that are of the greatest interest in terms of policy-making.

These were created in 1961. They are of two kinds:-

(i) Science Councils for Major Complex and Inter-Branch Scientific and Technical Problems attached to the State Committee for Science and Technology

Thirty three science councils for major complex and inter-branch scientific and technical problems are attached to the State Committee for Science and Technology. They cover such subjects as: automation and mechanisation of productive processes; electronic technology; energy and electrification; the study and development of the natural resources; study of the oceans and seas and utilisation of their resources; catalysis and its industrial utilisation; chemistry of natural and biologically important compounds; chemistry of natural and synthetic biologically active compounds; the introduction of chemical processes into agriculture; new materials based on glass; new welding processes and welding design; polymers in the national economy; use of rare metals in the national economy; development and introduction of mathematical methods, of computer techniques and of automatic systems of control, and of processing information into the national economy;

---

*(Zaleski et al; *Science Policy in the USSR*, P 228, OECD Paris 1969)

rare elements; synthetic polymer based materials; creation of new metallurgical processes.

The function of these complex councils of the State Committee for Science and Technology is to supervise work on the major problems included in the State Committee's plans. These problems invariably involve the participation of several research and development establishments under different authorities, and it is the job of the councils to overcome barriers between these establishments and to draw the threads of their work together. The councils cover between them all the complex problems in the plan. It has been questioned whether the councils have the necessary powers to carry out their work effectively. The councils of the state committee are reputed to be more influential than those of the academy in this respect. Their recommendations about the lines of development of the problem or problems for which they are responsible are usually accepted by the Council of Ministers. Nevertheless they lack the power of allocating resources, and it has been argued that their status would be considerably enhanced if they were given a budget.

(ii) Science Councils for Major Complex Problems in the Natural and Social Sciences under the supervision of the USSR Academy of Sciences

Twenty seven science councils for major problems in the natural and social sciences come under the supervision of the USSR Academy of Sciences (nineteen for the natural sciences and eight for the social sciences). They include such subjects as; solid state physics; physiology; photosynthesis; cybernetics; complex biogeocoenotic study of animate nature and the scientific base of its rational utilisation and conservation; complex research of the Earth's crust and the upper mantle; cosmic rays; international space co-operation; scientific principles of the application of chemicals in agriculture; radio biology; radioastronomy.

The science councils for complex problems of the USSR Academy used to be controlled by the presidium of the Academy but in 1964 they became attached to the newly created sections and divisions. They are composed of leading scientists, manufacturers, industrial designers and specialists from scientific research establishments in industrial enterprises and higher educational institutions. Their role is: to analyse the existing state of research work; to help to determine the lines along which it should move; to co-ordinate research work in their respective fields; to examine the most convenient organisational forms for research throughout the country; to arbitrate conflicts; to indicate means for executing research projects; and to set research targets within the academy system. Their sphere of influence includes the entire academy system (Republican academies and agricultural and medical academies), as well as the higher educational establishments.

The status of the Complex Problems Councils is not clear. Recommendations made by them are supposed to have a decisive influence on the presidium, departments and institutes of the academy, as regards research developments, material supplies, personnel and editing work. However, their members work on a voluntary basis and they have no permanent staff. They do, however, serve the important purpose of drawing scientists into the higher circles of decision-making, thus enabling them to help formulate and implement policy on a national scale.

In addition to the councils for complex problems, there are a large number of ordinary science councils scattered throughout the academy system. They generally have a consultative and co-ordinating function, the principle being to involve the entire scientific community in the solution of problems under consideration.

(II) PLANNING

The development plans for science and technology in the USSR are an integral part of the State Plan for the country as a whole.

The Plan for scientific research and the introduction of scientific and technological achievements covers the most important scientific research work on complex problems, important individual problems in science and technology, theoretical problems of the natural and social sciences, the application of advanced technological processes, the mechanisation and automation of production, and the manufacture of new machines, equipment, instruments and materials. It also includes tasks assigned by the government to scientific organisations of the republics or ministries.

The plan is composed of special sections:

> Research and experimental design work.
>
> Introduction of the achievements of science and technology into the national economy.
>
> Ensuring the material and technological aspects of scientific work and measures for applying advanced technology, the mechanisation and automation of production processes and control systems.
>
> Financing scientific research.
>
> Capital investments in the development of science.
>
> Training scientific personnel.

(i) The Plan for Scientific Research considers:

(a) Natural sciences - theoretical and experimental studies and research, primarily in the fundamental sciences (mathematics, physics, chemistry, and biology), which will lead to further scientific and technological developments:

(b) Social sciences - the study of the socialist society constructing communism, theoretical generalisations which can be drawn from this study and the creation of scientific concepts concerning the development of socialist production;

(c) Technology - theoretical research and studies which will contribute to deciding future development and progress in all branches of the national economy and industry;

(d) Agriculture - research is planned which will help to identify agricultural production;

(e) Medicine - research is planned into fundamental medical problems such as methods for controlling the life processes, the question of hygiene at work and in food, and preventive medicine.

(ii) The Plan for Experimental Design Work deals with the creation of models for new machines, instruments, materials and goods, which will increase technical and economic efficiency;

(iii) The Plan for Introducing the Achievements of Science and Technology deals with the question of introducing into industry the results of scientific research and experimental design work which has already been completed and tested; utilising the advanced experience of the USSR and other countries to improve the technology of production, increase productivity, cut out repetitive work, increase output, improve quality, cut production costs and speed up construction. Plans for introducing new techniques are drawn up for each branch of the national economy and each industry. They include calculations on the economic effectiveness of the measures envisaged.

(iv) The Plan Providing for the Material and Technical Execution of Scientific Research and Experimental Work is concerned with the question of meeting demands for equipment, apparatus, special instruments, machines, mechanisms, installations, materials and goods.

(v) The Finance Plan deals with allocations for scientific research. It covers important research problems of national significance, together with other more local scientific problems worked out by the state committees, ministries, departments and union republics. The plan is based on the number of workers in scientific establishments and expenditure on scientific research (without taking into account capital investments).

(vi) The Plan for Capital Investments allocates money for the construction of scientific research establishments, laboratories, and experimental stations.

(vii) The Plan for Training Scientific Personnel deals with the question of training qualified scientific staff for the national economy and, in particular for the most important branches of science and those undergoing rapid development.

(A) Breakdown of the Plans

The plans concerning scientific and technological research consist basically of an input side, made up primarily of wages, and an output side, concerned with scientific research targets. Attempts are made to evaluate the research targets. They are divided in order of importance into 'problems', which in turn are divided into 'research projects', which are again subdivided into 'works',

The *Problems* are of a scientific/theoretical, scientific-technical/experimental and technical-economic/applied character. They are divided into: fundamental problems (physics, chemistry, biology, mathematics, human sciences); branch problems (improvement of production); territorial problems (development of productive forces); inter-branch problems; inter-regional problems.

The *Research Projects* maintain the same main subdivisions, namely scientific, branch and territorial projects.

The *Works* consist of new technological processes, designs for new machines, the preparation of technical documentation for the introduction of new technology into production, etc. Problems which fall into the category of '*exploratory work*' are subdivided into exploratory work of a theoretical nature and project exploratory work. The latter cannot be evaluated according to either expense or economic effect and so research organisations tend to present their work as 'exploratory work'.

So-called 'eternal' subjects - for example, energy and electrification are at present divided by the State Committee for Science and Technology into 'research project tasks for five to ten years', with sub-divisions into 'tasks' and 'stages' to be executed in one to two years.

(B) Planning Process

The planning process can be divided into four main stages; transmission of directives to lower organs; presentation of draft plans to higher authorities; approval of the plan and its transmission; plan adjustments. These stages, however, are only approximate. In practice, the higher and lower organs often make their plans at the same time, and the directives and approvals are then replaced by bargaining and mutual agreement.

The basic elements essential for drawing up the long-term plans of scientific research establishments are:

(i) Proposals of the Council of Ministers of the USSR and the Republican councils of ministers;

(ii) Main orientation of the development of science and technology prepared by the State Committee for Science and Technology together with the Academy of Sciences;

(iii) Long-term plans for scientific research work of the USSR ministries and departments.

When the plans are being drawn up, wide use is made of authors' certificates, of information on the position and direction of development in the appropriate branches of science and technology both in the USSR and abroad, and of suggestions from scientific workers. Advisory bodies, such as the science councils, and scientific societies give guidance to the State Committee for Science and Technology, the Academy of Sciences, the Republican academies and the USSR and Republican ministries when they are drawing up their plans.

The long-term plan fixes the basic direction of the economic development of the country and reflects the related interests of science and production.

Draft Plans. The draft plans for the development of scientific research are based on the principal trends in the development of science and technology, as enumerated and sanctioned by the Council of Ministers. After these have been communicated to its management, each enterprise works out a draft plan for its work for the next specified number of years. Draft plans are worked out at all levels, from individual enterprises up to the one for the whole country.

Union Republics. The draft plans are elaborated by the Republican ministries and departments to include the most important measures in the plans for the technical development of the enterprises coming under their jurisdiction. The plans also include tasks important for the development of specialised branches of industry as a whole and also tasks provided for in government directives. When the plans are being worked out, proposals from higher and specialised educational institutions concerning the execution of work in collaboration with industrial enterprises or on contract from them are taken into account.

The draft plans worked out for the specialised branches of industry are presented by the State Planning Committees attached to the councils of ministers of the Union Republics.

The Republican councils of ministers elaborate the draft plans for scientific research and the introduction of scientific and technological achievements into the economy of the republics. These cover material and technological provisions for, and the financing of, scientific research, capital investments for the development of science, and the training of scientific personnel. When elaborating these plans account is taken of the proposals put forward by the scientific research, design and planning organisations, individual enterprises, scientific councils, the ministries and departments of the republic concerned, as well as recommendations from the Academy of Sciences, the State Committee for Discoveries and Inventions, ministries and departments on an all-union level.

The republic plans include tasks laid down for that republic in the overall plan, tasks deriving from resolutions of the Council of Ministers, and also any important work done for the republic or executed under its authority, but carried out in the republic itself.

In addition to putting forward proposals for their own republic's plans, the various organisations within the republic make suggestions for the draft plans for the branches of the national economy and industry.

Industrial and specialised ministries. The plans for scientific research and experimental design work within the specialised branches of industry and the national economy are elaborated by their particular ministry. They include tasks sanctioned by the all-union plan, which are important to that particular branch of industry, carried out by research organisations either directly subordinate to that ministry or under the authority of another department or organisation connected with it. The most important tasks contained in the republic plans are also included.

The All-Union Plan. The Union republics and the industrial and specialised ministries send their draft plans to the State Committee for Science and Technology, Gosplan, the State Committee on Construction, the Ministry of Agriculture, the Academy of Sciences and the Ministry of Health. The State Council for Construction, Ministry of Agriculture, Ministry of Health and Academy of Sciences then work out and present to the State Committee for Science and Technology and to Gosplan, draft plans for scientific research on questions on building, agriculture, hydrology, medical sciences and technology, pharmacology, and the natural and social sciences, taking into account the proposals put forward by the Union republics and ministries.

The USSR Academy of Sciences draws up the draft plans for scientific research in the Republican academies of science, based on proposals put forward by the Republican councils of ministers. These are then put before the State Committee for Science and Technology and the Gosplan. Certain sections of the draft plan - the introduction of advanced technology, the mechanisation and automation of production processes, the manufacture of new types of industrial products, the removal of out-of-date goods from production - are presented with the necessary technical and economic justifications, and are then co-ordinated with the plans for production.

All work included in the draft plan must be approved by the services under whose authority the executants or co-executants are placed.

The State Committee for Science and Technology and the Gosplan - on the basis of these draft plans presented by the Republican councils of ministers, the ministries and departments - present the USSR Council of Ministers with the draft plan for scientific research and the introduction of the achievements of science and technology into the economy of the USSR. After the Council of Ministers has approved it, this then becomes part of the national plan.

(C) Difficulties

A number of difficulties arise when research problems are integrated into the National Economic plans:

(i) Research and development targets are not easily divided into the traditional five-year, annual or quarterly planning periods. For this reason the targets tend to be elaborated only for the annual plans. But even here they are not usually fulfilled and it is difficult to impose rigid plan indicators.

(ii) It is impossible to enforce centralised decisions by aggregating the indicators as is done for production targets. For example, in the case of steel production it is possible to have a few aggregate indicators in the National Plan, and to give more detailed targets for quality, distribution, etc. in the Ministerial and Enterprise Plans. This is impossible for research and development, and here the government can only select items of particular importance and supervise the execution of 'complex' problems of national interest.

(iii) Research and development targets do not always fit into the standard classification of the National Economic Plans. Many inter-branch and inter-regional research works are executed jointly by different administrative bodies, which makes overall surveillance very difficult.

(iv) No bridge exists to link research and development plans with production targets. The 'Scientific Research Work and Introduction of Science and Technology into the Economy' section is grouped into a special chapter of the plan. Other indicators connected with research and development such as investment, employment, costs, formation of cadres, etc. are dispersed throughout the plan. Indicators concerning financial and credit resources for research and development are not even included in the plan, but are put in separate documents annexed to it. In addition the 'Scientific Research Work' section includes only targets of all-union interest, whereas the other sections of the State Economic Plan and the annexed documents include the financing, material supplies, investment and so on for *all* research and development targets: all-union, republican and local.

(D) Perspectives

Up to 1949 research and development plans were elaborated only by the scientific research establishments and by the testing-design organisations. The National Economic Plan included targets such as the introduction of new technology, mechanisation, automation etc., in the plans for investment and production. The first annual plan for the introduction of new technology was approved by the government in 1949. From then up to

1965 the sections dealing with science and technology became increasingly comprehensive. However, this expansion did not necessarily signify a continuous improvement in research planning. The general movement towards giving industrial enterprises increased autonomy affected the research and development sector. In the 1965 economic reforms, new regulations were introduced concerning the central planning of research and development. The number of chapters of the Central Plan for Research and Development has been reduced, as also has the number of problems, subjects and works. The plan for 1967 includes only "the most important scientific and technical problems of great interest to the National Economy".

(III) FINANCE

(A) Allocation and distribution

With the aim of maintaining standards of development in scientific research, annual and long-term State plans for financing such research, including the necessary capital investments, have been drawn up within the framework of the national annual and long-term plans since 1962.

Table 1
General Expenditure on Science in the USSR, 1959-1967
(in thousand million roubles
Percentage increases are shown in brackets: 1959 = 100)*

Year	Total	Scientific research	Capital investment
1959	3.3 (100.0)	2.8 (100.0)	0.5 (100.0)
1960	3.9 (118.2)	3.3 (117.9)	0.6 (120.0)
1961	4.5 (136.4)	3.8 (135.7)	0.7 (140.0)
1962	5.2 (157.6)	4.3 (153.6)	0.9 (180.0)
1963	5.8 (175.8)	4.9 (175.0)	0.9 (180.0)
1964	6.4 (193.9)	5.4 (192.9)	1.0 (200.0)
1965	7.1 (215.2)	6.0 (214.3)	1.1 (220.0)
1966 (planned)	7.7 (233.3)	6.5 (232.1)	1.2 (240.0)
1967 (expected)	9.0 (282)	7.2 (257)	1.8 (360)

* Source: *Science Policy and Organisation of Research in the USSR*, P 57 (UNESCO Paris 1967)

Between 1959 and 1966, expenditure on scientific research and on the capital construction of laboratories and experimental bases increased nearly two and a half times from 3,300,000,000 roubles to 7,700,000,000 roubles. Out of these figures, scientific research took 2,800,000,000 roubles in 1959 and 6,500,000,000 roubles in 1966, while construction of scientific units took 500,000,000 roubles and 1,200,000,000 roubles. The average annual rate of increase of expenditure on science during these years was roughly 15 per cent. Table I illustrates the growth of this expenditure. By 1969 the total expenditure reached 10,000,000,000 roubles.

Two points should be noted about the figures given here for scientific research. First, they include expenditure by industrial and construction enterprises through contracts with research organisations and similar bodies. This amounts to about 30 per cent each year. Secondly, the figures do not include money spent on scientific research carried out in laboratories and design offices of factories, plants and construction enterprises. This expenditure is balanced by the value of the production of these factories etc., which makes them self supporting. In any case it is only a very small fraction of the total expenditure on science.

The budget plans fix the amount to be spent on scientific research and capital construction of scientific units in each ministry and department of the USSR and each union republic; they also decide where the finance is to come from. In addition, the scientific research budget contains indices for general expenditure on scientific research, the wage fund and the number of workers in scientific research establishments. Budget estimates for research institutes - whether they are on the State budget or self supporting bodies - have to conform with the relevant section of the all-union and republic budgets. The estimates are drawn up by the ministries, taking into account the total volume of general expenditure and the staff quota existing in the finance plan for that particular period.

(B) General Expenditure

General Expenditure on scientific research includes:

(1) Wages. The wages bill takes up some 37 per cent of the total expenditure on science, including capital investments. If capital investments are not included, the bill then fluctuates, according to the particular scientific field, from 30 to 80 per cent of the scientific research budget.

Wages expenditure includes the payment of wages according to fixed official rates; the payment of increments for long service, for working in distant localities and so on, (these rates are established by the State); the payment of differences between special and official rates; the payment of fees to active and corresponding members of the academy; and the payment of premiums awarded in the special premium system for fulfilling and overfulfilling the plan.

(2) Apart from wages, general expenditure on scientific research includes the cost of making experimental models, pilot machines and special laboratory testing equipment; the acquisition of material and equipment for scientific research work (reagents chemicals, utensils, paper, drawing sets, metals, medicines for scientific work, seeds, etc); the purchase, delivery, and maintenance of experimental animals and livestock; the organisation of scientific congresses and meetings, and of scientific expeditions; the cost of experimental work carried out for the scientific institution by an independent organisation; the preparation of maps, diagrams, sketches, models and other such material needed for displays and exhibitions; the cost of electric power and other sources of heat needed for technical experiments or for heating greenhouses, vivaria, hot-houses and so on; the publication of scientific works; the safety regulations in scientific research establishments.

(C) Capital Investment

The All-Union Capital Investment Plan for the Development of Science includes figures for the capital investment and construction and installation work planned for each ministry and department in the USSR and for each Union Republic. It includes allocations for the construction of new research institutes, and for the expansion and reconstruction of existing ones. The same applies to experimental enterprises, workshops and scientific stations.

Expenditure on the *purchase of equipment, instruments, and apparatus* for scientific research institutions, including allocations for these purposes included in the capital investment plan, has increased 3.3 times from 1959 to 1966. The proportion of this expenditure in the general budget has similarly risen from 5.5 per cent to 15.1 per cent during the same period. This relatively high rate of increase in comparison with other science expenditures is a result of the increasing complexity of scientific experiments.

In 1966 expenditure on research into complex scientific and technical problems amounted to 80 per cent (according to the plan) of the general expenditure on scientific research. This is compared with 72 per cent in 1965 and 56 per cent in 1963. Expenditure on non-guided research represents some 12 per cent.

Expenditure on unforeseen research, that arises in the course of the year, amounts to some 2 per cent of the general expenditure. This is covered by reserve funds provided for this purpose in the State budget and by resources freed when work is cancelled. The amount of cancelled work dropped from 1 per cent in 1963 to 0.3 per cent in 1965. This was a result of the increasing role of the single state plan for scientific research.

(D) Sources of finance

The main sources for financing Soviet investment expenditures are :

(i) Budgetary subsidies which finance most of the centrally planned investments (some 53 per cent in the Plan for 1968);

(ii) Enterprises' own resources (from profits, amortisation allowances, sale of equipment, and so on) which provide finance for some of the centralised investments and for a major part of decentralised and kolkhoz investments.

(iii) Bank credits which reinforce the enterprises' 'own' resources and the financial resources of the kolkhoz and population.

The pattern of Soviet investment falls into three stages. First, the construction and assembly work, followed by the acquisition of equipment, instruments and apparatus. Then thirdly, there are other expenditures which include project survey expenditures, boring for oil and gas, expenditures which do not increase the value of fixed capital, and so on.

There was strong criticism of the financing of investment by budgetary subsidies since it did not encourage an economic use of capital. The 1965 economic reforms introduced payments for the productive capital of enterprises. This encourages the allocation of long-term credits for centralised investments and increased the significance of the enterprises's own resources in the financing of decentralised investments. In 1967, the result was an increase in decentralised investments of 20 per cent compared with a 5 per cent increase for centralised investments. In addition, enterprises have two main resources of their own for encouraging workers, engineers and managers to introduce new technology into production.

(a) The Fund for the Development of Production, created in 1965, is administered by those enterprises operating under the new planning system. Its main function is to encourage modernisation, automation and the introduction of new products. It can also finance improvements which increase labour productivity, reduce costs, improve the quality of the goods produced or lead to a higher rate of profit. The bulk of the Fund is used to purchase capital equipment. The fund is financed by deductions from enterprise profits, 30 to 50 per cent of amortisation allowances used for replacement of capital and differentiated by branches of industry, and by receipts from sales by the enterprise of unused and superfluous equipment.

(b) Credits for Mechanisation, Automation and Introduction of New Technology are generally allocated by the State Bank (for industrial enterprises) and by the Bank for Construction, or Stroibank (for construction and assembly, project design and geological prospecting organisations). The money should not be used for building new enterprises, for reconstructing existing ones, or for financing projects included in the

Investment Plan. For sums up to 50,000 roubles the local Gosbank director can allocate the credits, but larger sums depend on the president of the State Bank and recommendations from the Republican councils of ministers and union ministries.

Credits are normally given for periods of three years on the basis of project estimates evaluating the economic efficiency of the proposed measures. Provisions for wages cannot exceed 40 per cent and construction and assembly expenditures cannot exceed 25 per cent of the total cost. As yet these credits represent a very small percentage of the total government allocations for new technology. In the 1964/5 plan allocations for the acquisition of equipment (used in new technology) in industry was some 15,000,000,000 roubles, while the amounts for credits was in the region of 500,000,000 roubles.

(c) The Fund for Consumer Goods is another less important source for extra-budgetary financing of new technology. The money comes from profits obtained by the sale of consumer goods made out of the remnants of the enterprises' main production process, and also from profits accruing from raw materials and semi-finished products produced from these raw materials. Up to 70 per cent of the fund can be used to increase the number of workshops for consumer goods, to improve the quality of such goods, to prepare new prototypes and their production, and to build housing.

(E) Special funds

Extra-budgetary resources also exist for financing research and development. They consist of various funds administered by ministries, central administrations and enterprises.

(i) The Special Fund for Financing Scientific Research is used for branch and inter-branch scientific research and for important testing-design work. Part of the fund goes direct to the scientific institutes, while another part is used to pay for scientific research contracts and testing-design work. Payments made by enterprises into the fund are included in their production costs. After the 1965 economic reforms, the amount deducted from the new Ministerial Fund varied between a lower limit of 1 per cent and an upper limit of 3 per cent of the cost of finished production. These funds are particularly important for financing industrial research, and are included in the National and Ministerial Plans.

(ii) The Fund for Assimilation of New Technology was created in 1960 on a ministerial level for the purpose of reimbursing enterprises for any extra planned expenditures on design, testing and other work needed to achieve batch production. It is financed through deductions of planned cost on the factory level. These vary from 1 per cent of costs in the case of ball-bearing and motor industries to 3.5 per cent in the case of heavy engineering and machine tools. The total amount of the Fund was 300,000,000 roubles in 1967.

(iii) Commercial Surcharges are also used to finance some testing and scientific research work of enterprises specialising in batch production. They are used in situations where the enterprise starts the work on its own initiative without orders from other plants. The amount of the surcharge has to be approved by the Republican Ministry of Finance.

(iv) The Special Enterprises Funds. Enterprises financed directly by the State Budget also occasionally have access to "special funds". These come from rents, entrance fees to museums and exhibitions belonging to the particular establishments, and from incomes of laboratories and cabinets of the enterprises' scientific research and testing establishments. Such earnings can supplement the financial resources of scientific research and testing-design organisations allocated through the State Budget.

(IV) MANPOWER

(A) Scientific workers

In the Soviet literature, the word scientific worker is used in the broad 'wissenschaft' sense, where science embraces all scholarship and not just the exact and natural sciences.

This fact has to be kept in mind when looking at manpower figures for the Soviet Union. To try to avoid confusion, the term scientific worker rather than 'scientific' has been used when the text refers to people working in academic and research fields covering the humanities to the sciences.

(B) Number of Scientific Workers

In 1969 the total number of scientific workers was 883,400. The growth of their numbers over the past 55 years is shown in Table 2.

Table 2*
(numbers in 1,000s)

	1914	1940	1950	1960	1965	1969
Total number of scientific workers	10.2	98.3	162.5	354.2	664.4	883.4
At research establishments	4.2	26.4	70.5	200.1	390.4	-
At higher educational establishments	6.0	61.4	86.5	146.9	221.8	-

*Source: *Science Policy and Organisation of Research in the USSR*, P 61, (UNESCO Paris 1967)

Table 3*

	1940	1950	1960	Jan.1 1967	Doctors of Science	Candidates of Science
USSR	98,315	162,508	354,158	711,552	16,591	152,272
RSFSR	61,872	111,699	242,872	488,364	11,584	100,305
Ukrainian SSR	19,304	22,363	46,657	98,410	2,138	21,796
Byelorussian SSR	2,227	2,629	6,840	16,168	301	3,600
Uzbek SSR	3,024	4,541	10,329	17,827	344	4,383
Kazakh SSR	1,727	3,305	9,623	20,325	253	3,880
Georgian SSR	3,513	4,843	9,137	14,988	708	4,455
Azerbaijan SSR	1,933	3,364	7,226	14,068	418	3,592
Lithuanian SSR	633	1,402	3,320	6,541	82	1,466
Moldavian SSR	180	745	1,999	4,420	75	1,115
Latvian SSR	1,128	2,184	3,348	6,651	93	1,505
Kirgiz SSR	328	841	2,315	4,151	85	1,026
Tadzhik SSR	353	715	2,154	3,853	60	868
Armenian SSR	1,067	2,000	4,275	9,112	309	2,347
Turkmen SSR	487	656	1,836	2,940	44	713
Estonian SSR	544	1,221	2,227	3,734	88	1,112

*Source: V. Yelyutin *Higher Education in USSR*, P 80, (Novosti Press Agency, Moscow)

In 1969 women made up 39 per cent of the total of scientific workers.

The distribution of scientific workers in the Union Republics of the Soviet Union is given in Table 3 for the years 1940, 1950, 1960 and January 1, 1967.

(C) Number and type of scientists and technologists

Engineers, scientists and medical specialists account for more than three quarters of the scientific workers. The figures for 1961, 1965 and 1969 are given in Table 4.

Table 4*
(numbers in 1,000s)

	1961 total	%	1965 total	%	1969 total	%
Total of scientific workers	404.1	100	664.5	100	883.4	100
Engineers	151.8	37.6	298.8	45	390.9	44.2
Sciences of which	119.4	29.5	171.5	25.8	223.8	25.3
Physical + Math.	35.1	8.7	63.9	9.6	89.0	10.1
Chemistry	32.3	8.0	33.5	5.0	44.0	5.0
Geology-mineralogy	12	3.0	16.4	2.5	19.6	2.2
Biology	16.2	4.0	27	4.1	36.4	4.1
Agriculture and Veterinary	23.8	5.9	30.6	4.6	34.8	4.0
Medicine and Pharmaceutics	34.2	8.5	36.7	5.5	47.7	5.4

*Source: *Narodnoe Khoziastvo SSR v 1969 g*, P 695 (Moscow 1970)

Table 5*

	total	%
Total of scientific workers in VUZy	221,800	100
Engineering	53,454	24.1
Sciences of which	53,232	24.0
Physical + Math.	26,172	11.8
Chemistry	9,094	4.1
Geology-mineralogy	2,440	1.1
Biology	7,098	3.2
Agriculture and Veterinary	8,428	3.8
Medicine and Pharmaceutics	25,285	11.4

*Source: Zaleski et al: *Science Policy in the USSR*, Table 12, P 155 (OECD Paris 1969)

The figures in Table 4 are average figures for scientific workers in both research organisations and higher educational establishments. In fact there is a higher percentage of scientists and technologists among the total workers in research organisations, and a smaller percentage in the VUZy. Table 5 gives the figures for the VUZy in 1965.

The total number of scientific and technological academic and research personnel for the year 1969 was 883,400. The number of scientists and technologists working in the economy as a whole is, of course, somewhat bigger than this figure. The total number of engineers, for example, was 1,630,800 in 1965. Of these, the largest number – 530,010 – were employed in industrial enterprises; next came project and design organisations with 313,114, scientific and research establishments with 226,681, construction organisations with 127,202, organs of state and economic management and so on with 115,787 and educational establishments with 109,264. The remainder were engaged in transport and communication enterprises (79,909), agriculture enterprises, repair and technical equipment stations (13,046), geological survey organisations (42,401), and other unidentified areas (73,386). (Based on Zaleski et al: *Science Policy in the USSR*, OECD 1969, Table 5, P 145).

The Centre for Russian and East European Studies of the University of Birmingham, as part of its contribution to the OECD report on *Science Policy in the USSR*, made revised estimates for the total number of people engaged in research and development in the Soviet Union. For 1966 it gave a range of 1,655,000 (cautious estimate) to 2,291,000 (more generous estimate). Of these between 476,000 (cautious estimate) and 670,000 (more generous estimate) were graduates. These figures covered scientists, engineers and medical research workers. It has to be noted that these figures were full-time equivalents, giving the number of man-years estimated to be devoted to research and development. The actual number of people employed, many of whom would be working part-time, would be much larger.

(D) Number and analysis of people with graduate and post-graduate education

The numbers of scientific workers with academic positions is given in Table 6.

*Table 6**
(numbers in 1,000s)

	1950	1960	1965	1969
Total	162.5	354.2	664.6	883.4
Doctors of Science	8.3	10.9	14.8	21.8
Candidates of Science	45.5	98.3	134.4	205.4
Academician, corres. member of academy, professor	8.9	9.9	12.5	16.9
Assistant professor	21.8	36.2	48.6	64.9
Senior research assistant	11.4	20.3	28.7	37.3
Junior research assistant	19.6	26.7	48.9	48.4

*Source: *Narodnoe Khoziastvo SSR v 1969 g*, P 694 (Moscow 1970)

Table 7 gives the numbers of scientists, engineers, and medical doctors among the higher degree holders for the year 1965:

*Table 7**

	Candidates	Doctors
Total	134,427	14,757
Engineering	34,818	3,043
Physical, life and agricultural sciences of which	44,132	5,902
Physics and Mathematics	12,151	1,637
Chemistry	7,632	843
Geology and mineralogy	4,184	763
Biology	10,557	1,647
Agriculture and Veterinary Medicine	9,308	1,012
Medicine and Pharmacy	18,248	3,204

*Source: Zaleski et al: *Science Policy in the USSR*, P 149 (OECD Paris 1969)

The importance of engineering among higher degree holders (making up 26 per cent of the total in 1965) is even more emphasised at the graduate level. In 1965 they numbered 157,400, or 39 per cent of the total 403,900 graduates. In the same year graduates specialising in agriculture (including agronomy, animal husbandry, forestry and veterinary medicine) numbered 33,900 or 8.4 per cent of the total; mathematics, physical and biological science graduates numbered 15,400, or 3.8 per cent of the total; and graduate teachers in mathematics and other sciences numbered some 25,000, or 6.2 per cent of the total. Thus the engineering, natural and medical sciences accounted for 58.3 per cent of the total graduate output in 1965.

(E) Trends

Between 1961 and 1965, the total number of people employed in the Soviet economy rose by 16.7 per cent; during the same period, the total number of higher education graduates employed in the national economy rose by 27.9 per cent, while the total number of scientists employed rose by 64.4 per cent. In other words, the growth in the number of scientists was roughly four times as fast as that of the working population. At the same time, the actual share of scientists in the total working population remained small; it rose from 0.6 per cent in 1961 to 0.9 per cent in 1965.

Within the scientific manpower force, the engineer plays a key role. In 1965 engineers accounted for 45 per cent of all scientists employed in the national economy. During the 1960s, the emphasis in engineering was on the electrical engineering and electrical instrument-making industries. In particular, a large number of graduates enrolled in electronics. In the physical and life sciences, physics and mathematics courses have the largest numbers of graduates, followed by chemistry. The latter gives an indication of the importance attached to the development of a strong chemical industry.

3 Structure of the Research Bodies*

There are essentially three groupings within Soviet research. Firstly, there is the academy system, which includes the Academy of Sciences of the USSR, its affiliates and institutes, and the republican academies of sciences. Secondly, there is the ministerial system which includes a large number of specialised research institutes, mainly concerned with fields connected with production. Thirdly, there is the university system, including the universities, institutes of higher education and the research facilities connected to them.

(I) NETWORK OF THE ACADEMY OF SCIENCES OF THE USSR

3(I) ACADEMY OF SCIENCES OF THE USSR

(A) Historical Background

The Academy of Sciences (in 1925 given the name the Academy of Sciences of the USSR) is the oldest scientific establishment in the country. It was founded in 1725. It is ruled by a statute which is periodically renewed to conform with changing conditions in science and society. The statute defines the main tasks of the Academy, its organisational structure, the rights and obligations of its members and their election procedure.

In 1917 the organisational structure of the academy corresponded basically to the 1836 statute. It consisted of three departments: physical and mathematical sciences; Russian language and literature; and history and philology. The academy was headed by a president, vice-president and a permanent secretary. It called meetings and conferences for the discussion of scientific and organisational questions.

The October Revolution opened up a new era in the organisation of science, which then became one of the most important of government affairs. The Academy of Sciences agreed to become an organising centre; it concentrated its efforts on a study of the country's natural resources and on finding the best means to exploit them. It also helped create a series of specialised research institutes, and organised and led a number of scientific expeditions.

---

* Much of the information in this chapter is taken from two important sources: Zaleski et al: *Science Policy in the USSR* (OECD, Paris, 1969); and *Science Policy and Organisation of Research in the USSR* (UNESCO, Paris, 1967)

In 1925 the Central Executive Committee of the USSR and the Council of People's Commissars of the USSR (now the Council of Ministers) passed a resolution "On the recognition of the Russian Academy of Sciences as the highest learned establishment in the USSR", following which it became the All-Union Academy, or the Academy of Sciences of the USSR. During its 200th anniversary celebrations that same year, the significance of the Academy as an all-union scientific centre was emphasised. In addition it had a duty to help the union and autonomous republics in the development of their economy and culture.

The changes brought about by the revolution made it necessary to draw up a new statute. This was done by a special commission and adopted by the General Assembly of the Academy of Sciences on February 5, 1927. The principal tasks of the academy, as laid down by the statute were:

(a) to develop and improve scientific disciplines under its authority, benefiting them with new discoveries and methods of research;

(b) to study the natural resources of the country and contribute to their utilisation;

(c) to adapt scientific theory and the results of scientific experiment and observation for practical application in industry and the cultural and economic construction of the USSR.

The number of departments within the academy was reduced to two: the Department of Mathematics and Natural Sciences, and the Department of the Humanities - which took in the previous History-Philology and Language-Literature Departments. There was a new procedure for proposing candidates to membership of the academy. The right to put forward candidates was granted to groups of scientists, individual scientists, and public organisations. The election of candidates had to be preceded by a commission comprising academicians and scientists of the country.

The statute of the academy soon became subject to modification and change as conditions in the country and the academy's role within them changed and altered. After the adoption of the first five-year plan, for example, the academy's statute was altered to introduce the idea of the close relationship between the work of the academy and socialist reconstruction. The Academy of Sciences, as the leading scientific organisation of the USSR, had to contribute to the working out of a single scientific method based on materialist ideology, which would systematically orientate the whole system of scientific knowledge to satisfy the needs of the socialist reconstruction of the country.

The second five-year plan, however, called for a completely new statute, adopted in 1935. Its aim was to try to expand the scientific research work of the academy to help in the creation of a modern technological industrial base in the country. The 1935 statute divided the academy into three departments (for social sciences, for mathematics and natural sciences, and for technology) each subdivided into groups corresponding to a particular scientific specialisation. However, only three years later this structure was again modified. The academy's general assembly abolished the groups and instead formed eight departments: physical and mathematical sciences; chemical sciences; geological sciences; biological sciences; technological sciences; history and philosophy; economics and law; literature and language. The divisional structure has remained basically the same since then.

During the Second World War, science was confronted with entirely new problems, and the activity of the academy was correspondingly reorganised to meet the needs of war. Many academic establishments were evacuated from Moscow and Leningrad. The number of scientific research establishments within the academy system rose during this time, and new academicians and corresponding members were added to the staff. Scientific research developed widely in the union republics, their scientific personnel grew and a number of republican academies of sciences were formed.

After the war the government recentralised the network of institutes which had proliferated during the war. During the post-war period the government spent some time in organising new institutes and in extending the network of bases and affiliates attached to the academy. Collaboration between the All-Union Academy, the Republican academies and specialised scientific research centres was consolidated. The academy's information links were extended, and its role as the central scientific research centre of the country was reinforced.

In 1947 the affiliates which had achieved republican academy status were placed under the supervision of the newly created Council for the Co-ordination of Scientific Activity of the Republican Academies of Sciences (this was attached to the presidium of the USSR academy). In 1955 the affiliates of the academy, were also placed under this council's supervision.

At the end of the 1950's the academy was still the most important organisation responsible for research, in terms of the number of institutions under its control, numbers of staff at its disposal, and budget. However, there was now a problem of duplication of research for the institutes under the academy and those in the industrial sector. In 1959 a new statute was introduced, with the aim of redefining the academy's role in planning and organising scientific activities on a national scale. The academy was directed to establish closer contacts between its research organisations and industrial enterprises. It was at this time that the Siberian Branch was being set up.

The Statute also provided for annual elections of new members. Formerly these had been held irregularly. Elections to available vacancies were determined according to fields of speciality to be approved by the academy's presidium. When the government created new vacancies, the elections were to take place at that time.

Two years later, in 1961, the role of the academy with respect to the over-all leadership of scientific activity came under review. It was emphasised that the academy should not be allowed to become a national Ministry of Science, but should concentrate on key problems of fundamental research. Shortcomings in the academy system resulted from its preoccupation with industrial research projects, taking time away from theoretical problems especially in key fields like cybernetics and biology. To reduce the academy's role in applied research, the industrially orientated and branch specialised institutions of the academy were transferred to the appropriate state committee (now ministry); and the size of the academy's Department of Technical Sciences was greatly reduced.

In 1963 a special decree relieved the academy of all its remaining direct involvement in applied research and development. Co-ordination became the concern of the newly created State Committee for Co-ordination of Scientific Research. The academy retained responsibility for developing research in: the key directions of the natural sciences and for long-range research directly connected with production; fundamentally new possibilities for technical progress; research in the social sciences; and the study and dissemination of developments in world science. The affiliates of the academy were returned to the control of the presidium. Its departments were expanded to 15 according to fields of speciality and grouped under three main sections: physical and technological sciences and mathematics; chemical and technological sciences and biology; and social sciences. Since then the organisation has been slightly altered to make four sections and sixteen departments, but the Academy is basically ruled by the 1963 statute.

(B) Objectives and Functions

The USSR Academy of Sciences carries out a number of roles. It is a community of leading scientists working in different fields of science and technology. It is also directly involved in education, supervising graduate work at many of its institutes, some of which can award academic titles and degrees.

Its essential function, however, is to carry out a certain part of the total research work being done throughout the Soviet Union. The 1961 and 1963 reforms placed the academy in charge of all research in the natural and social sciences. It also has to determine basic trends in these sciences and to co-ordinate studies in these fields.

The most important areas of research are:

(a) *Theoretical research*: mathematics, physics, chemistry biology, and sciences of the Universe and Earth in areas of fundamental research.

(b) *Applied research*: electrification, mechanisation and automation of production, "chemicalisation" in the key branches, new synthetic materials, radio-electronics, new sources of energy, new methods of energy conversion.

(c) *Research analysis and planning*: study of the achievements of world scientific progress, research to advance technical progress, and recommendations for the introduction of new technology into the economy.

To enable it to carry out such research, the academy was given the following *functions*:

(a) approval of plans for basic research and control of scientific research in the natural and social sciences wherever conducted;

(b) presentation to the Council of Ministers of plans for financing for materials and technical supplies, and for investments in its own establishments and in the scientific establishments of the Union and Republican Academies;

(c) planning and implementation, together with the State Committee for Science and Technology, of international scientific relationships;

(d) execution, in conformity with the State Economic Plan, of research plans in scientific establishments;

(e) training of scientific personnel;

(f) reports on scientific achievements and their dissemination throughout the country.

(Source: Zaleski et al: *Science Policy in the USSR*, PP 207-208, OECD Paris 1969)

(C) Structure

(1) Departments. There are sixteen departments in the Academy of Sciences of the USSR (see Fig. 3) grouped under four sections.

(a) Section of Physico-Technological and Mathematical Sciences - includes, Department of Mathematics; Department of General Physics and Astronomy; Department of Nuclear Physics; Department of Physico-Technical Problems of Energetics; Department of Mechanics and Cybernetics.

(b) Section of Chemico-Technological and Biological Sciences - includes Department of General and Technical Chemistry; Department of Physical Chemistry and Technology of Inorganic Materials; Department of Biochemistry, Biophysics and the Chemistry of Physiologically active Compounds; Department of Physiology; Department of General Biology.

(c) Section of Earth Sciences - includes, Department of Geology, Geophysics and Geochemistry. Department of Oceanology, Physics of the Atmosphere and Geography.

(d) Section of Social Sciences - includes, Department of History; Department of Philosophy and Law; Department of Economics; Department of Literature and Linguistics.

Each department groups together members of the Academy with the corresponding specialisations.

PRESIDIUM OF THE ACADEMY OF SCIENCES OF THE USSR

- USSR Academy of Sciences Siberian Division
  - Buriyat Branch
  - East Siberian Branch
  - Far Eastern Branch
  - Yakut Branch
- Section of Earth Sciences
  - Department of Geology, Geophysics and Geochemistry
  - Department of Oceanology, Physics of the Atmosphere and Geography
- Section of Chemical-technological and Biological Sciences
  - Department of General Biology
  - Department of Physiology
  - Department of Biochemistry, Biophysics and the Chemistry of Physiologically Active Compounds
  - Department of Physical Chemistry and Technology of Inorganic Materials
  - Department of General and Technical Chemistry
- Section of Social Sciences
  - Department of Literature and Linguistics
  - Department of Economics
  - Department of Philosophy and Law
  - Department of History
- Section of Physico-technical and mathematical sciences
  - Department of Mechanics and Cybernetics
  - Department of Physico-technical Problems of Energetics
  - Department of Nuclear Physics
  - Department of General Physics and Astronomy
  - Department of Mathematics
- All-Union Institute of Scientific and Technical information
- GIPRONII
- Bashkir Branch
- Daghestan Branch
- Kola Branch
- Karelia Branch
- Komi Branch
- Urals Branch

**Figure 3
Organisation chart of the Academy of Sciences of the USSR**

Source: Millionshchikov, M.D.: *Research Centres of the Academy of Sciences of USSR* (Moscow, 1969)

Control over the activities of the departments is exercised solely by the presidium. Each department is formally governed by a general assembly consisting of those full and corresponding members of the academy working in that particular department. Day-to-day control is in the hands of the academic secretary, his deputy and the bureau elected by the general assembly of the department.

According to the 1963 decree, the departments are the scientific and organisational centres for their own particular branch of science. They are responsible for the development of their own fields in the country as a whole and also for research in those fields within the establishments of the academy.

(2) General Assembly. The General Assembly of the Academy of Sciences of the USSR is the supreme body of the Academy of Sciences of the USSR. It is made up of the regular and corresponding members. In 1967 these totalled 557, of which only 174, the full members, were entitled to vote.

The general assembly has a number of functions. It discusses questions concerning the development of science in the country; it decides the direction of scientific research in the natural and social sciences; it makes decisions on important organisational questions; it ratifies the reports of the presidium, and receives reports from the academy departments, branches, establishments and individual members; and it also elects the full, corresponding and foreign members of the academy.

The general assembly holds sessions at least twice a year. The Annual General Assembly usually takes place at the beginning of March each year.

(3) Presidium. The decisions of the general assembly are carried out by the Presidium of the Academy of Sciences of the USSR which directs the activities of the academy between sessions of the general assembly. The presidium has direct control over a number of academy bodies which means that in effect it is the central executive body of the academy.

Membership of the presidium is made up of the president, vice-presidents, the chief learned secretary, academic secretaries of the departments and other members, making a total of thirty three in all. This number is fixed by the general assembly. Elections for the members of the presidium, including the officers, are held every four years. Only full members of the academy are eligible. The presidents of the Republican academies have the right to attend meetings of the presidium, but can only exercise full rights when the discussion concerns affairs of their academies.

The presidium exercises wide control over the academy system. It controls the departments in a number of ways; through their academic secretaries, who are automatically members of the presidium; by approval of new establishments, science councils and new personnel; by approval of research plans; by controlling all international contacts of the departments. With respect to the Republic academies, the presidium controls their relationships with other bodies and supervises their work through the Council for Co-ordination of Scientific Activities of the Republican academies of Sciences. All affiliates of the academy are subordinate to the presidium, as are some of the more important scientific establishments and science councils. The presidium also controls the main libraries and editing facilities of the USSR Academy including the publishing house *Nauka*.

(D) Membership

The Academy of Sciences of the USSR comprises full members or academicians, corresponding members and foreign members. The number of these in 1969/70, was 229 academicians, 419 corresponding members, and 65 foreign members. Each full and corresponding member of the academy is a member of one of the departments, according to his own particular specialisation, and he presents an annual report on his activities....

The actual number of scientists working within the academy network is much greater than this, being of the order of some 34,000.

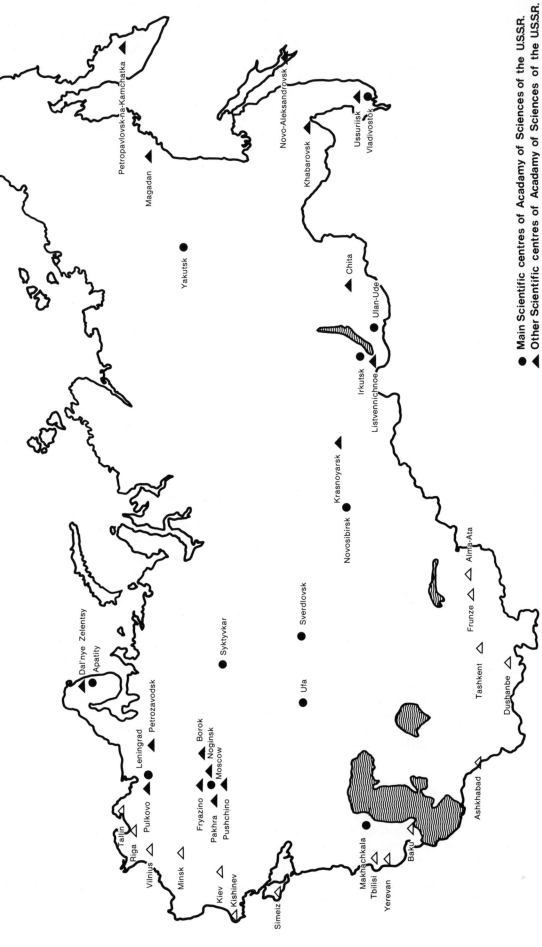

Figure 4 Source: Millionshchikov, M.D., *Research Centres of the Academy of Sciences of the USSR*, (Moscow, 1969)

(E) Scientific Establishments

Scientific establishments within the Academy of Sciences of the USSR include scientific research institutes, laboratories, observatories, experimental stations, libraries and museums. At the end of 1969 there were 229 of them.

(i) Scientific Research Institutes form the backbone to the scientific research activities of the Academy of Sciences of the USSR. Institutes may be under the direct authority of the presidium or that of an academic department. The presidium determines the nature and direction of their scientific work following proposals put forward by the appropriate department and with the approval of the relevant section. The section is in charge of the general direction of activities of the institute, whereas the department is responsible for its scientific and methodological work.

In general, an important research institute is chosen to carry out a specific project and is held responsible for its execution. In order to do this, the institute is responsible for: the organisation of its work taking into account the recommendations of the appropriate scientific councils engaged on the problem in question; carrying out research and experimental work in accordance with the plans; taking measures, in conjunction with other interested bodies, to utilise the results of research in the national economy; agreeing with other scientific research establishments on the co-ordinating of research activities; assisting the scientific establishments of the Republic academies in the development of scientific research and the training of scientific personnel; training research workers; organising debates, lectures, and conferences on the scientific questions under study at the institute; the publication of research results; the diffusion of scientific knowledge and information on the research accomplished by the institute, and the results achieved; taking care of the general direction of work on the periodical which is the press organ of the institute. (Source: *Science Policy and Organisation of Research in the USSR*, P 47, UNESCO, Paris 1967).

Each institute is headed by a director, one or more assistant directors for each scientific specialisation, the scientific secretary of the institute and an assistant director for administrative and financial affairs. The directors are elected for a term of four years by the general assembly of the appropriate department. The election is then sanctioned by the presidium of the academy. The director is responsible for the organisation of work within the institute. He selects the staff and allocates the budget. He has the power to make certain alterations or innovations to the planned projects. For example, he can purchase locally produced materials without the authorisation of the material supplies authorities, and sell (also without their authorisation) materials and equipment which are no longer needed. The director is also assisted by the scientific council of the institute. This is a consultative body comprising the institute's leading scientists, and party and trade union representatives. It is nominated by the director. The assistant directors and the scientific secretary of the institute are nominated by the presidium on the basis of proposals submitted by the director.

Research work is carried out in the research departments, which are further subdivided into laboratories and research groups. The director is responsible directly for the most important of these bodies as well as for experimental stations, observatories, libraries, archives and information sections. Institutes which train research workers have an education department.

(ii) Leading Institutes are specialised research establishments whose research project plans are larger and more precisely defined than those of a regular institute which are ordinarily preoccupied with specific problems geared to local needs. Leading institutes combine research and development activities and often have laboratories for working on production problems attached to them. Examples of leading institutes are the Institute of Cybernetics and the Institute for the Study of Properties of Materials attached to the Ukrainian Academy of Sciences, the Institute of Physics of the Byelorussian Academy of Sciences and the Byurakan Astrophysical Observatory of the Armenian Academy of Sciences.

3(2) SIBERIAN BRANCH

The Siberian Branch has the same status as a department of the Academy of Sciences of the USSR. Its members are full and corresponding members of the academy, it has to submit its research plans to the presidium for approval, and it controls its own research institutes and other scientific establishments. However, in reality, it has more power than the normal academy department, resembling more closely the status of a Republican academy. It is headed by a presidium, not a bureau as in the other departments. The President of the Siberian Branch is automatically a vice-president of the USSR academy, and his first vice-president is automatically a member of the academy's presidium. Its financial resources are allocated from the budget of the Russian Federal Republic. Unlike other departments, it has control over the disposal of credits allocated to it by the presidium of the academy.

The Siberian Branch was created in 1957, since when it has developed very rapidly, becoming a powerful centre for the physico-mathematical, chemical, geological and engineering sciences. Its creation coincided with the reorganisation of industry carried out in 1957. The aim was to encourage the geographic dispersal of the academy's resources and to develop original research centres in Siberia and the Far East. Among the new centres which have already been developed or are still in the planning stages are the complex research centres at Novosibirsk, Krasnoyarsk, Yakutsk, Irkutsk and Vladivostock, and the regional research establishments of Ulan Ude, Magadan, Novo Aleksandrovsk on Sakhalin Island, and Petropavlovsk-na-Kamchatka. All these centres are affiliates or institutes set up by the Siberian Branch. In 1966 there were, under the Siberian Branch, some 45 institutes (nearly 25 per cent of the total number of the USSR Academy network), employing a total of 5,500 scientists, of whom 18 were full members and 45 corresponding members of the academy.

3(3) NOVOSIBIRSK CENTRE AT AKADEMGORODOK

The Novosibirsk Centre at Akademgorodok is the head-quarters of the Siberian Branch. It is situated on the River Ob, some 25 miles from Novosibirsk. The Novosibirsk Centre was founded in 1957 by a decree of the Council of Ministers. A total of 219 million new roubles were spent on its construction between 1958 and 1963, roughly 35 to 40 per cent of all the funds allocated for investments in the USSR Academy during that period. Contrary to usual procedure, the money was given on a lump sum basis with no stipulations from Gosplan as to how it should be allocated.

The Novosibirsk Centre concentrates a large proportion of the total scientific manpower and research facilities of the Siberian Branch. At the end of 1964, it had more than a third of the scientific establishments, employed 70 per cent of the total personnel and 58.8 per cent of all scientists, and it accounted for all the Siberian Branch's full and corresponding members of the USSR academy, as well as 64 per cent of the doctors of science and 60.7 per cent of the candidates of science.

At the end of 1967 the Novosibirsk Centre comprised 20 institutes employing 16 academicians, 28 corresponding members of the Academy of Sciences, 95 doctors and 950 candidates of science. The total personnel was 16,780. When the Centre was established the initial 15 institutes reflected the desire of the government to bring together inter-disciplinary fields of science in order to facilitate the solution of complex problems. They covered mathematics and calculus, applied and theoretical mechanics, hydrodynamics, automation, physics of heat, electronics, nuclear physics, solid state physics and semi-conductors, catalysis, kinetics and combustion, chemistry or minerals, organic chemistry, cytology and genetics, geology and geophysics, and economics and statistics. Since then additional institutes have been created.

Each of the original institutes was set up by an eminent scientist with administrative experience and is now headed by either a full or a corresponding member of the academy. The aim was to attract highly qualified personnel to work at the centre.

A series of workshops and experimental factories were established alongside the scientific research institutes to encourage the integration of technical advances into industry. Units with large-scale model equipment were built for the Institutes of Catalysis and Organic Chemistry. Large-scale experimental plants were constructed for the Institutes of Theoreticalaand Applied Mechanics, Chemical Kinetics and Combustion, and Nuclear Physics. There is an experimental factory for special precision machinery and equipment. In addition, nearly all the institutes have their own workshops.

There are also plans to construct a series of design bureaux and experimental factories on the outskirts of Akademgorodok. These would be under the supervision of the appropriate ministry and department, with industry supplying the finance. But the scientists working at the Novosibirsk Centre would be responsible for the work done at the design bureaux. An example of this kind of co-operation between research and industry is provided by the Novosibirsk Institute of Hydrodynamics, whose design bureau works on orders from a number of ministries and developed a hydraulic press now being used by a number of factories. The factories pay royalties to the institute for the use of the press.

Because of the scarcity of qualified scientific personnel in Siberia and the Far East, the University of Novosibirsk was founded in 1959. It has close links with Akademgorodok; members of the Centre hold chairs in the University, and students are employed in the design bureaux and experimental factories of Akademgorodok. The emphasis is on research and the aim is to produce qualified personnel for the whole area.

3(4) IRKUTSK CENTRE

The nucleus for development in Eastern Siberia will be the Irkutsk Centre on the Angara River. The Centre is intended to be a supplement to Akademgorodok, and so institutes are only being built for fields not already represented at the Novosibirsk Centre.

The Irkutsk Centre began taking shape between 1958 and 1960. It now includes eight institutes; the Siberian Institute of Power, the Geochemical Institute, the Institute of Organic Chemistry, the Institute for the Study of the Earth's Mantle, the Limnological Institute, the Institute of Siberian and Far Eastern Geography, the Institute of Plant Physiology and Biochemistry, and the Institute of Terrestial Magnetism. Together they employ some 800 scientists with a total staff of some 2,742 persons.

In addition to the eight institutes, the centre has six laboratories, for research and experimentation, seven institutes of higher education, a large polytechnic institute and there are plans for a university. Staff at the Irkutsk Centre total about 4,500 of whom more than 100 are doctors and some 750 are candidates of science.

3(5) AFFILIATES OF THE USSR ACADEMY OF SCIENCES

New affiliates of the academy have been created when it has been necessary to redistribute the academy's regional network. They exist in the autonomous republics and districts of the Russian and other Republics. They group together the scientific research institutes and other scientific establishments of the USSR Academy in their area.

The legal status of the affiliates is similar to the Republican academies and the Siberian Branch, but their administrative powers are less. They come under the supervision of the USSR academy, whose general assembly elects their president for a period of three years on the basis of the presidium's proposals. The affiliate presidents have no *ex officio* duties in the USSR academy's presidium. The direction of the affiliate is in the hands of its presidium, which consists of the president, the vice-presidents and members.

At present there are ten affiliates: Dagestan, Kola, Komi, Ural, Bashkir, Karelian,

Buryat, Eastern Siberia (centred in Irkutsk), Far East (named Komarov) and Yukutsk. The last four were founded recently and were placed directly under the administrative jurisdiction of the Siberian Branch.

(i) Buryat Affiliate. Following the principal of concentrating research effort into key fields, the Buryat Affiliate was created in 1966. The fields it covers are biochemistry, the biology of micro-organisms, the study of minerals in the Baikal region, pedology, geochemistry, the chemistry of carbons, the synthesis of monomers and polymers contained in nitrogen, radio physics and solid state physics.

(ii) Far East or Komarov Affiliate. The Komarov Affiliate is situated in Vladivostock. It has four institutes covering geology, biologically active substances, biology and soil science and sea biology, and three departments for chemistry, economic studies and history. In addition it administers the Komarov Mountain Taiga Station, a Sun Station and the Suputinsky Reservation.

(iii) Yakutsk Affiliate. The Yakutsk Affiliate has institutes of geology, biology, space physics research and aeronautics, and linguistics, literature and history. It also has two departments for economics and energy.

(iv) Bashkir Affiliate. The Bashkir Affiliate has three institutes - organic chemistry, biology, and mining and geology.

(v) Daghestan Affiliate. The Daghestan Affiliate is situated in Makhachkala on the Caspian Sea. It administers two institutes for physics and history, linguistics and literature. It also has a department of economics.

(vi) Kola Affiliate. The Kola Affiliate, centered in the Murmansk region, has institutes covering geology, the chemistry and technology of rare elements and minerals, mining and metallurgy, marine biology and geophysics in the polar regions. It also administers a polar alpine botanical garden.

(vii) Komi Affiliate. The Komi Affiliate is situated in Syktyvkar in the Komi autonomous republic, north east of Moscow. It has institutes of geology, biology, and departments of chemistry, and energy and water conservation.

(viii) Ural Affiliate. The Ural Affiliate is centered at Sverdlovsk. It has institutes of plant and animal ecology, geophysics, geology and geochemistry, chemistry, and electro-chemistry. It also administers the Ilman State Reservation, the Kungar 'Ice Caves' Station and the Salekhard Station in the Tyumen region.

(ix) Karelian Affiliate.. The Karelian Affiliate has institutes of geology, forestry and biology. It also administers a department of water problems.

(II) ACADEMIES OF SCIENCES IN THE UNION REPUBLICS

There are at present fourteen Republican academies of sciences, controlling some 340 research establishments, employing around 28,000 scientific workers. The Russian SFSR is the only republic not to have its own national academy, but its needs are served by the USSR Academy.

(A) Historical Survey

Before the October Revolution there were practically no scientific establishments in the regions which now form the Union Republics. Since 1917, however, a network of scientific establishments has been set up, grouped around the academies of sciences of the Union Republics. There are now fourteen Republican academies of sciences. The first to be founded was that of the Ukraine in 1919. Since then they have been founded successively in the Byelorussian SSR in 1929, the Lithuanian SSR in 1941, the Georgian SSR in 1941, the Uzbek and Armenian SSRs in 1942, the Azerbaijan SSR in 1945, the Latvian, Estonian and Kazakh SSRs in 1945, the Tadzhik and Turkmen SSRs in 1951, the Kirgiz SSR in 1954 and the Moldavian SSR in 1961.

Some academies were created out of scientific establishments already existing in the republic. For example, this happened in the Ukraine, Byelorussia, Lithuania, Latvia, and Estonia. In other republics, where there were very few native scientists and little scientific tradition, the all-union bodies helped to organise scientific establishments.

In 1930, a network of affiliates and bases of the academy began to be formed in the Union Republics. The transcaucasian affiliate of the USSR Academy of Sciences (Zafkan) was set up in Tbilisi, Georgia, with a department in Baku. In 1932, the Kazakh and Tadzhik bases of the USSR academy were established, later becoming academy affiliates. In 1935 Zafkan was divided into three, making the separate Georgian, Armenian and Azerbaijan affiliates. The Kazakh affiliate was formed in 1938, the Uzbek in 1940 and the Tadzhik, Turkmen and Kirgiz affiliates in 1941.

Scientists in the academy affiliates tended to concentrate on the study of the natural resources of the republics, on the history, language, and literature of the different peoples of the republic, and also on the training of young scientists.

This meant that at the beginning, the institutes coming under the administration of the affiliate usually covered the humanities, botany, zoology, geology, and chemistry. Organisations for the study of physics, mathematics and technology were not developed until the Second World War or after. However, as time went on, the affiliates were unable to meet the growing demands upon them - the result of the economic development and spread of education in the Union Republics. New specialised establishments were needed with larger experimental bases. In addition, the number of qualified scientists was growing all the time. All these factors created the prerequisites for the reorganisation of the affiliates of the Academy of Sciences of the USSR into individual academies of sciences of the Union Republics.

(B) Organisation

The internal organisation of each Republic academy is similar to that of the Academy of Sciences of the USSR. Formal power is concentrated in the general assembly while real power lies in the presidium, and research is carried out in the scientific establishments headed by the various departments. However, there are less departments in each Republican academy compared with the USSR Academy.

The Republican academies receive the principal part of their budget through the Republican council of ministers. (They also receive money in payment for contract work, but this never represents more than 25 per cent of their budget and is usually considerably less.) This method of financing gives them a certain autonomy with regard to the USSR academy, Gosplan and the State Committee for Science and Technology. However, since the 1963 reforms which gave the USSR academy the right to exercise general guidance over all research in major problems of the natural and social sciences, the supervisory role of the USSR academy over the Republican academies has increased. This is exercised through the USSR academy's presidium and departments of the Republics. The activities of the Republican academies are directed by the USSR academy's presidium together with the councils of ministers. These two bodies also have to approve the plans for the most important research projects in the natural and social sciences. The USSR academy departments examine reports from the Republican academies on their scientific activities, the organisation of scientific manpower, and the practical application of research results. They then submit their recommendations to the presidium of the USSR academy. They also look at suggestions concerning the creation of new scientific establishments in the Republican academies.

In addition, all vacancies for full or corresponding membership of the Republican academies can only be filled with the permission of the USSR academy, whose recommendations must be taken into account. All nominations for directors of research institutes have to be approved by the Academy of Sciences of the USSR. However, the presidents of the Republican academies are present at presidium meetings of the USSR academy, and they have a deliberative vote on matters concerning their own academies.

The USSR academy also exercises indirect supervision through the Council for the Co-ordination of Scientific Activities of the Republican Academies of Sciences.

(C) Research Policy and Orientation

The Republican academies of sciences plan and co-ordinate research in the fields of the natural and social sciences, train postgraduates as highly qualified scientific workers, and direct and execute fundamental research work in the natural and social sciences.

They are the main centres of research within the Republics, and have given rise to a large number of specialised institutes, including ones for applied research. This results from the fact that certain branches of industry still do not have enough of their own specialised institutes. The juxtaposition, within the republican academy system, of specialist institutes for different fields of science and technology encourages the solution of complex scientific and technological problems.

Since 1963 attempts have been made to promote the development of particular branches of the economy in different areas of the country and to bridge the barriers between research and production (see the Siberian Branch, Section 3(2)). With respect to the Republican academies, each academy is supposed to concentrate on a special field or fields chosen on the basis of the economic and cultural development of its Republic, and taking into account the needs of the national economy.

(III) ACADEMIES DEPENDING ON USSR AND REPUBLICAN MINISTRIES OR STATE COMMITTEES

There are a number of specialised academies which are not subordinate to the Academy of Sciences of the USSR, but are administered by a Ministry, State Committee or Republican council of ministers.

The academies and their respective administrative bodies are as follows:

1. *All-Union Lenin Academy of Agricultural Sciences* - under the Ministry of Agriculture of the USSR.

2. *Republican Academies of Agriculture of the Ukraine, Byelorussia, Uzbekistan, Kazakhstan, Georgia and Azerbaijan* - under the Ministry of Agriculture of the Republic.

3. *Academy of Construction and of Architecture of the USSR* - under the State Committee for Construction of the USSR.

4. *Academy of Construction of the Ukrainian Republic* - under the State Committee for Construction of the Ukrainian Republic.

5. *Academy of Arts of the USSR* - under the Ministry of Culture of the USSR.

6. *Academy of Medical Sciences of the USSR* - under the Ministry of Health of the USSR.

7. *Academy of Pedagogical Sciences of the USSR* - under the Ministry of Education of the USSR.

8. Academy of Municipal Economy of the RSFSR - under the Council of Ministers of the RSFSR.

In 1960 these specialised academies had 946 full and corresponding members, working in 478 scientific establishments which employed a total of 20,077 scientists. They represent the third largest sector of the academy system.

(A) Structure

The over-all structure of a specialised academy generally resembles that of the USSR or Republican academies with a general assembly, presidium, departments and scientific establishments. They have full, corresponding and honorary members (except for the Academy of Municipal Economy of the RSFSR, which has only an administrative staff). In some instances members of the USSR academy are also members of a specialised academy. The general assembly has less powers than those of the USSR or Republican academies. The presidium is not required to inform the assembly of important decisions that have been made. Separate libraries and editing offices are usually attached to the presidium. In the Academies of Arts and of Pedagogical Sciences, the presidium also controls higher educational establishments within its field. The number of departments within the specialised academies varies from three to six.

(B) Finance

Financial arrangements are made in accordance with the orientation of the academy's research: fundamental research is financed out of the State budget, applied research is financed by the appropriate Ministry or department. Work done on the basis of contracts is financed by the contractee. A large part of the resources of the specialised academies comes from their own work. Forty per cent of these are set aside for theoretical work and purchase of equipment.

(C) Research Policy

The specialised academies differ from other sectors in the nature of research they conduct and in their organisational network. Practical research predominates since they were created to promote scientific and technical progress in one particular branch of the economy. As a general rule, they work in areas of research not covered by the USSR academy. There is a tendency for close vertical integration within the branch, and an emphasis on strong administrative direction.

(IV) MINISTERIAL SYSTEM

(A) Historical Background

The ministerial system covers all scientific establishments not subordinated to the academic or higher educational systems. The bulk of research carried out within the ministerial system is in the applied fields, being connected with the various industrial ministries.

The system emerged from the industrialisation drive of the 1930s, with the central government directing the development of technology and technologically based industries from above. This facilitated the rapid introduction of Western technology into existing Soviet industries, and the establishment of new industries utilising technology adapted from advanced Western firms. It also proved successful in mobilising resources for major research and development projects concerned with the military and later the space programmes.

The 1957 reforms attempted to decentralise the ministerial system by dissolving a number of the industrial ministries and placing their scientific establishments under regional economic councils. At the same time a number of key sectors, such as the defence, aviation, radio engineering, shipbuilding and chemical industries, retained centralised control of their scientific establishments, and, in fact, at least a half of the industrial scientific establishments remained under some central state authority (state committee, central agency or Gosplan). In the period after the 1957 reforms the ministerial system developed rapidly. The number of scientific establishments grew from 2,018 in 1956 to 3,285 in 1961; and the number of scientists increased from 74,250 in 1956 to 180,910 in 1961.

In 1961, a clear cut distinction was drawn between the academic and ministerial systems with regard to the scope of their research. Before that year, the Academy of Sciences of the USSR had had a sector dealing with the technological sciences, which had led to a certain amount of duplication of effort between the academy institutes and the branch institutes. It had also meant that there was a dispersal of the academy's resources over a wide variety of problems, thus detracting from more fundamental research in the natural and social sciences. In 1961 the relevant technological institutes were transferred from the jurisdiction of the academy to that of the relevant ministry, department, or economic council leaving the academy with overall control over research in the natural and social sciences.

Duplication of effort and lack of co-ordination still existed after 1961. Many scientific and design institutions were under the authority of the economic councils in different economic areas. Establishments with similar interests worked under different economic councils and there was no overall guiding centre. In 1963 branch-specialised research institutions were transferred from the jurisdiction of the Republican academies of sciences to that of the State committees and departments responsible for research in the branches concerned. In 1965 the industrial ministries, which had been abolished in 1957, were restored, and almost all research and development within the ministerial sector came under central government control again. The regional economic councils were abolished. However, the economic reforms introduced at the same time in 1965 increased the number of scientific establishments operating on an economic accounting basis, and gave more autonomy to enterprises in their planning and management.

The traditional pattern of the ministerial system is a network of specialised research establishments operating separately, both geographically and organisationally, from the industrial enterprises. Although both the scientific establishments and the industrial enterprises may come under the authority of the same ministry, the lines of command within the ministry are separate. This traditional pattern is now changing with the need to achieve closer liaison between research and innovation into industry.

(B) Ministries

Each Soviet ministry is divided into a number of different departments:

The *Technical Department* is the one mainly responsible for research and development. It acts as a clearing house for plans for new technology put forward by the *Production Department* (these usually have their own technology division). These plans are later presented to the ministry for confirmation. The Technical Department also guides, plans and supervises the work of research institutes under the authority of the ministry. It has an inventions and rationalisation bureau and a patents bureau. The production departments have similar bureaux or appoint specialists to the job.

The *Scientific and Technical Council* of the ministry is a non-executive body. Its function is to look at plans for the introduction of new technology, for automation and mechanisation, and to discuss the main directions of technical development within the field covered by the ministry. Its special concern is to establish a scientifically based unified technical policy for the industry, and to introduce the latest achievements of Soviet and foreign science and technology. The council also organises bureaux and sections to deal with the most important problems of industry and industrial production. It is headed by the minister or his first deputy and its members consist of scientists, industrial specialists, leading workers, inventors and rationalisers, representatives from the economic administration and leading members of the party, soviet, trade union and komsomol organisations.

(C) Scientific Establishments

(i) Branch or Departmental Research Institutes. The titles of branch or departmental research institutes varies according to their importance. Major research institutes are usually called "All-Union", "Central" or "State" research institutes. They often have branches or departments in other parts of the country.

There are also different titles referring to the type of research being carried out:

(a) Research institutes dealing with product design
are called project-design (NIPKI) or design (NIKI).

(b) Research institutes dealing with process design
are called project-technological (NIPTI) or design-technological (NIKTI) or technological (NITI).

(c) Research institutes dealing with plant design
are often called project (NIPI).

(ii) Design Bureau. The design bureau is the main type of organisation responsible for designing new products. As with research institutes there is considerable variation in their titles. The major design bureaux operate independently, with their own accounting system. They have their own experimental and testing facilities and are responsible for the manufacture and testing of prototypes as well as for design. Other smaller bureaux are attached to research institutes or factories, and with them the prototype is tested at the factory.

(iii) Head Institutes. The role of head institutes - the most eminent institute in their particular field - was enhanced during the period when the industrial ministries were abolished (1957-65) and when there was little centralised scientific and technological control within branches of industry. They helped to co-ordinate research and development and to eliminate duplication of research effort. Since 1965 it has been suggested that they should be responsible for complex problems.

The power of a head institute over other organisations in its field varies. They have the right to review projects submitted to them and to accept or reject them. A few are able to take decisions about future work within their field, and have the formal powers to implement these recommendations. Some have established a dialogue with industry to help determine the main lines of research in their field.

(iv) Factory Centres. The factory centre is an arrangement intended to overcome the barriers to innovation which exist between research and industry. It brings together the research, development and production processes into a single unit. In order to achieve this the research institutes are put under the control of the factory. In this situation the branch research institutes tend to concentrate on more theoretical work where results are difficult to predict, while the factory carries out all the production oriented research and is responsible for all the stages in the development of a new product or process.

Not many large-scale factory centres exist at the present. The Elektrosila works in Leningrad is one successful example.

(v) Research Complexes. There are a number of large new centres where various research institutes work in related scientific fields. These are broadly termed 'research complexes'. The best example of such an establishment is the Novosibirsk complex at Academgorodok (see Section 3(3)).

(vi) Research Corporations. The research corporations are responsible for setting up permanent links between the research institutes or design bureaux and groups of factories requiring their services.

There is also a long-term aim to create specialist groups of factories and scientific establishments under the authority of industrial ministries which would develop, manufacture and sell new products and processes to the rest of industry.

(V) UNIVERSITY SYSTEM

(A) Historical

Before the Revolution the universities and institutes of higher education were the main centres for scientific research, apart from the Academy of Sciences of the USSR. Lobachevsky, Mendeleev, Pavlov and Timiriazev were all attached to universities. However after 1917 their main concern, and in some cases their sole concern, became teaching and research was reduced to a secondary role. The authorities did not regard this as an ideal situation and there have been decrees since 1936 on the need to integrate the higher educational establishments (VUZy) more closely into the research and development effort of the country. The point has become greatly stressed over the past decade or so. New regulations brought out in 1961 reaffirmed that research together with training was the main aim of VUZy. A number of research institutes have been placed directly under the jurisdiction of universities or the ministry of education. However, one of the main problems that remains is the heavy teaching load of the VUZy staff which makes it difficult for them to carry out original research as well.

(B) Structure and Organisation

There are two kinds of higher educational establishment in the Soviet Union:

(i) Universities. The universities are divided into faculties for the humanities and natural sciences. In addition to research facilities within their faculties, some universities maintain separate research institutes or laboratories which are considered to be distinct scientific institutions. Research in the universities is primarily directed towards the basic sciences. (See Chapter 4, Section F (i)).

(ii) Institutes of Higher Education. The institutes of higher education concentrate on specific fields and are primarily occupied with applied research.

(C) Administration

The administrative structure for supervising VUZy is complex. Higher and specialised secondary education is primarily directed by the USSR Ministry of Higher and Specialised Secondary Education. However, other ministries and state agencies are endowed with administrative and supervisory functions, and these in their turn are further controlled by the Union ministry and Republican councils of ministers.

3(6) USSR MINISTRY OF HIGHER AND SPECIALISED SECONDARY EDUCATION

The Union Ministry of Higher and Specialised Secondary Education[*] works jointly with the Republican councils of ministers and the Republican ministries of higher and specialised secondary education. The Union ministry has general supervisory powers over all VUZy, as well as direct control over those educational establishments under its direct administration.

Decisions within the ministry are taken by the *Collegium* (Board) presided over by the *Minister*. Questions concerning the activities of VUZy are considered by a variety of *Ministerial Councils*:

---

* See also Section 4 (iii)

The *Scientific-Technical Council* is responsible for general policy determination. It plans and co-ordinates the research activities of all higher educational establishments. and is responsible for expediting the practical application of their research results It contains a number of sections concerned with different fields.

Other Councils are responsible for specific fields of research. For example, the Scientific-Methodological Council for Theoretical Mechanics and the Inter-Departmental Scientific Council for Programmed Planning.

Supreme Attestation Commission. This is headed by the Minister and had some 200 members in 1965, of which more than half were full or corresponding members of the Academy of Sciences of the USSR. It is a powerful body in the control of research. It selects the higher educational establishments which are to engage in advanced training in research. It reviews all candidate degrees, and decides on the award of the higher degree of Doctor of Sciences. It is responsible, in fact, for all higher degrees, and not just those awarded by VUZy. It makes the appointments to all senior positions (professorships and associate professorships), and must ratify the appointment of junior personnel (teachers and instructors). It can revoke certification of advanced degrees or advanced ranks. It keeps files on all advanced academic teaching and research personnel.

The Union Ministry exercises joint control with the State Committee for Science and Technology over the creation and reorganisation of VUZy, the planning of scientific research, the organisation of scientific information and the approving of members of VUZy councils and of scientific establishments of the ministries. It approves the designs and inventory of general educational and laboratory equipment. It directly controls various research establishments and its orders concerning educational methodology, scientific research and ideological work in higher and secondary educational establishments are obligatory for all ministries supervising such establishments. It organises research work in those VUZy under its jurisdiction. This role gives the ministry a powerful say in financial allocations.

3(7) REPUBLICAN MINISTRIES OF EDUCATION

The Republican ministries of education have a similar role to the Union ministry. They have more VUZy under their direct supervision and take a more active role in the day-to-day planning of research. They elaborate prospective and current scientific research plans for the most important research projects and present these for approval to the respective Republican council of ministers. They organise research within their competence, approve plans for all research in the republic, control the execution and assist in the practical application of research results in the economy.

They also nominate rectors and deputy rectors of VUZy and the deans of faculties. They keep files on teaching personnel, determine the categories of the VUZy, plan their investments, employment, cost, material supplies and allocate financial resources.

(D) Finance

On the basis of Soviet evaluations of the cost of training students and aspirants one can get a general idea of expenditure for training research personnel. According to selective inquiries made by the USSR Ministry of Higher and Specialised Secondary Education, the average yearly cost of student training in all VUZy was as follows: 1,066 roubles for daily, regular courses; 280 roubles for evening courses; and 85 roubles for extension-correspondence teaching. In universities, the respective cost was: 970 roubles, 290 roubles, and 80 roubles. One year of aspirant training costs 2,600 roubles.

The average expenditure on research and laboratory equipment for VUZy were, in 1964, five times lower than those for academy institutes. The press is filled with criticisms of the lack of equipment, space and financial resources. In 1967 the Minister of Higher and Specialised Secondary Education, V.P. Eliutin, pointed out that the investments in VUZy were totally inadequate for implementing the plans for the formation of specialists. (Source: Zaleski et al: *Science Policy in the USSR*, P 318, OECD Paris 1969).

Government encouragement of VUZy research and its lack of finance adequately to back this up perhaps represent the slightly two-edged position VUZy find themselves in with regard to their role and the relationship they should find between teaching and research.

(E) Research Institutes and Laboratories of VUZy

The organisation of research institutes and laboratories has been the traditional way of stimulating research in VUZy. Although the majority of these research units were attached to VUZy departments, the most significant were those which had the status of separate scientific institutions. Since 1958-59 there has been a concentrated effort to develop VUZy research by creating special problem research laboratories and branch-of-industry laboratories.

Problem Research Laboratories are attached to a section of a higher educational establishment; their main aim is to strengthen links with industry, and they are designed to work on specific problems or to aid a department in its research.

Branch-of-Industry Laboratories are subordinated to an institute of higher education or to a department. They are designed to solve specific problems of interest to a ministry or enterprise. The latter provide the finance for the research.

(F) Personnel

The distribution of scientists in higher educational establishments by branches of sciences over the past 20 years is shown in Table 8.

Table 8*
Distribution of Scientists in Higher Educational Establishments by Branches of Science - 1950-1965

	1950 Official Data		1960 Official Data		1965[1] Obtained by Extrapolation of Overall Trends
	Number	%	Number	%	%
Engineering	17,575	20.3	35,271	24.0	30.0
Sciences	19,612	22.7	35,351	24.1	23.9
Physico-Mathematical	6,755	7.8	17,272	11.7	14.0
Chemistry	4,180	4.8	6,054	4.1	2.9
Geology-mineralogy	1,058	1.2	1,679	1.1	0.9
Biology	3,400	3.9	4,719	3.2	3.1
Agriculture and Veterinary	4,219	4.9	5,627	3.8	3.0
Medicine and Pharmaceutics	11,609	13.4	16,786	11.4	7.0
Pedagogy	6,583	7.6	10,584	7.2	6.2
Social Sciences	23,762	27.5	36,869	25.1	23.5
History and Philosophy	7,536	8.7	10,634	7.2	5.5
Economics	2,993	3.5	7,748	5.3	6.3
Philology	11,866	13.7	16,781	11.4	10.8
Geography	1,367	1.6	1,705	1.2	0.9
Humanities	3,948	4.6	5,569	3.8	3.0
Art	2,847	3.3	3,833	2.6	2.1
Architecture	387	0.4	611	0.4	0.3
Jurisprudence	714	0.8	1,125	0.8	0.6
Others (unknown)	3,453	4.0	6,485	4.4	6.4
TOTAL	86,542	100.0	146,915	100.0	100.0

*Source: Zaleski et al: *Science Policy in the USSR*, P 317, (OECD Paris 1969)

1. The figures represent a tentative distribution pattern made on the basis of the changing annual numbers of scientists (*naucyne rabotniki*) in Scientific Research Institutes, VUZy, and other unidentified Institutes.   Sources: Years 1950 and 1960: *Vyssee obrasovanie v SSSR, op. cit., 1961,* P 204.

Year 1965: Joseph Kozlowski, *The Scientific and Engineering Manpower Research Resources of the USSR*:   1961-1966, Part II of the present publication.

So far as research activities are concerned, a number of difficulties face the scientific personnel working in VUZy.   First, the ratio of auxiliary and technical workers to scientific workers is low.   One example cited by the Minister of Higher and Secondary Education in the Uzbek Republic is of only four technical workers for every 10 teaching persons in the Uzbek VUZy, compared with 1.5 technical workers for each scientist in USSR Academy establishments.   Second, the teaching load is often very heavy leaving little time for independent research.   According to estimates made by Alexander Korol, in 1960, 50 per cent of VUZy scientists were engaged in research and devoted 30 per cent of their time to this research.   This figure is equivalent to 22,000 scientists engaged full time in research, and they were distributed according to special field as shown in Table 9.

*Table 9**

	Number of scientific workers	
Physico-mathematical	3,000	12 (per cent)
Chemical	1,000	4 "
Agricultural and veterinary	1,000	4 "
Biological	1,000	3 "
Geologo-geographical	1,000	3 "
Medical and pharmaceutical	2,000	11 "
Technical (engineering)	5,000	24 "
Social Sciences	8,000	39 "

*Source: Zaleski et al: *Science Policy in the USSR*, P 330, (OECD Paris 1969)

Various attempts have been made to encourage teaching staff to carry out research. Rectors of VUZy can authorise faculty members to supplement their regular work with research in enterprises, branch research institutes, and so on.   In 1963 the government created posts for scientists in VUZy involved in research in order to reduce teaching pressures while they were preparing for their dissertations.   Only candidates of the sciences engaged in serious research work and able to write a doctoral dissertation were eligible.   The maximum time span during which they could be relieved of teaching obligations was two years, and the total number participating could not exceed 1,000 persons.

(G) Training

The higher educational establishments train scientists and research personnel through:

- general courses or seminars for improving the professional quality of faculty personnel;
- research programmes for the higher degrees of Doctor and Candidates of Sciences;
- participation of students in the research projects carried out in the higher educational establishments.

People preparing doctoral dissertations are attached to the departments of a VUZy. The nature and quality of research on which doctoral dissertations are to be based, are therefore influenced by the work of the department, and this is not always concerned with very suitable problems or dissertations in the basic sciences. Apart from teaching requirements, the department may have prior commitments to practically oriented, contract research designed to satisfy an immediate need and which is only of limited interest for fundamental research. Attempts have been made to start long-term research within departments by encouraging them to undertake complex research projects in co-operation with other research groups.

Students participation in research is seen primarily as a way of influencing the general formation of future research workers and of enabling faculty staff to directly influence the nature of their work. There is usually a close association of students in faculty research work. The numbers of students' design, project-design, technical bureaux and research laboratories are on the increase - 150 in 1960-61 and 250 by 1967. Their research is closely associated with production, and administratively they are either a subsection of a student scientific society of a VUZ or are productive units attached to a concrete scientific project. They carry out research work at the request of a department laboratory or industrial enterprise. The government also sponsors a number of annual competitions to encourage student research.

(VI) SCIENTIFIC SOCIETIES[*]

(A) Historical

Societies of scientists or scholars working in one field or in several allied fields play a prominent part in the formulation and discussion of many major problems of science.

The first association of scientists in Russia was the Free Economic Society founded in 1765. The development of higher education, research facilities and of industry in the 19th century gave rise to a number of new scientific and technical societies. For example, the Russian Technical Society was founded in St. Petersburg in 1866, followed by the Russian Chemical Society two years later. Today this still exists as the Mendeleev Chemical Society.

In the initial years after the October Revolution these early societies continued to function as before. It was only in 1929 and 1930 that any real attempt was made to reorganise them; particularly, so as to bring their activities into closer relationship with the country's practical life. The *USSR Association of Scientists and Engineers for Assistance in the Building of Socialism*, founded in 1927-28 on the initiative of leading scientists, played a major role in this process. It helped to dovetail scientific and practical activity, to introduce discoveries and inventions in production and to popularise scientific knowledge among working people.

---

[*]The main source for this section is: Zvorykin, A.A. "*The Organisation of Scientific Work in the USSR,*" *Impact of Science on Society*, PP 102-4 (UNESCO, vol.XV, no.2 1965)

(B) Organisation and Policy

At present the establishment of societies is generally covered by the legislation of the respective Union republic. Members of the public, of State or public organisations, or of economic units are entitled to be founders or members. The rules of scientific societies define the aims and principles of their activities, and are adopted at their general meetings or congresses. Large societies have several branches in different cities and local sections in institutes or enterprises. The congress, conference or general meeting of the society is its Supreme policy making body. Between sessions, the elected board, council or presidium headed by the president or chairman, carry out the administration. Societies are supported by membership dues or state subsidies, or both.

There are a large number of scientific societies in the USSR. They include the Geographical Society, the Mineralogical Society, the Moscow Society of Naturalists, the Astronomical-Geodesical Society, the Botanical Society, the Entomological Society, various medical societies, and 21 scientific and technical societies which correspond to the various branches of industry.

These 21 industrial branch scientific and technical societies had more than 2,500,000 members between them, and some 57,000 local sections operating in industrial and construction enterprises and collective farms throughout the country in 1966. Their activities are co-ordinated by the *USSR Council of Scientific and Technical Societies*, which is elected at the national congresses of the 21 societies. The function of the industrial branch scientific and technical societies is to explore new potentialities for the acceleration of scientific and technological progress, and to aid programmes for research and the application of science and engineering to production. The societies exercise a degree of public control in their respective fields of interest and also arrange for scientific consultations. In October 1962, the Council of Ministers adopted a decision "on the improved utilisation in the national economy of recommendations and proposals made by scientific and technical societies". The societies were called upon to discuss plans for research and help to apply science and engineering in the national economy. Due care was to be taken to ensure that adequate equipment was provided for scientific and technical societies, as well as public laboratories, design offices and teams.

Between 1960 and 1963 the scientific and technical societies worked out more than 450,000 recommendations and proposals, and held more than 25,000 contests attracting a total of about 310,000 entries. They guide the activity of some 48,000 public science and engineering teams, 18,000 public design and technological offices, 19,000 economic analysis offices and teams, and 1,390 public research institutes, teams, laboratories and other groups. Between 1963 and 1964 these societies carried out some 400,000 investigations. In addition some 15,000 of the societies' local sections act as technological production councils in their respective enterprises.

The scientific and technical societies take an active part in discussing and drawing up current and future State plans for new techniques and promoting their introduction into industry

4 Education

(A) Historical Introduction

Compared with much of the Western world pre-revolutionary Russia was generally an educationally backward country. The 1897 census showed that only 28.4 per cent of the population over the age of nine years was literate. And in the outlying republics of the empire the situation was much worse. Among the Uzbeks, Tadzhiks, Kirgiz and Turkmens of Central Asia not more than 2 - 3 per cent of the people were literate, and many had no written language of their own. The 150 million strong population of the empire had only 150,000 teachers between them, and about 8,000,000 children went to school. This did not mean, of course, that the actual education received by these 8,000,000 was of poor quality. The nineteenth century had in fact seen a tremendous development in educational standards and the universities particularly had become internationally recognised centres of learning. Practically all Russian scientists of this period worked within the university framework, there being little tradition of the West European gentleman scientist. However, even then the facilities were highly inadequate and generally available only to the more privileged sections of society. And again most of the outlying republics of the empire had no universities or colleges for higher education, nearly all the higher education facilities being concentrated in Moscow and St. Petersburg (now Leningrad).

After the revolution, the government policy was to provide education for the whole population. The first step was a big literacy campaign and then in 1930 universal and compulsory four-year schooling was introduced for all children aged eight years and over, and seven-year compulsory schooling for those living in urban districts. In non-Russian areas teaching was provided in the native language; education was separated from the church.

In the same period steps were taken to develop higher education. The courses were free and the state provided students with stipends. Special "workers' faculties" were set up to give workers crash courses to prepare them for college and university entrance.

The war naturally wrought a lot of havoc on the economy including the schools, but recovery was fairly quick. The school-entering age was lowered to seven in 1944 and then five years later universal seven-year compulsory schooling was introduced. Ten years later it was increased to eight years and during the past five-year plan this has again been increased to universal compulsory ten-year schooling.

The post war period has also seen a further development of the higher education facilities. Key professions such as engineering, construction, transport, communications, forestry, agriculture and teacher training have been especially developed. And the geographical location of higher educational institutes has widened out over the whole country, so that both Siberia and all the non-Russian republics have their own facilities.

(B) Administration

The education system is administered by the State. The Supreme Soviet lays down the basic educational principles and the legislation it passes is binding on all the Union republics. However, special features of the population may be taken into account from area to area.

The Council of Ministers of the USSR is the supreme executive and administrative organ of State authority in education. It issues regulations on educational matters within the terms of the laws passed by the Supreme Soviet. The republican Councils of Ministers carry out the same function on the republican level.

There are two Ministries which carry out the day-to-day administration of education. The *Ministry of Education* is mainly concerned with general education, while the *Ministry of Higher and Specialised Secondary Education* is concerned with higher education.

(i) The Ministry of Education and its Republican counterparts are responsible for pre-school institutions such as kindergartens and playgrounds, for general secondary schools, teacher training institutes, and extra-curricular institutions for children. They work out curricula, approve textbooks, and are responsible for the general direction of educational and teaching work. The Republican ministries guide and control local educational organisations, such as the local Soviets, their executive committees and local public education bodies, who control education within their area.

The day-to-day administration and educational direction of the schools, kindergartens and other institutions is carried out by their heads or directors. Their rights and duties are defined by standard regulations approved by the central committee of the teachers' trade union, and by the appropriate Republican Ministry of Education.

(ii) Voluntary Organisations. After the school reform carried out by Khrushchev in 1958 a process started for increased participation by the public in education. A number of voluntary organisations grew up. One of these is the "public education council". These councils are attached to a government education department and occur at all levels. Five commissions dealing with general education, polytechnical and industrial training, out-of-class and out-of-school instruction, schools for working youth, and pre-school education carry out the practical work of the council. Members of the councils usually represent a fairly wide cross-section of the public including professional and public organisations, industrial enterprises, parents' associations, and the Communist Party. The main aim is to keep an eye on the quality of instruction in general education in their area and to give advice to teachers.

The "council of principals" is another public organisation, created to direct, supervise and co-ordinate a group of schools within one region. They have numerous duties including advising teachers, studying teaching experience, promoting teacher training, arranging summer holidays for the pupils and looking after their health. "Councils of principals" are able to take organisational decisions, unlike the "public education councils" which are essentially advisory bodies.

Other public organisations include "pedagogical reading sessions", "parents'" universities, and "subject groups" (to discuss teaching methods).

(iii) Ministry of Higher and Specialised Secondary Education. The Ministry of Higher and Specialised Secondary Education* is responsible, as its name indicates, for the overall administration of the higher education network and for specialised secondary schools.

The ministry, together with the Ministry of Finance, is responsible for opening,

---

* See also Section 3(7)

reorganising or closing specialised secondary schools. It approves their curricula and teaching methods, including lessons, lectures, practical studies, and laboratory work..

Within the field of higher education the USSR Ministry of Higher and Specialised Secondary Education co-ordinated the work of all higher educational colleges all over the country, seeing that scientific and educational standards are maintained. It is also directly responsible for some 30 major higher educational establishments (VUZy). The main functions of the ministry are: to draw up and approve syllabuses, to plan the publication of textbooks and other teaching and student aids; to approve textbooks for publication; to plan and co-ordinate research; to co-operate with Gosplan in drawing up plans for the development of university education (this covers enrolment, graduates, training of scientists and so on; to organise exchanges with foreign countries.

Each Union republic has a ministry or committee in charge of higher educational institutes located on its territory. This supervises all aspects of college administration such as finance, enrolment, teaching and research, allocation of jobs, and postgraduate courses.

A large number of higher educational institutes come under the aegis of the specific ministry with which they are concerned. For example, transport colleges come under the Ministry of Transport, agricultural colleges under the Ministry of Agriculture.

(iv) Structure and Administration of VUZy. Each higher educational institute is managed by a *rector* who has three assistants in charge of the studies, scientific work and maintenance. The rector presides over the *academic council* which is composed of study and research directors, the deans, department heads, some of the teachers, representatives of the social organisations within the college and also representatives from enterprises and other organisations working within the same field as the institute.

The academic council considers current developments both in relation to research content and teaching methods, examines the experience of the different departments, and decides promotions.

The rector approves the plan of research and sees that the results of the institute's work are made available to industry.

A faculty may offer courses in one or several related professions. It is headed by a dean, who is elected by the professors responsible for the major subjects taught in that department. The dean supervises the teaching and research carried out in the various departments of the faculty. He organises the teaching and education process, sees to the timetable, is responsible for discipline and arranges the students' practical work.

The department is the primary unit of the institute. It carries out work in teaching, method development and scientific research in one of several related subjects. The department head must be a professor. He supervises the work of the laboratories and study rooms, delivers lectures in the main courses, advises the other professors, assistant professors and teachers on their work, supervises student and postgraduate work and sees that members of the department improve their qualifications. The staff of the department include professors, assistant professors, teachers, junior teachers and laboratory and study room attendants.

(C) Educational System

Although the Soviet Union has established a uniform system of education on the principal of equal opportunities for all, there are numerous variations within that system. On the first step of the educational ladder are the pre-school institutions for children under seven. These are not compulsory and as yet facilities tend not to satisfy the demand. The second step is the compulsory schooling period and it is here that there are a number of variations in the types of schools available. The establishments providing a general education, eight-year schools, complete secondary school, incomplete secondary schools, boarding schools, secondary schools for children with physiological defects, language schools, music schools, schools with a bias towards mathematics, biology, chemistry and so on. Then there are also schools and colleges which provide a vocational-technical education and secondary specialised schools. The third step within the system provides higher education, and includes the universities, institutes of higher education and colleges.

(i) Pre-school system. The network of pre-school institutions has grown enormously since the revolution. In 1920 there were facilities for 250,000 children in kindergartens and creches. By 1959 the figure had risen to nearly 3,000,000. Ten years later it was 9,000,000 and by the end of the 1965-1970 plan period it was hoped to have increased this to 11,500,000.

Kindergartens accept children from the ages of three to seven years. Most operate for nine to twelve hours daily. There is no charge except for a contribution towards the cost of the child's food.

The headmistress of a kindergarten must be a trained teacher with at least five-years' experience and special medical training. The teachers themselves must have completed their secondary education and done a teacher training course.

(ii) Secondary school system

(a) The *Eight-year School* provides universal compulsory mixed education for pupils ages 7 to 15-16 years. It is the official title for incomplete secondary general schooling and it gives the pupils the fundamentals of general and polytechnical knowledge. Eight-year schooling was first introduced in 1958, and in 1963 it became compulsory replacing the previous scheme of seven-year schooling.

Eight-year schooling consists of a primary and secondary stage. During the last five-year plan the curriculum was reorganised so that the primary stage was reduced from four forms to three, while the secondary stage was expanded to cover forms four to eight. The aim behind this was to adapt the education to the demands of a highly industrialised society. The natural sciences are now introduced in the second and third form and the amount of mathematics taught in primary forms has also been increased, with a certain amount of algebra and geometry being first presented in the fourth and fifth years.

(b) The *Ten-year School* provides complete secondary general schooling. With the introduction of compulsory ten-year schooling during the last five-year plan, these schools will gradually become general, replacing the eight-year schools.

Most schools within the secondary system are day schools, but in recent years the number of boarding schools has increased. Before 1956 these existed mainly for children needing special care. But they are now becoming more generally popular. The number of children going to boarding schools has increased from 180,000 in 1958 to more than 4,000,000 in 1968. The plan is to build sufficient schools to meet demand. Parents contribute towards their children's maintenance according to their means. In most boarding schools the children return home at the weekends. They are particularly useful in the rural areas.

There are also "prolonged day" schools where the children remain from 8.0a.m. to 6.0p.m. doing their homework at school.

Other types of schools within the secondary system are those which provide special teaching on a particular subject such as languages, dancing, music, science or mathematics. These schools provide the normal eight or ten-year schooling but at the same time they are given extended teaching in their particular speciality. In language schools much of the teaching may go in the language in which the school specialises. Dancing and music schools are usually open to pupils from the age of seven who show a particular talent for the subject. The mathematics and science schools have been a more recent innovation and they normally cater for pupils in the higher grades (from fourth or fifth up).

All in all, there were some 204,000 secondary schools in the 1968/69 school year, with 49,000,000 pupils attending them. Some idea of the growth of school education during the past 50 years in the country as a whole and in the individual Republics can be gained from Table 10.

Table 10*
Number of pupils 1914-15, 1940-41 and
1966-67 (thousands)

	1914-15	1940-41	1966-67
USSR Total	9,656	35,552	48,168
Russian Federation	5,684	20,633	26,186
Ukrainian SSR	2,607	6,830	8,468
Byelorussian SSR	489	1,737	1,769
Uzbek SSR	18	1,325	2,592
Kazakh SSR	105	1,158	2,865
Georgian SSR	157	767	928
Azerbaijan SSR	73	695	1,199
Lithuanian SSR	118	380	562
Moldavian SSR	92	440	763
Latvian SSR	172	242	343
Kirgiz SSR	7	334	657
Tadzhik SSR	0.4	315	613
Armenian SSR	35	333	553
Turkmenian SSR	7	252	455
Estonian SSR	92	121	215

* Source: Mikhail Prokofyev: *Education in the USSR*, P 53 (Novosti Press Agency)

(c) Vocational and Technical Secondary Schools provide pupils with training in a specific profession or trade. There are two types: the vocational-technical school where the curricula is based on that of the ordinary secondary school together with a particular training: and the technical schools which take on pupils who have completed their secondary training and give them a training of one to three years depending on the trade or profession. Apart from classroom studies the students have to carry out practical work in the school workshop, local factories or farms. They are paid for this.

In addition to the full-time day vocational-technical schools, there are evening or shift schools and evening departments at day schools for training people working full-time during the day.

In 1968 there were 4,800 vocational-secondary schools throughout the country, in which 2,000,000 people were enrolled on a full-time basis. The number benefiting from the part-time facilities was in the region of 10,000,000.

The network of vocational and technical schools developed at the beginning of the 1920s, from the need to provide skilled workers for the growing industry. As the economy of the country reaches higher levels, so the requirements for technical workers increases. Thus continued and possibly increased demands are placed on the vocation and technical schools. There are plans to widen the network of schools and to attach a vocational-technical school to every medium or large enterprise, and to attach several schools to some of the larger enterprises.

(d) Secondary-specialised schools provide a specialist training for students who have completed the equivalent of eight-year schooling. They are open to people between the ages of 14 and 30 who have passed the entrance examinations. The schools provide full-time day tuition, evening tuition and correspondence courses. The specialisations and trades that they cover are very varied. The vocational secondary schools, or technicums, specialise in subjects such as industry, economics, transport, communications, construction, agriculture and so on. These schools train medium grade qualified specialists. It is not clear from the literature whether they are independent of, or overlap with, the vocational-technical secondary school described in the previous section. It is possible that they offer slightly higher qualifications. Secondary-specialised schools, equal in status to the technicums, specialise in such subjects as medicine, art, music, mathematics, physics and chemistry. All secondary-specialised schools have their own laboratories, workshops and offices. They are attached to, or maintain close contact in some way with, enterprises or organisations whose work corresponds to the speciality they are teaching. The students follow a four-year course at the end of which they sit for state exams, or defend a thesis for a diploma.

(D) Teachers and Teacher Training

In 1968/69 there were 2,563,000 teachers working within the general secondary school system, and 160,000 teachers working in the specialised-secondary schools.

There are 208 teacher training colleges with an enrolment of nearly 1,000,000 students. Another 481,500 teachers and instructors qualify at evening classes or through correspondence courses. About 300,000 students attend specialised pedagogical secondary schools. And finally, many university graduates enter the teaching profession.

The course at a teacher training college is three years for a student who has completed eight-years' schooling and two years for a student who has completed the full ten-year secondary education. The courses cover educational theory, psychology and teaching methods for a particular subject. Most teachers train to teach at least two subjects as in the smaller rural schools they may often have to cover more than one subject.

Of teachers of primary classes (the first three forms), 92.5 per cent have completed either a specialised pedagogical school or the primary school faculty in a training college. Of teachers of secondary forms (fourth to eighth years), 82 per cent have complete or incomplete higher education. And the majority of ninth and tenth form teachers are college or university graduates.

There are extensive facilities for taking refresher courses. Each administrative region has a special teachers' refresher college and every Union republic has its own central refresher institute. Once every five years teachers have to take a refresher course in their own field. Teachers also account for nearly 75 per cent of the student attendance at the correspondence and evening departments of pedagogical colleges.

A large amount of literature is published on teaching methods and pedagogical research generally. A *Teachers Newspaper* is published in Moscow and numerous other specialised magazines are put out at republican or all-union level. Pedagogical societies set up in each of the union republics help to exchange information and experiences.

The work schedule of a primary grade teacher is fixed at 24 hours a week, and that of a secondary grade teacher at 18 hours a week.   There are supplementary payments for any teaching done over these basic amounts.

(E) Curricula

A new standard curriculum is being introduced into the secondary schools at present. 1971 is the planned date for its complete introduction. The aim of the new curriculum is to meet more satisfactorily the requirements of the highly technological and industrialised society. The biggest change is the reduction of the primary course from four to three years thus increasing the pace of education. There are additions in the mathematics and Russian language programmes; and the fundamentals of the natural sciences are now included in the second and third form syllabuses. The second or middle stage has been expanded to include the fourth to eighth forms. This makes it possible to add new material to the various subjects taught. For example, priority is given to mathematical subjects, with the elements of algebra and geometry being introduced in the fourth and fifth form. New data have been added to various science subjects to keep up-to-date with changing knowledge. However, the natural sciences-humanities ratio is remaining much as it was previously with the humanities taking up some 40 per cent of tuition time, the natural sciences and mathematics some 37 per cent. An important aim of the new curriculum is to encourage initiative and the ability of pupils to think for themselves, rather than to acquire just a vast amount of factual knowledge.

Table 11 shows the new standard school curriculum, form by form. Appendix A gives a more detailed description of the curricula of science subjects.

*Table 11**
Standard school curriculum*

Subject	Number of hours per form										Total hours per week	
	1	2	3	4	5	6	7	8	9	10	1967-68	1968
1. Russian Language	12	10	10	6	6	3	3	2	2/0	-	53	57
2. Literature	-	-	-	2	2	2	2	3	4	3	18	19
3. Mathematics	6	6	6	6	6	6	6	6	5	5	58	59
4. History	-	-	-	2	2	2	2	3	4	3	18	20
5. Social science	-	-	-	-	-	-	-	-	2	2	2	2
6. Natural science	-	2	2	2	-	-	-	-	-	-	6	2
7. Geography	-	-	-	-	2	3	2	2	2	-	11	12
8. Biology	-	-	-	-	2	2	2	2	0/2	2	11	11
9. Physics	-	-	-	-	-	2	2	3	4	5	16	17
10. Astronomy	-	-	-	-	-	-	-	-	-	1	1	1
11. Draftsmanship	-	-	-	-	-	1	1	1	-	-	3	4
12. Foreign language	-	-	-	-	4	3	3	2	2	2	16	20
13. Chemistry	-	-	-	-	-	-	2	2	3	3	10	11
14. Pictorial art	1	1	1	1	1	1	-	-	-	-	6	7
15. Singing and music	1	1	1	1	1	1	1	-	-	-	7	8
16. Physical culture	2	2	2	2	2	2	2	2	2	2	20	22
17. Manual work	2	2	2	2	2	2	2	2	2	2	20	58
Total of compulsory lessons	24	24	24	24	30	30	30	30	30	30	276	330
optional lessons	-	-	-	-	-	-	2	4	6	6	-	-
Grand Total	24	24	24	24	30	30	32	34	36	36	-	-

*Source: Mikhail Prokofyev: *Education in the USSR*, P 55 (Novosti Press Agency)

(F) Higher Education

There are two basic kinds of higher educational establishments (VUZy): universities and institutes of higher education. The latter are sometimes further subdivided into specialised higher educational branch institutions, and polytechnical institutes.

The majority of the higher educational establishments are State bodies, maintained by the State. However, there are a small number that belong to co-operative or private organisations; examples of these would be the Trade Union School, the Co-operative Institute and the Patrice Lumumba Friendship University. The VUZy provide three types of teaching: full-time, part-time evening, and correspondence. Usually all three types of training are found in each VUZy.

(i) The *universities** occupy a special place in higher education. Historically they are the oldest higher educational organisations and so they have long-established teaching and research traditions. Unlike the other higher educational establishments their teaching covers all branches of the arts, natural and social sciences and mathematics. The special feature of the universities is that they give a wide general education. The student must not only know his own special subject but also have a good command of allied and general subjects. Most courses last $4\frac{1}{2}$ to 5 years, and the last year is usually devoted to preparing a thesis.

In 1969 there were 46 universities with 500,000 students (compared to 13 universities and 43,000 students in 1913). Since then the number of universities has increased further to 49. Every republic now has its own university, and universities have also been established in a number of autonomous republics, including the Moldavian, Daghestan, Yakut, Bashkir, and Kabardino-Balkarian. Recently established universities include the ones at Vladivostock, Novosibirsk and Donetsk.

(ii) The *institutes of higher education* provide a more specialist training. The *polytechnical* colleges usually have a number of faculties and provide training for a considerable number of different professions or specialised trades. The Leningrad Polytechnical Institute is the largest establishment of this kind. It has nine faculties: physics and metallurgy, mechanical engineering and machine construction, power machinery, engineering, electrical engineering, engineering physics, hydraulic engineering, radio engineering and so on. It gives instruction in 48 special trades to some 1,000 students. There are 890 professors and lecturers and 200 laboratories.

Similar establishments include the Urals Polytechnical Institute (13 faculties and 36 specialised trades), the Kharkov Polytechnical Institute (15 faculties, and 38 specialised trades) and the Kaunas Polytechnical Institute (7 faculties and 18 specialised trades).

(iii) The *Specialised institutes* provide training for a specific subject, either a branch of industry, teacher training, law economics, music, drama, art, agriculture, medicine and so on. Those catering for branches of industry usually have less faculties than the polytechnical colleges, though the division between the two types of institute is sometimes nominal.

The figures for institutes of higher education are as follows:

    228 higher technical colleges with almost two million students;
     98 agricultural colleges with some 400,000 students;
     90 medical colleges with some 250,000 students;
    206 teacher training colleges with some 820,000 students;
     24 economic institutes:
     48 art institutes with some 50,000 students.

These latter include 22 music academies, 12 drama institutes, several fine arts academies, plastic arts institutes, decorative and applied art institutes, industrial art

---

* See also Chapter 3, Section V.

schools and an institute of literature.

Most courses in institutes of higher education last five years; the medical institutes are an exception, with a training period of six years.

(iv) Evening and Extra-Mural Education. Part-time education began in the 1930s when there was a desperate need for specialists. The network now consists of 29 correspondence and evening colleges and more than a thousand departments with an enrolment of more than 2,382,000 students (more than half the total number of college students in the country).

A recent development in part-time education has been the organisation of factory training and state farm colleges. These colleges are both attached to a factory or farm and to a higher educational establishment. In the factory training colleges, instruction is organised on the basis of one week's full-time study followed by one week working in the factory with classes in the evening. The students train to become skilled engineers or technicians needed by the factory or some associated enterprise. The farm colleges work on the same basis. The advantages of these training methods are that the student combines study with useful work and also becomes well acquainted with all the processes connected with his future work. The Likhachev Automobile Plant, the Leningrad Metal-Working Factory, and the Rostov Farm Machinery Works are examples of factories which have colleges of this type attached to them.

(v) College Teaching Staff. There are 263,200 teachers and lecturers in higher educational establishments.

A continuous effort is being made to improve the quality of the teaching staff. "Special courses to improve the qualifications of teachers" in mathematics, physics, chemistry and theoretical mechanics were started in 1967. Since then further courses have been opened for pedagogical institute teachers, and for specialised and general engineering teaching staff. The idea is to provide short concentrated courses (of a few months duration) on the latest teaching methods and latest achievements of the particular subject. Another trend is towards shorter two or four week seminars.

(vi) Staff Grades. A number of different grades of staff exist in universities and colleges. Starting at the top, there are the *professors*, who have a Doctor of Sciences degree or who have conducted extensive teaching or research work. On the next step down the ladder are the *docent* or *senior scientific workers*, titles usually conferred on people who have their master or candidate of sciences degree, and who are conducting appropriate teaching or research work. At the bottom of the scale there are the *assistant professors, lecturers* or *junior scientific workers*. These are graduates and they work under the guidance of a professor or docent. The title of junior and senior scientific workers is usually used for college staff, while the others are reserved for university staff.

(vii) Degrees. There are two higher postgraduate degrees: the *Master or Candidate of Sciences* and the *Doctor of Sciences*.

The graduate working for a candidate of sciences degree is called an *aspirant* and his training is called *aspirantura* training. Applicants for admission have to be under 35 years of age for a full-time course or under 45 years for a part-time course. They must have completed a higher educational course in the subject they are specialising in, and usually they have had at least two years research, teaching or industrial experience in that particular field.

Full-time aspirants are given grants, and benefit from all library, laboratory and other facilities. Aspirants completing the degree on a part-time basis get 30 days extra holiday a year from their employment as well as one day a week off on half pay. And in the last year this may go up to two days a week without any salary reduction.

The full-time course normally takes three years, and the part-time one four years. During the first two years of training, the aspirants have to complete their reading requirements, carry out experimental work, fulfil two to three months worth of teaching and seminar requirements and learn to read a second language. They are regarded as junior members of their department. At the end of the two years the aspirant usually takes the *Minimum Candidate Examination* which he has to pass as part of his training. The third year is devoted to the preparation of his research and dissertation.

Aspirants who successfully defend their theses and obtain their candidate of sciences degree are then given the title of docent or senior scientific workers.

The degree of doctor of sciences is conferred on people who already hold the candidate of sciences degree or the grade of professor and who have successfully defended a doctoral dissertation. The requirements for these dissertations are quite strict. They must "represent independent research work, contain theoretical analyses and conclusions and suggest conclusions of significance for the general development of science, as well as for immediate practical application". (Zaleski et al: *Science Policy in the USSR*, P 339, OECD Paris 1969).

The Higher Attestation Commission on the proposal of the Academic Council of the Higher Educational Establishment awards the doctorate degrees. Only a small number of VUZy - 283 in 1960 - are able to accept doctoral theses and propose them to the Attestation Commission.

A person who successfully defends his dissertation and is awarded the degree of doctor of sciences is then given the title of professor.

(G) Research

The USSR Academy of Pedagogical Sciences is the main research body in the field of education. It is a specialised Academy which comes under the USSR Ministry of Education. It controls eleven research institutes, two school laboratories, ten experimental schools and 127 " support " schools scattered throughout the Russian Federated Republic. In 1965 a central commission and 15 specialised commissions were attached to the Pedagogical Academy and the USSR Academy of Sciences with the job of controlling curricula. The commissions are also responsible for new programmes and plans which have to be approved by the Ministry of Education.

The academy's function is to help in the development of public education, disseminate pedagogical ideas, study the problems of pre-school education, the theory and history of pedagogy, psychology, and teaching methods. It also helps draw up any legislative measures in the field of education.

There are a number of higher educational institutes which have been placed under the direct authority of the Ministry of Higher and Secondary Specialised Education. They will become centres carrying out the generalisation and development of teaching methods, the compiling of text-books and manuals, and the training and re-training of scientific teaching personnel for all higher educational institutions.

Journals: There are over 30 journals published which are concerned with educational questions. They include *Byulletin Ministerstva Vyshego i Srednego Spetsial'nogo Obrazovaniya SSSR* (Bulletin of the Ministry of Higher and Secondary Specialised Education of the USSR); *Vestnik Vyshei Shkoly* (Herald of the Higher School); *Voprosy Psykhologii* (Problems of Psychology); *Doshkol'noe Vospitanie* (Pre-school education); *Sovetskaya Pedagogika* (Soviet Pedagogy); *Srednee Spetsial'noe Obrazovanie* (Secondary Specialised Education); and *Shkola i Proizvodstvo* (School and Industry).

(H) Finance

State expenditure on education increases each year. The 1966 budget for example, allocated 18,700,000,000 roubles or 17.8 per cent of the whole budget to education, science and culture. Of this, general secondary education received 5,000,000,000 roubles, an increase of 6.4 per cent over 1965. The vocational-technical, secondary specialised, and higher educational establishments received 3,300,000,000 roubles, an increase of 8.5 per cent over the 1965 allocation. Spending on science, education and culture was proportionately more than on military and defence purposes in 1966, the latter only getting 12.8 per cent of the budget.

The state bears the major responsibility for financing the education programme. Contributions from other sources are relatively minor. In 1964 they amounted to 7.2 per cent of all expenditure. Such contributions include payments from parents towards the maintenance of their children in kindergarten, creches, pioneer camps, boarding schools, or similar establishments; payments from professional unions and from factories and other industrial or state enterprises towards the running of their own creches and kindergartens; and thirdly, another source of outside finance is social security.

In higher education about 45 per cent of expenditure goes on the wages of the professors, lecturers and other personnel, and 30 to 35 per cent on student scholarships. Other categories of spending include around 5 per cent on teaching, practical training, scientific research work and acquisition of books, some 3 per cent on equipment and furniture, 7 per cent on investment in construction and 2.5 per cent on repairs to buildings and installations. (Source: Zaleski et al: *Science Policy in USSR*, P 364, OECD Paris 1969).

The Ministry of Education has a strong say in the allocation of the state budget; the VUZy themselves cannot decide how to apportion the money they receive from the state, but have to use it according to the Ministry allocations. However, with the recent encouragement of VUZy to improve their research standards and undertake more contract work, since 1966 they have been given the power to allocate 75 per cent of any income from such contract scientific research work, to their own expansion and improvement. These allocations are in addition to those made from the state budget.

Secondary compulsory education is free for all children. Parents of children going to boarding schools or to pre-school institutions or to holiday and pioneer camps usually contribute towards their maintenance according to their means. Students at vocational schools are supported by the State which provides free board and uniforms and free hotel accommodation if the student comes from a distance. Students at secondary specialised schools and higher educational establishments get a small grant. This is not usually enough to cover all their living expenses and has to be supplemented by their parents or by working during the vacations. Supplementary grants are given to students with very good marks, and there are also a number of larger scholarships founded in honour of various eminent people. Some students may get a grant from the factory or farm where they previously worked.

(I) Policy and Plans

The basic principles of Soviet education as set out by Mikhail Prokofjev, the Minister of Education are:

- *(a)* State supported education system – this ensures a uniformity of organisation and curricula so that training levels can be maintained throughout the country.

- *(b)* Equal opportunities for all peoples – this entitles the non-Russian nationalities within the USSR to receive their education in their own language. In some cases this has meant writing the language down for the first time.

- *(c)* Equality of sexes – all schools, pre-school establishments and higher school institutes are co-educational. In addition, teachers receive the same salary whatever the sex.

- *(d)* Unified school system – means that there are no dead-end schools which rule out further education. Each educational stage is linked to the next.

(e) School and other educational and training institutions are completely separated from the church - this, of course, contrasts greatly with the situation before the revolution when many schools were run by the priests.

(f) Broad contacts between school and public - every school has a parents committee elected at the beginning of the academic year. The trade unions, Young Communist League and other public organisations take an active part in the work of the schools and other educational institutions.

(J) Training of Scientists

The compulsory secondary schooling provides a general education up to school leaving age around 17 to 18 years. There is no specialisation before this point, every pupil following a general course including both the arts and the sciences. The recent curricula reforms have altered the syllabus of the secondary school to introduce some mathematics and elements of the natural sciences earlier than before. It has also brought the content of the science teaching up to date. However, the pupil following this normal general schooling does not begin to specialise in science, if that is his choice, until the higher school level. However, recently established special science schools are an exception to this rule. They have been organised during the past ten years or so and provide specialised teaching in subjects such as mathematics, physics, chemistry and biology. The aim is to encourage talented scientists in the same way as dancing or music schools cater for budding dancers and musicians.

The special mathematics schools were the first to be formed, and the first one of its kind was organised in Akademgorodok, the science city near Novosibirsk. Here pupils gain entry by competing in the maths olympiad. Each year a test paper compiled by members of the Institute of Mathematics is published in local newspapers throughout the country. It is aimed at boys and girls of 14 to 15 years of age, though the standard of the problems posed is pitched higher than average for that age. The Institute usually receives about 10,000 replies to this initial test paper. It whittles these down to 5,000 candidates who are invited to interviews and further written tests at a number of centres throughout the country. At this point the field is drastically cut to some 600 who are then invited to come to Akademgorodok for a month. All their expenses are paid. They attend a series of classes and tutorials and at the end of the month they are tested solely on what they have learnt during that time. This is meant to reduce to a minimum the influence of their different backgrounds, standards of teaching and so on. Two hundred boys and girls are selected from the 600 and invited to join the special mathematical school in Akademgorodok. Here they finish their final two years of school. They continue with all the general subjects but at the same time follow a rigorous and extremely advanced (for their age) mathematics course. By the time they have finished school their standard in mathematics is equivalent to that of a second year university student. The members of the Mathematics Institutes and other scientists at Akademgorodok believe that this advanced grounding in mathematics is absolutely necessary if they are going to get a steady flow of highly qualified and talented recruits to both mathematics and mathematics-based sciences. They are now expanding the scheme to other science subjects.

Once at the higher school level there are two basic methods for training highly qualified personnel - the universities and the technical colleges. The universities give their students a broad, general scientific training, but do not prepare them fully for work in specialised scientific research institutes or design offices. The technical colleges, in contrast, do not provide a broad enough training, while training the specialist. Both approaches have their advantages and disadvantages. The Moscow Institute of Physics and Technology (MFTI) was founded in 1946 as a new type of higher educational institution which would train specialist physicists combining the breadth of the university with the practical approach of a technical college education. At MFTI there are two stages of training. During the first stage the students follow a general scientific course as in a university. The second stage, which begins in the second or third year, then consists of training in a particular field together with independent research carried out at a research institute. The students visit design offices and scientific research institutes, where they attend lectures on specialised subjects. They take part in seminars, and in the third year they are included in the scientific research programme. Each one is given a specific problem to solve. This system at MFTI makes it possible to train quite quickly highly qualified specialists for the most up-to-date and rapidly developing fields of physics and technology. There is some scepticism about the possibility of deciding on the scientific potential of the student at such an early stage. However, the graduates from MFTI appear generally to have been very successful, with 97 per cent working in scientific research institutes and major design offices. (Data based on A. Dorodnitsyn et al: 'The Training of Physicists', *National Lending Library Translations Bulletin*, P 839, 1967).

Young people living in the country and wishing to specialise in subjects related to agriculture or forestry are to be given certain advantages in admission to VUZy, according to a 1969 Novosti report. The idea is to encourage the flow of specialists to the country areas, and to stem the flow of young people away from them to the towns. Agriculture is experiencing certain difficulties with regard to skilled specialists. (Source: *Soviet News*, 22.7.69).

Appendix

THE REVISED CURRICULA OF THE 10-YEAR SCHOOL*

"The first, primary stage, has been reduced from four to three forms, raising the pace of school education. Additions are made to the mathematics and Russian language programmes. Fundamentals of natural science are now included in the 2nd and 3rd form syllabuses. This has been done to raise the general level of primary education.

The second, middle stage, has been expanded to include the 4th to 8th forms. By this means the volume of information can be increased without changing the tempo of schooling, certain topics can be shifted from the upper forms' programmes and new facts and ideas added in a form comprehensive to the middle group of pupils. To this end much new material, theoretical theses and scientific methods have been included which enrich the students' knowledge, develop their intellect and outlook, help them apply their knowledge in practice both at school and at work upon leaving it.

Priority is given to mathematical subjects in view of their great educational value. In the new programme for the 4th and 5th forms mathematics is represented by arithmetic with elements of algebra and geometry. Included in the section "arithmetic" are the concepts of plurality elements. These are explained with the help of various concrete examples and appropriate exercises.

Systematic algebra and geometry courses begin in the 6th form. Thus the three stages of secondary education are forms 4 and 5, 6-8 and 9,10. The shifting of the foundations of algebra and geometry to the 4th and 5th forms and their correlation with arithmetic greatly enhance the level of knowledge acquired. It grows further in the middle forms and rises sharply in the 9th and 10th forms where the pupils begin to study finite processes characteristic of higher mathematics. One of the topics here - "Algebra and Elements of Analysis" - includes the derivative, the integral and the fundamentals of the theory of probability as well as information on electronic computers.

The physics course has likewise been augmented with new scientific data. The key theory of modern physics is the theory of the structure of matter. Physics considers matter in two forms - substance and field - to study which it is essential to be familiar with the modern theory of atomic structure, elements of statistics, quantum mechanics and the basic points of the theory of relativity.

The pivotal questions of the physics course are :

---

*Source: Prokofjev, M.: *Public Education*, P 36 (Novosti Publishing House, Moscow)

6th form - molecular structure of substances; movement and forces; pressure of liquids and gases; work and power; the concept of energy.

7th form - thermal phenomena; heat transfer and work; changing aggregate states of matter; thermal engines; electricity; structure of the atom; intensity of current, tension, resistance; work and power of electric current; electromagnetic phenomena.

8th form - mechanics; foundations of kinematics; variable motion; Newton's laws of motion, their application; forces in nature; addition of force; work and energy.

9th form - molecular physics; fundamentals of the kinetic theory of gases; internal energy of ideal gas; properties of vapours; properties of solids and liquids; fundamentals of electrodynamics; electric fields; direct current; electromagnetic field; electromagnetic induction; production, transmission and utilisation of electricity.

10th form - oscillations and waves ; mechanical oscillations and waves; alternating current; electromagnetic oscillations and waves; optics; light waves, geometrical optics; radiation and spectra; action of light; light quanta; fundamentals of the theory of relativity; physics of the atomic nucleus; atomic nucleus; elementary particles; nuclear energy.

The chemistry course in the secondary school unites the fundamentals of inorganic and organic chemistry. An emphasis on theory raises its level. The theoretical foundations of chemistry are explained more deeply at an earlier stage. The molecular-atomic theory is introduced at the beginning of the course and serves as the basis for understanding chemical phenomena, reactions and calculations. One year earlier than they used to before pupils begin to learn the periodic law and periodic system of elements as well as the electrolytic dissociation theory. Deeper study is carried out on the structure of matter. All this gives the chemistry course in the 7th and 8th forms the character of general chemistry, while in the senior forms pupils have a chance to study the electron theory.

A deeper study is made of heteropolar and homopolar bonds, donor-acceptor and hydrogen bonds. Material on the energy of chemical processes is supplemented by the concept of the energy effect of reactions and its quantitative expression. This enables the student to characterise the durability of compounds (with substances consisting of two elements as examples) and show that chemical processes are governed by the conservation of energy laws.

7th form - elementary chemical concepts of oxygen, oxides, combustion, hydrogen, acids, salts, water, solutions.

8th form - basic classes of inorganic compounds; calculation of chemical formulae and equations; Mendeleev's periodic law and periodic system of elements; structure of matter; halogens; oxygen sub-group.

9th form - electrolytic dissociation theory;
nitrogen and phosphorus; carbon and silicon;
metals.

10th form - introduction to organic chemistry;
theory of mechanical structure of organic
compounds; basic classes of organic compounds.

In the school biology course the general trend is determined by the latest advances in cytology, genetics and selection. In structure this course differs from the physics and chemistry courses since it consists of four different subjects - botany, zoology, human physiology and general biology. Botany, zoology and human physiology are augmented by new elements of cytology, ecology, biocenology, genetics and the theory of evolution. In the course of studying botany, zoology, and human physiology, pupils gradually accumulate knowledge about the cell as an elementary living system forming the basis of the structure, life, and growth of organisms, about the cell's structure and its chemical components, about its proteins which perform fermenting, transporting and motive functions and which are the basic building material of the cell's nucleus, cytoplasm and organs. The general biology course supplies detailed information on the structure and functions of the desoxyribonucleic acid which forms the material particles of chromosomes - the genes, the carriers of hereditary function. The second nucleic acid, ribonucleic, is also studied, as well as the role of both in the synthesis of protein.

Since all functions of the cell are accompanied by expenditure of energy, pupils learn about energy exchange in cells, adenosine triphosphoric acid, this universal energy substance, its chemical composition and structure as well as synthesis and decomposition in the process of energy exchange.

Rich in material, the biology course gives pupils systematised knowledge about the basic laws of living nature.

The profound upheavals in the world demand that the youth receive an all-round ideological and political training. This is done in school when studying all subjects, primarily the humanities.

The new history programmes reflect basic teaching requirement - that the entire course be interpreted as one process, governed by laws, with comprehensive explanations of facts, cause and effect, and the events and historical characters adequately described.

Included in the new literature programme is a carefully selected and comprehensive list of the works of 19th century Russian classical writers and Soviet authors and outstanding masterpieces of world literature. These are compulsory titles for study in class and at home. In some cases the teacher and pupils in senior forms are given the right to choose one or several works out of the compulsory list. A number of books are given for compulsory reading in connection with the appropriate topic. These are not discussed in detail in class, but the students work on them independently following the teacher's instructions.

A new Russian language programme has also been drawn up. Its systematic course now begins in the 4th, not 5th form on the basis of approximately the same material as before. The programme includes material on phonetics, lexicology, word formation, grammar, oral and written practice, elements of style and some general information about the language.

The raising of the level of teaching foreign languages was done in two directions. In the first place the makers of the new programmes supply not only a list of grammatical phenomena but also structures, or typical phrases, to illustrate them. This helps form habits of speech. Secondly, the qualitative and quantitative indicators of the ability to speak, understand spoken speech, read and write have been reconsidered in accordance with the latest psychological and methodological studies".

5 The Natural Sciences

(A) History

The development of the natural sciences in the Soviet Union since the revolution can be divided into three stages: the first ten years when the basic areas of scientific research were organised and consolidated; the 1930s when there was an all-out attack and support for scientific development; and the present period when science has become an important factor in the technological and cultural life of the country.

Pre-revolutionary Russian science was centered in the few universities and institutes of higher education, and its development was extremely uneven. For example, it lagged badly in the field of physics and Russian physicists contributed little to the tremendous discoveries about the nature of the atom at the beginning of the twentieth century. On the other hand there were a number of outstanding scientists covering a large number of scientific fields, for example, Mendeleev and Butlerov in chemistry, Zhukovsky and Chaplygin in aerodynamics, Sechenov and Pavlov in physiology of the nervous system and Dokuchaev in the soil sciences.

In the 1920s immediately after the revolution, there were two immediate tasks: to provide the large number of science teachers needed to cater for the wide development of science that was planned; and to build up existing scientific institutes and develop new ones particularly in branches of science relevant for the various branches of technology that the country was trying to develop. For example, the production of electric lamps, optical instruments, radio transmitters and the development of the chemical industry all demanded scientific knowhow.

The 1930s, the period covering three five-year plans, was one of very rapid quantitative development in science. The number of people studying at higher educational institutes rose from 112,000 in 1914, to 177,000 in 1928 and 667,000 in 1941. At the same time the number of higher educational establishments has risen from 91 in 1914 to some 800 in 1941. The scientific centres expanded out from the European part of Russia, where they had mostly been centred in the 1920s, to the various republics and eastern regions of the country. Geological mapping and prospecting had covered nearly two-thirds of the Soviet Union by 1940, compared with the 10 per cent known before the revolution, and had led to the discovery of some of the country's vast resources. A lot of original scientific research was also carried out in this period. Thus by the beginning of the 1940s the natural sciences had become an integral part of the whole economy of the country. After the Second World War, apart from the need to rebuild the disrupted institutes and scientific organisations, the main task facing the scientists was the development of atomic energy - both for military and peaceful purposes. The successful exploding of the first atomic bomb in 1949 was followed by the rapid development of rocketry and cosmonautics in general.

The organisational network and manpower of science grew in this period. The 1,800 scientific establishments existing before the war had grown to 4,724 by 1965; various new research institutes were founded; the number of scientific workers increased from 100,000 before the war to 930,000 in 1970; and the average number of students was 4,600,000 in 1970, compared with 667,000 in 1941.

(B) Mathematics

There are a number of major mathematics' scientific research centres. The Steklov Mathematics Institute, the Siberian Mathematics Institute, and the Institute of Applied Mathematics, are all attached to the Soviet Academy of Sciences, as are the various computing centres. Then there are the mathematics departments of the Moscow, Leningrad and Novosibirsk universities, as well as institutes and computing centres in the Republican academies. The major mathematical schools are in Moscow, Leningrad, the Ukraine, Georgia, Armenia, Uzbekhistan, and Lithuania.

They cover a wide range of mathematical ideas. Their areas of research include the theory of numbers (I.M. Vinogradov); a school of algebra founded by D.A. Grave in the 1920s in Kiev; mathematical logic; topology; a school founded by P.S. Uryson, P.S. Alexandrov and L.S. Pontriagin; differential geometry (for example the work of A.D. Alexandrov and N.V. Yefimov and A.V. Pogorelov on the differential geometry of a formation as a "whole"); the theory of functions of a complex variable (Zhukovsky, S.A. Chaplygin, N.I. Muskhelishvili and M.A. Lavrentiev); the theory of the approximation of functions in a complex area (M.V. Keldysh and M.A. Lavrentiev); the development of functional analysis; the theory of ordinary differential equations and the theory of derivative differential equations; the probability theory; and statistics.

The 1950s and 1960s saw the development of computers and with it the accompanying techniques. This was accompanied by an increasing influence of mathematics both on other sciences (chemistry, geology, biology and linguistics, for example, apart from the more traditional physics and mechanics) and the economy.

Journals: The main journals are: *Matematicheskii Sbornik* (Mathematical Symposium); *Izvestiya Akademii Nauk SSSR. Seriia Matematika* (Proceedings of the USSR Academy of Sciences. Mathematics Series); *Uspekhi matematicheskikh nauk* (Progress in Mathematics); *Trudy Matematicheskogo Instituta im. V.A. Steklov* (Transactions of the Steklov Mathematics Institute); *Prikladnaya matematika i mexhanika* (Applied Mathematics and Mechanics); *Ukrainskii matematicheskii zhurnal* (Ukrainian Mathematics Journal; *Sibirskii matematicheskii zhurnal* (The Siberian Mathematics Journal).

(C) Mechanics

Pre-revolutionary Russia had two significant mechanics schools: the Moscow school, formed around N.E. Zhukovsky and S.A. Chaplygin, which brought a mathematical approach to technical problems; and the engineering school which grew up at the beginning of the twentieth century around V.L. Kirpichev, and one of whose outstanding representatives was S.P. Timoshenko. However, both these schools had only limited possibilities because of the low level of industrial growth in Russia at the time. After the revolution, the Central Aerodynamic Institute, founded in 1919, became the nucleus for scientists specialising in mechanics. By the 1930s it had become a leading theoretical and applied science research centre. A smaller centre grew up in Leningrad at the beginning of the 1930s within the structure of the USSR Academy of Sciences and the university - the Leningrad Polytechnical Institute. An Academy Mechanics Institute was formed in Moscow in 1939, and in Novosibirsk in 1958. Since the 1930s scientific centres have grown up in other towns as well.- at first in Kazan, Kiev, Saratov and Kharkov, and then in Yerevan, Riga, Tashkent, Tbilisi and others.

The main areas of research are the following:

(i) Mechanics of liquids and gases: work on the mechanics of liquids and gases includes aerodynamic problems of supersonic and hypersonic speeds; magnetohydrodynamics; and hydraulics (the main centre here is the *All-Union Institute of Hydraulic Engineering* in Leningrad which co-ordinates all work on hydraulics throughout the country.

(ii) Mechanics of deformed solid bodies: including elasticity; plasticity; rheology; construction. Important centres of research are the Institutes of Mechanics in Kiev and Moscow, Leningrad University, the Scientific Research Institute of Mechanics attached to Moscow University, and the Voronezh State University.

(iii) General mechanics.

Journals: *Prikladnaya matematika i mekhanika* (Applied Mathematics and Mechanics); *Izvestiya Akademii nauk SSSR. Seriia otdeleniya tekhnicheskikh nauk*, and *Seriia Mekhanika zhidkosti i gaza* (Proceedings of the USSR Academy of Sciences. Series of the Department of Technical Sciences and Series of the Mechanics of Liquids and Gases); *Mekhanika tverdogo tela. Inzhenernii zhurnal* (Mechanics of Solid Bodies. Engineering Journal); *Prikladnaya mekhanika* (Applied Mechanics); *Zhurnal prikladnoi mekhaniki i tekhnicheskoi fiziki* (Journal of Applied Mechanics and Technical Physics); *Magnitnaya gidrodinamika* (Magnetic Hydrodynamics); *Mekhanika polimerov* (Polymer Mechanics).

(D) Physics

The university laboratories were the main centres of physics in pre-revolutionary Russia. The development of physics in the Soviet period has been characterised by the founding of major research institutes usually with a close interest in one particular field. The period 1918 to 1919 saw the founding of the *Physics and Biophysics Institute* in Moscow under the directorship of P.O. Lazarev, the *Physico-Technical Institute* under A.F. Ioffe and the *State Optical Institute* under D.S. Rozhdestvenskii in Leningrad. Later on the *Lebedev Physics Institute* and *Vavolov Institute of Physical Problems*, both attached to the USSR Academy of Sciences in Moscow, the *Ukrainian Physico-Technical Institute* in Kharkov, the *Siberian Physico-Technical Institute* in Tomsk, and a number of institutes within the Republican academies were organised. After the Second World War a number of new institutes were created both within and outside the Academy system. These included the *Institute of Nuclear Research of the Academy of Sciences* at Dubna, which in 1956 formed the base for the *Joint Institute for Nuclear Research* with members from 11 socialist countries, and the *Semiconductor Institute of the Academy of Sciences*. The physics faculties of the universities have also been expanded, the most important being the physics institutes of Moscow and Leningrad Universities.

The main areas of research are the following:

(i) Theoretical physics: the most prominent workers in theoretical physics before the war were A.A. Friedman and V.A. Fok. Since the war N.N. Bogoliubov, L.D. Landau, I.E. Tamm and their pupils have worked on the quantum theory. Research centres include the *Institute of Theoretical Physics*, Moscow and the *Institute of Theoretical Physics*, Kiev.

(ii) Atomic nucleus, elementary particles and cosmic rays: a large number of physicists are employed in particle physics and its associated fields. These include names such as I.E. Tamm, P.L. Kapitsa, A.I. Alikhan, A.I. Alikhanian, B.V. and I.V. Kurchatov, V.I. Veksler, B. Pontecoryo, and G.I. Budker. Their research apparatus includes a laboratory at high altitude in Armenia where cosmic rays are studied using a mass spectrograph to determine the particle mass. There are powerful accelerators at Dubna (10 GeV) and Serpukhov (70 GeV). Development is underway on a 1,000 GeV machine using a collision technique. There are a number of experimental atomic reactors, including ones for thermonuclear reactions. Other centres of research are the *Institute of Physics*, Kiev and the *Institute of Theoretical and Experimental Physics*.

(iii) High-temperature plasma physics: research in high-temperature plasma physics became particularly important in connection with the problem of thermonuclear reactions. Sakharov and Tamm worked on this directly after the Second World War. Since then L.A. Artsimovich and M.A. Leontovich have been two of the leading experimenters in the field of controlled thermonuclear power. In 1964, M.S. Josse developed new methods for compressing certain forms of unstable plasma, which helped increase the life-time of the contained plasma. Their research is carried out in the Kurchatov Institute of Atomic Energy, Moscow. Other centres of fusion research include the *Institute of Electro-Physical Apparatus*, Leningrad, the *Ioffe Physical Technical Institute*, the

Physico-Technical Institute, Kharkov, the *Lebedev Institute of Physics*, Moscow, and the *Institute of Nuclear Physics*, Novosibirsk and the *Physico-Technical Institute* at Sukhumi, Georgia.

(iv) Optics, atomic and molecular physics, spectroscopy: this is a very important field in relation to industry and it developed on the basis of the *State Optical Institute* in Leningrad fouded in 1919. Leading workers in this field at that time included S.I. Vavilov, and D.S. Rozhdestvenskii. P.A. Cherenkov, I.E. Tamm and I.M. Frank were awarded the Nobel Prize in 1958 for their discovery of the Cherenkov effect and their work on Cherenkov counters of fast charged particles. The other main research centres are the *Institute of Physics and Mathematics* in Minsk, the *Leningrad Institute of Precision Mechanics and Optics*, the *Institute of Physics* at Kiev and the Moscow, Leningrad, and Byelorussian Universities.

(v) Wave theory and radiophysics: the main contribution, both theoretical and technical in wave theory and radiophysics, came from L.I. Mandelshtam, N.D. Panalevski, and others. Centres of research today include the *Institute of Physics of the Atmosphere* in Moscow, *Moscow State University*, the *Radiophysics Institute* in Gorky.

(vi) Low Temperature Physics: research in low temperature physics has been carried out in two main centres: under P.L. Kapitsa in the *Institute of Physical Problems* in Moscow and in the *Cryogenics Laboratory of the Physical Technical Institute of the Ukrainian Academy of Sciences* in Kharkov (under B.G. Lazarev). L.D. Landau and others worked on the theory of movement of quantum liquids during the war, for which Landau was later awarded the Nobel prize in physics.

(vii) Solid state physics: A.I. Ioffe was one of the Soviet Union's leading scientists in the field of solid state physics. Main centres of research today include the *Lebedev Physics Institute, Moscow State University*.

(viii) Physics of dielectrics: again A.I. Ioffe and his school were among the main contributors to this field.

(ix) Semiconductor physics: as in other industrialised countries semiconductor physics has become of considerable importance in recent years. Main centres of research include the *Institute of Semiconductors* in Leningrad, *Leningrad University* and the *Institute of Semiconductors* in Kiev.

(x) Electronics: N.G. Basov and A.M. Prokhorov were awarded the 1964 Nobel prize in physics for their work in developing laser quantum light generators. They are attached to the *Lebedev Physics Institute*. Other research centres are the *Radiophysics Institute* in Gorky, the *Institute of Radio Engineering and Electronics* in Moscow, and *Moscow State University*.

(xi) Magnetism: research into ferro-magnetism takes place in the *Institute of Physics of Metals* at Sverdlovsk, at *Moscow State University* and the *Physical-Technical Institute* at Minsk, among others. Other centres of research into magnetic phenomena include the *Institute of Terrestial Magnetism, Radio Research and the Ionosphere* and the *Siberian Institute of Terrestial Magnetism, the Ionosphere and Radio Waves*.

(xii) Crystallography: Soviet physicists have made contributions to the field of mathematical crystallography and theory of symmetry and to the field of crystal growth. Research centres include the *Institute of Physical Chemistry* in Moscow, where experiments are being carried out on the growth of artificial diamond crystals, and the *Moscow and Leningrad Universities*.

(xiii) Acoustics: N.N. Andreev was head of a well-known school of acoustics dealing with problems ranging from the theoretical to the practical. Research centres today include the *Acoustics Institute* in Moscow, the *Institute of Radio Reception and Acoustics* in Leningrad and the *Moscow and Leningrad Universities*.

Journals: The main journals of physics are *Akusticheskii Zhurnal* (Journal of Acoustics);

Atomnaya Energiya (Atomic Energy); *Zhurnal tekhnicheskoi fiziki* (Journal of Technical Physics); *Zhurnal eksperimentalnoi i teoreticheskoi fiziki* (Journal of Experimental and Theoretical Physics); *Izvestiya Akademii nauk SSSR. Seria fizicheskaya* (Proceedings of the USSR Academy of Sciences. Physics Series); *Kristallografiya* (Crystallography); *Optika i spektroscopiya* (Optics and Spectroscopy); *Pribory i tekhnika eksperimenta* (Instruments and Experimental Technique); *Radiotekhnika i electronika* (Radiotechnology and Electronics); *Uspekhi fizicheskikh nauk* (Advances in Physical Sciences); *Fiziki metallov i metallovedeniia* (Physics of Metals and Metallography); *Fizika tverdykh tela* (Solid State Physics); *Yadernaya fizika* (Nuclear Physics).

(E) Astronomy

The main astronomical institutes and observatories include the following: the Pulkovo Observatory and the Engelhardt Observatory near Kazan, both built before the revolution; the Institute of Theoretical Astronomy in Leningrad, the Shternberg State Astronomical Institute attached to Moscow University (GAISh), the Crimean Astrophysical Observatory, the astrophysical observatories at Abastuman (in Georgia), Byurakan (in Armenia), Zelenchuk (North Caucasus) and Shemakha (in Azerbaijan), the Institute of Astrophysics of the Tadzhik Republic, the Astrophysical Institute of the Kazakh Republic with a mountain astronomical observatory, the Mountain Astronomical Observatory of the Pulkovo Observatory (near Kislovodsk), the Astronomical Observatory of the Estonian Academy of Sciences in Tyraver (near Tartu), the Astrophysical Laboratory of the Latvian Republic, and a latitudinal station in Kitab near Samarkand.

During the 1960s the instrumental resources continued to extend and improve. "The 260cm reflector named after Academician Shain, and installed at the Crimean Astrophysical Observatory in 1960 is so far the largest Soviet telescope. It has been substantially complemented by the two-metre reflector of the new Shemekha Astrophysical Observatory (near Baku). A 2.6 metre reflector is under construction for the Byurakan Observatory. The world's largest six-metre reflector is being built for the newly established astrophysical observatory in the north Caucasus. The number of radio telescopes is increasing. In addition to the 22 - m parabolic radio telescope of the Serpukhov Station of the Lebedev Physical Institute, established in 1958, an identical radio telescope has recently been installed near Simeiz at the station of the Crimean Astrophysical Observatory. It has fully automated control. Large cross-shaped radio telescopes have been built and others are under construction at the Serpukhov Station of the Lebedev Physical Institute and the Byurakan Observatory. A T-shaped radio telescope of the Radiophysics and Electronics Institute of the Ukrainian Academy of Sciences, designed for operation in the field of decametric waves, has been built near Kharkov. This telescope consists of 2,040 antennae vibrators. Each arm of the instrument is 900 metres in length. Advances have been made in the study of the X-ray range of wavelengths by means of instruments mounted aboard artificial Earth satellites and automatic interplanetary stations. Instruments are also sent up in balloons. (Bronchten, Vilaly: The Latest Achievements of Soviet Astronomy *Anglo Soviet Journal, XX1X, No. 2, P 3, 1969*). Pulkovo Observatory has a large coronograph (objective lens diameter of 53 cm), installed at its mountain Astronomical Station near Kislorodsk.

Planning and co-ordination of all astronomical research is carried out by the *Astronomical Council of the Soviet Academy of Sciences* and its various permanent commissions.

The main line of research is into the nature of the universe, but the past five years have seen interests shifting in two distinct directions. The first is concerned with the closest celestial bodies - the Moon, Venus and Mars, which are now becoming accessible to space probes. The second is concerned with the remotest bodies of the Universe - quasars, pulsars, quasistellar galaxies and the entire Universe in fact. Improved techniques of radio astronomy, X-ray astronomy and gamma-astronomy contribute to developments here. Work is also being done on the rotation of the Earth using atomic clocks; on compiling star catalogues (in connection with this, a long-term expedition has gone to Chile to study the stars of the Southern Hemisphere). The astronomical instruments industry developed considerably during the 1950s and 1960s.

Journals: The main journals of astronomy are *Astronomichuskii Zhurnal* (Journal of

Astronomy); *Astronomicheskii Vestruk* (Herald of Astronomy) and *Astrofizika* (Astrophysics).

(F) Geodesy

The *Chief Administration for Geodesy and Cartography*, attached to the Council of Ministers, has been the most important institution for encouraging the development of geodesy and cartography in Russia. It is responsible for directing all geodetic activity; for the production of basic geodetic works and topographic surveys; the publication of special and specialised maps and atlases; and for carrying out scientific research work in the fields of geodesy, cartography and so on. Apart from the Chief Administration for Geodesy and Cartography, there is a *Central Scientific Institute of Geodesy, Aerial Surveying and Cartography* in Moscow, and *Training Institutes for Engineers of Geodesy, Aerial Surveying and Cartography* in Moscow (1930) and Novosibirsk (1939). There are also a number of departments of geodesy in the higher educational establishments.

The whole territory of the Soviet Union is covered by a geodetic-astronomy network and a complete network of triangulations. A gravi-metric survey of the whole territory of the USSR has been carried out. Research into the Earth's gravitational field is well developed. Investigations of the deformation of the Earth's core are also underway. Since 1957 artificial Earth satellites have been used for geodetic work.

Some of the more important achievements in cartography are the production of various maps and atlases, including the *Large Soviet World Atlas* (1937), the *Ocean Atlas* (three volumes (1950-63), the *Physical Geographical Atlas of the World* (1964), the *Atlas of the Peoples of the World* (1964), and the *Atlas of the Antarctic* (one volume - 1966).

Journals: The main journals of geodesy are *Geodesiia i Kartografiia* (Geodesy and Cartography); and *Iz. VUZ. Geodesiia i Aerofotos'emka* (Proceedings of Higher Educational Establishments, Geodesy and Aerial Photography).

(G) Geophysics

Geophysical studies are carried out in a number of institutes and observatories, including the Institutes of Applied Geophysics, Physics of the Atmosphere, Earth Physics, Oceanology and Acoustics, all of which are attached to the USSR Academy of Sciences; the Institute of Geology and Geophysics in the Siberian Department of the Academy; the Sea Hydrophysical Institute of the Ukrainian Academy of Sciences; and the State Oceanographic, Arctic and Antarctic Institute (GUGMS). A number of Republican Institutes and other scientific establishments also undertake geophysical research.

The main areas of research include: the mantle and the Earth's crust; seismic disturbances; the behaviour of rocks under high temperatures and pressures; the gravitational and magnetic fields of the Earth; physics of the atmosphere and upper atmosphere; and the oceans. Since the International Geophysical Year (1957/58) and the Year of the Quiet Sun, a number of expeditions to the Antarctic, Pacific and Atlantic have been organised. Scientific stations have been established in the Arctic since the 1930s; some of these are temporary and based on icefloes.

Journals: These include *Izvestiia Akademii Nauk SSSR, serii Fizika atmosfery i okeana, i Fizika Zemli* (Proceedings of the USSR Academy of Sciences, Series of Physics of the Atmosphere, Oceans and Land); *Meteorologiia i gidrologiia* (Meteorology and Hydrology); *Okeanologiia* (Oceanology).

(H) Geology

Geological traditions in the USSR date back to the eighteenth century, when the then Imperial Academy of Sciences first organised expeditions to study and explore the country. At the end of the nineteenth century, research in this field expanded under a Geological Committee which organised surveys in the European part of Russia, in the Urals, the Altai and Turkestan.

After the revolution, the need to find extra resources of minerals, fuels and other raw materials resulted in a rapid widening in the scope of geological research. In 1918 the Moscow Mining Academy was created, and later geology departments were opened in practically every university, polytechnic and industrial higher educational establishment. Today the work of planning and organising geological research is carried out by the *Ministry of Geology of the Soviet Union* through regional administrations and republican ministries. In addition, a number of other ministries (for example, the gas, oil, coal, ferrous and non-ferrous metallurgy) have their own geological managements. There are about 80 scientific institutes and laboratories attached to the Ministry of Geology, the USSR Academy of Sciences, other ministries, and the Republican Academies which carry out scientific research. These include: the *Geological Institute of the Academy of Sciences*; the *All-Union Scientific Research Geological Institute (VSEGEI)* of the Ministry of Geology of the USSR; the *Scientific Research Institute of the Geology of the Arctic (NIIGA)*, the Institutes of Geology and Geophysics of the Siberian section of the Academy and of the Uzbek Academy of Sciences;. and the Institute of Geology of the Ukrainian Academy of Sciences.

Those dealing with problems of mineral resources and raw materials include: the *All-Union Institute of Mineral Raw Materials* (VIMS), the *Siberian Scientific Research Institute of Geology and Mineral Raw Materials* (SNIIGIMS); the *Kazakh Institute of Mineral Raw Materials* (Kazakh IMS); and the *Caucasus Institute of Mineral Raw Materials* (KIMS).

Those dealing with geochemical problems include the *Institute of Geochemistry and Analytical Chemistry im. V.I.Vernadskii* (GEOKhI); the *Institute of Geology of Ore Mine Deposits, Petrography, Mineralogy and Geochemistry* (IGEM) within the Academy system; the *Institute of Mineralogy, Geochemistry and Crystallochemistry of Rare Elements*; and the *All-Union Scientific Research Geology Prospecting Institute* (VNIGRI).

Those dealing with geophysical problems are the *Institute of Earth Physics of the Academy of Sciences*; and *Institute of the Earth's Crust* of the Academy; and the *Institute of Applied Geophysics*.

A number of institutes are concerned with problems of oil and gas research. These include the *Institute of the Geology and Exploitation of Fuel Resources*; the *All-Union Scientific Research Geological Prospecting Oil Institutes* (VNIGNI and VNIGRI); the *Ukrainian Scientific Research Geological Prospecting Oil Institute* (UKRNIGRI); and *All-Union Scientific Research Institute of Natural Gases* (VNIIGaz).

Since the revolution, besides studies of the central area of the country, geological surveys have been undertaken of the Polar regions, Western Siberia, Eastern Siberia, the North-East, the Far-East, the Pamir, and all the Union Republics. In 1917 10.5 per cent of the territory had been geologically surveyed. By 1940 this percentage had grown to 65.8 per cent, and today the whole territory has been studied and mapped. Various types of geological, tectonic and other specialist maps of the USSR both as a whole and of particular regions have been published.

Apart from the development of purely geological sciences - stratigraphy, tectonics, mineralogy, petrography, lithology, study of mineral resources, hydrogeology and engineering geology - a whole complex of sciences connected with geology, such as geochemistry and geophysics, also have developed.

Journals: The main geology journals include the *Izvestiia AN SSSR, seriia geologicheskaya* (Proceedings of USSR Academy of Sciences. Geology Series); *Sovetskaya Geologiya* (Soviet Geology); *Geologiya rudnykh mestorozhdenii* (Geology of Ore Deposits); *Razvedka i Okhrana Nedr* (Prospecting and Protection of Mineral Resources); *Zapiski Vsesoyuznogo Mineralogicheskogo Obshchestva. Vtoraya seriia* (Notes of the All-Union Mineralogical Society, Second Series); *Byulleten' Moskovskogo Obshchestva Ispytatelei Prirody. Otdel geologicheskii* (Bulletin of the Moscow Naturalists' Society. Geology Section); *Litologiya i poleznye iskopaemye* (Lithology and Minerals); *Geologiya nefti i gaza* (Geology of Oil and Gas); *Geologiya i geofizika* (Geology and Geophysics).

(I) Geography

Geographical research in Russia as with geological research, dates back to the eighteenth century when the first scientific expeditions were organised to explore the country. After the revolution the network of geography faculties on the universities and of separate geographical institutes developed rapidly because of the need for research and study of the country's natural resources.

The 1920s and 1930s were characterised by a number of large expeditions organised by the USSR Academy of Sciences to study the geography of the various republics, and the more distant regions of the country. In particular, one of the more complex, was that to the Arctic to help develop the Northern Sea Route. In 1937 the first drifting station on an icefloe was established called *North Pole-1*. These expeditions were able to fill in regions which up to then had been blank spots on the maps. in the 1950s and 1960s, in connection with the International Geophysical Year, Soviet geographers carried out extensive research in the Antarctic and the Pacific Ocean.

A trend that has emerged under Academician Gerasimov is for the geographer to see himself concerned with the problem of the transformation of nature - the impact of industry on the countryside.

Geographical research is carried out both in the specialist institutes (such as the *Institute of Geography of the Academy of Sciences*, the *Institute of Geography of Siberia and the Far East* attached to the Siberian section of the academy), but also in Institutes whose main concern is some closely related subject. These include the *Oceanology Institute*, the *Botany Institute* the *Institute of Economics*, the *Institute of Ethnography* and other similar institutes all under the USSR Academy of Sciences. The *State Hydrological Institute*, the *Chief Geophysical Observatory*, the *Soil Institute*, the *State Oceanographical Institute*, the *Arctic and Antarctic Institute* are also all important centres of geographical research. The Republican academies and the geography departments (physical and economic) of some of the leading universities (Moscow, Leningrad, Kiev, Voronezh, Riga, and Tashkent) and polytechnical institutes also carry out work in the field. The *Geographical Society* has played an important part in encouraging and developing interest in geography in the country. It also helps to organise conferences.

The chief areas of research within physical geography are: general studies of the Earth; geomorphology; climatology; hydrology of the drylands; glaciology; geocryology; soil geography; geobotany and zoogeography; medical geography; oceanology.

Journals: The main journals dealing with research in geography are: *Izvestiia AN SSSR, seriia geograficheskaya* (Proceedings of the USSR Academy of Sciences. Geographical Series); *Izvestiia Vsesoyuznogo geograficheskogo obshchestva* (Proceedings of the All-Union Geographical Society); *Doklady Instituta geografii Sibiri i Dal'nego Vostoka* (Proceedings of the Siberian and Far East Institute of Geography); *Vestniki and Trudy* (Heralds and Transactions); of Moscow, Leningrad and other universities; *Voprosy geografii* (Problems of Geography).

(J) Chemistry

There was a well-developed chemical tradition in pre-revolutionary Russia, which had arisen around people like Zinin, Butlerov and Mendeleev. The development of chemistry since the revolution has been connected with the industrialisation of the country. New research institutes include the *Karpov Physical-Chemical Institute*, the *Institute of Applied Chemistry*, the *High-pressure Institute*, and the *Fertiliser Institute*. In recent years there has been an increased emphasis on the importance of chemical research. In 1966 there were 50 institutes of the Union and Republican academies of sciences working on chemical problems, as well as a number of institutes attached to ministries, the chemical industry, and higher educational establishments.

The main areas of research include:

(i) Physical Chemistry: Research in physical chemistry includes chemical kinetics, catalysis, reaction mechanisms, electrochemistry and chemistry of surface phenomenon. N.N. Semyonov, together with C.N. Hinshelwood from England, was awarded the 1956 Nobel prize in chemistry for his work on chain reactions.

(ii) General and Inorganic Chemistry: The Institute of General and Inorganic Chemistry directed by Academician I.I. Chernyaev is an important centre - the centre of the so-called 'Moscow' school in this field. The main achievements of the Moscow school are the development of the stereochemistry of platinum compounds, and of the chemistry of rhodium, iridium, uranium and transuranium elements. The Leningrad school of general and inorganic chemistry, on the other hand, is centered round Academician A.A. Grinberg. He works at the Leningrad Technological Institute and the Radium Institute in Leningrad, as well as being attached to the Institute of General and Inorganic Chemistry in Moscow. This Leningrad School has provided experimental proof of a number of concepts of Werner's theory, it has created the basis for a stereochemistry of palladium, and developed a theory of acidic-basic and oxidising-reducing properties of complex compounds. Other important areas of research in general and inorganic chemistry include the chemistry of silicates (Institute of Chemistry of Silicates in Leningrad), or rare elements and Fluorine (Institute of General and Inorganic Chemistry, Moscow), and of ozonides (Karpov Physical Chemical Institute, Moscow).

(iii) Analytical Chemistry: Fields of research include automation, analysis of difficult-to-determine elements, fluorescent methods, instrumentation, microtechniques and the use of radioactive isotope. Academician A.P. Vinogradov who directs the Institute of Geochemistry and Analytical Chemistry in Moscow, has contributed to the development of modern methods for analysing trace elements in minerals and meteorites. He is a key figure in analysing the Moon rock brought back to Earth by Luna 16.

(iv) Organic Chemistry: Research in organic chemistry is particularly widely developed in connection with the oil industry. The production and transformation of nitro derivatives is a well studied field (Institute of Petrochemical Synthesis, Moscow); Voronexh State University; the Scientific Research Institute of the Dairy Industry). Academician Semyonov at the Institute of Chemical Physics in Moscow has been working on the direct oxidation of hydrocarbons. The State Research Institute of Synthetic Alcohol and its products also does similar work.

(v) Elementary Compounds: The chemistry of elementary compounds is a wide-ranging field standing independently between organic and inorganic chemistry. Main centres of research are: the Institute of Elementary Organic Compounds (I.L. Krumyats, A.N. Nesmeyanov, M.I. Kabachnik, and R. Kh. Friedlin), the Karpov Physical-Chemical Institute (K.A. Kocheshkov), Moscow State University (Nesmeyanov and O.A. Reutov,) Kazan University and the Arbuzov Chemical Institute in Kazan (A.E. Arbuzov and B.A. Arbuzov).

(vi) Heterocyclic Compounds: Main centres of research include the Institute of Chemistry of Plant Substances, Tashkent (S. Yu. Yunusov) and the Physical Technical Institute of Tashkent (A.S. Sadykov), which specialise in the chemistry of cellulose, and the Institute of Chemistry of Natural Compounds in Moscow (M.M. Shemyakin and N.K.Kochetkov).

(vii) High Molecular Compounds: Main centres of research are the Institute of Chemical Physics (N.N. Semyonov), the Institute of Elementary Organic Compounds (V.V.Korshak, R.Kh. Friedlin and A.N. Nesmeyanov), the Moscow Institute of Fine Chemical Technology, (V.V. Korshak) and Moscow State University where an Interdivisional Specialised Research Laboratory of High Molecular Compounds was established in 1959 by V.A. Kargin, and the Institute of Physical Chemistry.

Journals: The main chemical journals include *Doklady Akademii nauk SSSR, seriia Khimiia* (Transactions of the USSR Academy of Sciences. Chemistry Series); *Izvestiia Akademii nauk SSSR, seriia Khimicheskaya* (Proceedings of the USSR Academy of Sciences. Chemical Series); also *Seriia Neorganicheskie materialy* (Inorganic Materials Series); *Zhurnal fizicheskoi khimii* (Journal of Physical Chemistry); *Zhurnal obshchei khimii* (Journal of General Chemistry); *Zhurnal neorganicheskoi khimii* (Journal of Inorganic Chemistry); *Kolloidny zhurnal* (Colloids Journal); *Zhurnal analiticheskoi khimii* (Journal of Analytical Chemistry); *Zhurnal prikladnoi khimii* (Journal of Applied Chemistry); *Zhurnal organicheskoi khimii* (Journal of Organic Chemistry); *Elektrokhimiia* (Electrochemistry); *Kinetika i Kataliz* (Kinetics and Catalysis); *Neftekhimiia* (Petrochemistry); *Zhurnal Strukturnoi khimii* (Journal of Structural Chemistry); *Uspekhi khimii* (Advanced in Chemistry); *Ukrainskii khimicheskii zhurnal* (Ukrainian Chemical Journal);

Zavodskaya laboratoriya (Factory Laboratory); *Vysokomolekuliarny soedineniia* (High Molecular Compounds); *Izvestiia AN Kazakhskoi SSR, seriia Khimicheskaya* (Proceedings of the Kazakh Academy of Sciences. Chemical Series); *Khimicheskaia promyshlennost'* (Chemical Industry); *Zhurnal Vsesoyuznogo khimicheskogo obshchestva im D.I. Mendeleev* (Journal of the Mendeleev All-Union Chemical Society).

(K) Biology

Biology was a well-established science in pre-revolutionary Russia, its best known practitioners ranging from Pavlov in the field of physiology, through A.O. Kovalevsky in embryology and K.A. Timiryazev in botany. Both Pavlov and Timiryazev lived through the revolution and helped provide for a continuity of tradition and development. As in the other sciences, the post-revolutionary period was one of expansion. N.N. Vavilov became director of the Bureau for Applied Botany, Genetics and Selection. A whole network of medical-biological institutes were set up, and the number of institutes and laboratories within the academy system grew rapidly. By 1966 some 120 scientific organisations were concerned with biological problems within the Soviet Academy system including 74 institutes. There has been a parallel development in the Republican academies of biological, medico-biological and agricultural institutes, laboratories and field stations. There are biology departments in many institutes of higher education and universities. In 1966 the total number of biological research establishments exceeded 220, and they employed more than 32,000 scientific workers.

Biological research received a set back in the 1940s and 1950s, because of the ascendancy of Lysenko and his ideas. Academician N.I. Vavilov was dismissed from his post and died in prison sometime in the early 1940s. Lysenko and his supporters occupied all the leading academic and propoganda positions and his influence spread right down to biology teaching in the schools. The fall of Lysenko coincided with the fall of Kruschev in 1964. Lysenko was dismissed from his position as director of the Academy of Sciences Institute of Genetics; many of his leading supporters also lost their posts; at the same time a number of good biologists, who had not supported Lysenko or had been supported by him, were awarded academic degrees without having to defend dissertations. Since then, new research institutes have been formed, the school biological curricula has been revised, and new textbooks published. However the process of reform is slow. There have been no whole-scale dismissals of Lysenko supporters from academic life, and a generation of biologists, which has grown up within the framework of Lysenko's ideas, now has to come to terms with the new situation.

The main fields of research include the following:

(i) *Genetics*: Institute of General Genetics, Moscow, Siberian Institute of Cytology and Genetics.

(ii) *Physiology*: Institute of Higher Nervous Activity and Neurophysiology, Moscow, the Pavlov Institute of Physiology and the Sechenov Institute of Evolutionary Physiology and Biochemistry, both in Leningrad; the Timiyazev Institute of Plant Physiology.

(iii) *Biochemistry*: Bakh Institute of Biochemistry, and the Institute of Biochemistry and Physiology of Micro-organisms in Moscow, the Institute of Protein in Akademgorodok.

(iv) *Microbiology*: Institute of Biochemistry and Physiology of Micro-organisms, Institute of Microbiology, Moscow.

(v) *Cytology and histology*: Institute of Cytology in Leningrad, and the Siberian Institute of Cytology and Genetics.

(vi) *Ecology*: the Severtsov Institute of Evolutionary Morphology and Animal Ecology.

(vii) *Helminthology, parasitology, entomology*:

(viii) *Bio-Oceanology*: Institute of Oceanology and Moscow State University.

(ix) *Palaeontology*: the Palaeontological Institute, Moscow.

(x) *Agriculture: crop selection and animal breeding*: All-Union Institute of Plant Breeding, the Academy of Agricultural Sciences, the Timiryazev Agricultural Academy are three of the leading institutes in this research field out of the main agricultural institutes throughout the country.

Journals: The main biological journals include *Zhurnal obshchei biologii* (Journal of General Biology); *Byulleten experimental'noi biologii i meditsiny* (Bulletin of Experimental Biology and Medicine); *Izv estiia Akademii Nauk SSSR, seriia biologicheskaya* (Proceedings of the USSR Academy of Sciences. Biological Series; *Byulleten Moskovskogo obshchestva ispytatelei prirody, otdel biologicheskii* (Bulletin of the Moscow Naturalists Society Biological Department); *Zoologicheskii Zhurnal* (Zoological Journal); *Botanicheskii Zhurnal* (Botanical Journal); *Palaeontologicheskii Zhurnal* (Palaeontological Journal); *Arkhiv anatomii, histologii i embriologii* (Archives of Anatomy, Histology and Embryology); *Fiziologicheskii zhurnal SSSR im. I.M. Sechenov* (Sechenov Physiological Journal); *Zhurnal vyshei nervnoi deyatel nosti im I.P. Pavlova* (Pavlova Journal of Higher Nervous Activity); *Biokhimiia* (Biochemistry); *Prikladnaya biokhimiia i mikrobiologiia* (Applied Biochemistry and Microbiology); *Mikrobiologiia* (Mircobiology); *Fiziologiia rastenii* (Plant Physiology); *Radiobiologiia* (Radiobiology); *Biofizika* (Biophysics); *Tsitologiia* (Cytology); *Genetika* (Genetics); *Rastitel'nye resursy* (Vegetative Resources); *Kosmicheskaya biologiia i meditsina* (Space Biology and Medicine); *Uspekhi sovremennoi biologii* (Advances in Contemporary Biology); Abstract journals are *Biologiia* (Biology); and *Biologicheskaya khimiia* (Biological Chemistry)

6 Space Research

(A) Introduction

The Russians had a strong tradition of research and interest in rocketry prior to the revolution. The 19th century saw the widespread use of military rockets, with General K.I. Konstantinov introducing winged and stabilised rockets in the 1860s. A number of other engineers working at the same time suggested ideas for aircraft using the principles of rocket flight.

Space technology research began at the end of the century with the work of K.E.Tsiolkovskii on the principles of rocket flight, I.V.Meshcherskii on the dynamics of bodies of variable mass, and N.E.Zhukovskii on aerodynamics. The Soviet consider Tsiokovskii to be the founding father of astronautics.

Tsander, Kondratiuk, Rynin, Perel'man, Fortikov and other disciples continued and expanded their work in the 1920s and 1930s. The Gas Dynamics Laboratory (GDL) and the Jet Propulsion Study Group (GIRD) were established to develop solid and liquid fuel rockets. 1932 saw the first launchings of rockets with experimental reaction motors working on benzene and liquid oxygen. In 1942 the first aeroplane with such an engine was produced. The Second World War brought the development of air-to-air and ground-to-ground (the famous Katiusha) rockets. After the war the Soviets acquired a lot of German rocket production and testing equipment plus several hundred personnel, all of which boosted their rocket development programme. By 1953 the president of the USSR Academy of Sciences was able to comment publicly "Science has reached a state when it is feasible to send a stratoplane to the Moon, to create an artificial satellite of the Earth ..." In 1957 the Soviet Union launched Sputnik 1, the first artificial satellite of the Earth. Sputnik 2 came the same year and Sputnik 3 was the only launching of 1958. 1959 saw three lunar probes to the Moon, and 1960 another three Sputniks. By 1961 the Soviets were ready to launch a man into space.

(B) Organisation

The Russians have revealed very little about the organisation of their space programme. In 1955 they mentioned that they had set up a permanent space organisation. They have never volunteered data on their space budget, organisational relationships, or names of space officials. We know that the Soviet Academy of Sciences has at least an advisory function in the programme, and that the Soviet strategic rocket troops conduct the actual launches. (Source: *Review of the Soviet Space Programme*, Report of the Committee on Science and Astronautics, p 79, Washington, 1967.)

(C) Launching Sites

The Soviet Union has three known launching sites: at Tyuratam, Kapustin Yar and Plesetsk (see Figure 5).

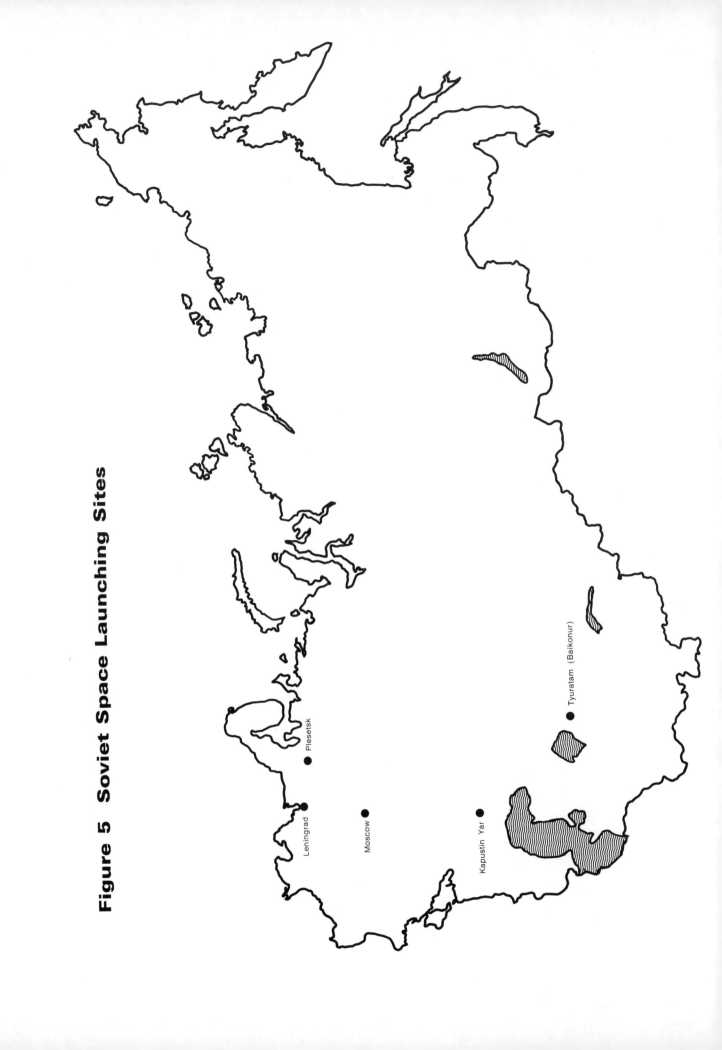

Figure 5 Soviet Space Launching Sites

(i) Tyuratam[*] lies in Kazakhstan, east of the Aral Sea. It is better known as the Baikonur Cosmodrome and was named as such by the Russians when Gagarin was placed in orbit in 1961. However, experts claim that the ground traces of numerous flights have consistently shown that the actual launching site lies in the vicinity of Tyuratam, 230 miles south-west of Baikonur village. Tyuratam has been used for ICBM tests, the early sputniks, all manned flights, all lunar and planetary attempts, all communications satellites and many other research and development flights. In many ways it is the Soviet equivalent of Cape Kennedy. And like Cape Kennedy, it is very likely that there are other launch pads in the same general vicinity as a result of proliferation of programmes and launch vehicles beyond the original one, while the first vehicle still continues to be used.

(ii) Kapustin Yar lies near the Volga River below Volgograd. It was known previous to 1957 as a launching site for vertical probes and intermediate range missiles. In 1962 it was first used for modest orbital flights of the Cosmos programme. Kapustin Yar plays a much more limited space role than Tyuratam. Every payload has had an inclination to the Equator of close to $49°$, and most of the payloads have been of modest size and spin stabilised.

(iii) Plesetsk is situated in the north of the USSR. The site was disclosed in 1966 by the team from Kettering Grammar School who keep a track of Soviet spacecraft. Since then Plesetsk has been found to be a nodal point for many flights and with enough variety of launch vehicles to indicate several different launch pads. It is probably also used for military flights.

(D) Programme

(i) Early Sputniks, 1957-58. Sputnik 1, launched on 4th October 1957, inaugurated man's exploration of space. It weighed 184 pounds and returned density, temperature, cosmic ray and micro-meteoroid data for 21 days. Sputnik 2, launched a month later on 3rd November, was much larger, weighing 1,120 pounds. It had the dog Laika on board, and carried out the first biomedical experiments in space during the seven days she remained alive. The satellite also carried out measurements of solar X-rays, ultraviolet radiation and cosmic rays. Sputnik 3, launched in 1958 (the only one for that year, see Table 12), was an extremely impressive orbiting geophysical laboratory. The payload was again increased, this time to 2,925 pounds and the conical spacecraft was the first Soviet one to use solar cells. Its many experiments included measurements of magnetic fields, micro-meteoroids, solar radiation, cosmic rays and atmospheric density and pressure. It returned data for two years, and the Soviets credit it with discovering the Earth's outer radiation belt.

(ii) Lunar Programme. The Soviet Union launched three lunar probes in 1959. Luna 1, launched on 2nd January, passed within 3,700 miles of the Moon and then continued into solar orbit becoming the first artificial satellite around the Sun. It is presumed that it had been intended to land on the Moon. This feat was in fact achieved nine months later with Luna 2, a 860 pound probe. It landed between the craters Archimedes, Aristillus and Autolycus, and detected no lunar magnetic field or radiation belt prior to impact.

Lunar 3, launched on the second anniversary of Sputnik 1, into a deep, highly elliptical Earth Orbit, passed behind the back of the Moon. From there it took pictures of the hidden side of the Moon, developed them automatically and transmitted them back to Earth. There was then a lull in the lunar programme until 1963, when attempts to soft land began. Luna 4 was launched in April 1963 but missed the Moon. Lunas 5, 6, 7 and 8, launched in 1965, also failed to make a soft landing. Finally Luna 9 was successful in 1966. It returned 27 pictures and answered the most crucial questions about the surface of the Moon. Its batteries ran out after two Earth days. In the same year three more probes, Luna 10, 11 and 12, were placed in orbit around the Moon. They took various measurements and Luna 12 returned some pictures. Luna 13 was another soft lander on the Moon's surface, launched at the end of 1966.

An 18-month lull followed until April 1968 when Luna 14 went into lunar orbit, studying Earth-Moon mass relationship and the Moon's gravitational field. A year later, in July 1969,

---

[*] The information in this section and in the following ones on Programmes and Launching Vehicles is primarily drawn from the *TRW Space Logs*, California, and the *Review of the Soviet Space Programme*, Washington, 1967.

Soviet Space Launches

1957

Sputnik 1
Sputnik 2

1958

Sputnik 3

1959

Luna 1
Luna 2
Luna 3

1960

Sputnik 4
Sputnik 5
Sputnik 6

1961

Sputnik 7
Venus Probe
Sputnik 9
Sputnik 10
Vostok 1
Vostok 2

1962

Cosmos 1
Cosmos 2
Cosmos 3
Cosmos 4
Cosmos 5
Cosmos 6
Cosmos 7
Vostok 3
Vostok 4
Cosmos 8
Cosmos 9
Cosmos 10
Cosmos 11
Mars 1
Cosmos 12

1963

Cosmos 13
Luna 4
Cosmos 14
Cosmos 15
Cosmos 16
Cosmos 17
Cosmos 18
Vostok 5
Vostok 6
Cosmos 19
Cosmos 20
Polyot 1
Cosmos 21
Cosmos 22
Cosmos 23
Cosmos 24

1964

Electron 1 ⎫
Electron 2 ⎭
Cosmos 25
Cosmos 26
Cosmos 27
Zond 1
Cosmos 28
Polyot 2
Cosmos 29
Cosmos 30
Cosmos 31
Cosmos 32
Cosmos 33
Cosmos 34
Electron 3 ⎫
Electron 4 ⎭
Cosmos 35
Cosmos 36
Cosmos 37
Cosmos 38 ⎫
Cosmos 39 ⎬
Cosmos 40 ⎭
Cosmos 41 ⎫
Cosmos 42 ⎬
Cosmos 43 ⎭
Cosmos 44
Cosmos 45
Cosmos 46
Cosmos 47
Voskhod 1
Cosmos 48
Cosmos 49
Cosmos 50
Zond 2
Cosmos 51

1965

Cosmos 52
Cosmos 53
Cosmos 54 ⎫
Cosmos 55 ⎬
Cosmos 56 ⎭
Cosmos 57
Cosmos 58
Cosmos 59
Cosmos 60
Cosmos 61 ⎫
Cosmos 62 ⎬
Cosmos 63 ⎭
Voskhod 2
Cosmos 64
Cosmos 65
Molniya 1
Cosmos 66
Luna 5
Cosmos 67
Luna 6
Cosmos 68
Cosmos 69
Cosmos 70
Cosmos 71 ⎫
Cosmos 72 ⎬
Cosmos 73 ⎬
Cosmos 74 ⎬
Cosmos 75 ⎭
Proton 1
Zond 3
Cosmos 76
Cosmos 77
Cosmos 78
Cosmos 79
Cosmos 80 ⎫
Cosmos 81 ⎬
Cosmos 82 ⎬
Cosmos 83 ⎬
Cosmos 84 ⎭
Cosmos 85
Cosmos 86 ⎫
Cosmos 87 ⎬
Cosmos 88 ⎬
Cosmos 89 ⎬
Cosmos 90 ⎭
Cosmos 91
Luna 7
Molniya 1
Cosmos 92
Cosmos 93
Cosmos 94
Proton 2
Cosmos 95
Venus 2
Venus 3
Cosmos 96
Cosmos 97
Cosmos 98
Luna 8
Cosmos 99
Cosmos 100
Cosmos 101
Cosmos 102 ⎫
Cosmos 103 ⎭

1966

Cosmos 104
Cosmos 105
Cosmos 106
Luna 9
Cosmos 107
Cosmos 108
Cosmos 109
Cosmos 110
Cosmos 111
Cosmos 112
Cosmos 113
Luna 10
Cosmos 114
Cosmos 115
Molniya 1
Cosmos 116
Cosmos 117
Cosmos 118
Cosmos 119
Cosmos 120
Cosmos 121
Cosmos 122
Proton 3
Cosmos 123
Cosmos 124
Cosmos 125
Cosmos 126
Cosmos 127
Luna 11
Cosmos 128
Cosmos 129
Molniya 1
Cosmos 130
Luna 12
Yantar 1
Cosmos 131
Cosmos 132
Cosmos 133
Cosmos 134
Cosmos 135
Cosmos 136
Luna 13
Cosmos 137

1967

Cosmos 138
Cosmos 139
Cosmos 140
Cosmos 141
Cosmos 142
Cosmos 143
Cosmos 144
Cosmos 145
Cosmos 146
Cosmos 147
Cosmos 148
Cosmos 149
Cosmos 150
Cosmos 151
Cosmos 152
Cosmos 153
Cosmos 154
Cosmos 155
Cosmos 156
Cosmos 157
Soyuz 1
Cosmos 158
Cosmos 159
Cosmos 160
Cosmos 161
Molniya 1
Cosmos 162
Cosmos 163
Cosmos 164
Venus 4
Cosmos 165
Cosmos 166 ⎫
Cosmos 167 ⎭
Cosmos 168
Cosmos 169
Cosmos 170
Cosmos 171
Cosmos 172
Cosmos 173
Cosmos 174
Cosmos 175
Cosmos 176
Cosmos 177
Cosmos 178
Cosmos 179
Cosmos 180
Molniya 1
Cosmos 181
Cosmos 182
Cosmos 183
Molniya 1
Cosmos 184
Cosmos 185
Cosmos 186 ⎫
Cosmos 187 ⎭
Cosmos 188 ⎫
Cosmos 189 ⎭
Cosmos 190
Cosmos 191
Cosmos 192
Cosmos 193
Cosmos 194
Cosmos 195
Cosmos 196
Cosmos 197
Cosmos 198

1968

Cosmos 199
Cosmos 200
Cosmos 201
Cosmos 202)
Cosmos 203)
Zond 4
Cosmos 204)
Cosmos 205)
Cosmos 206
Cosmos 207
Cosmos 208
Cosmos 209
Cosmos 210
Luna 14
Cosmos 211
Cosmos 212)
Cosmos 213)
Cosmos 214)
Cosmos 215
Cosmos 216
Molniya 1
Cosmos 217
Cosmos 218
Cosmos 219
Cosmos 220
Cosmos 221
Cosmos 222
Cosmos 223
Cosmos 224
Cosmos 225)
Cosmos 226)
Cosmos 227
Cosmos 228
Cosmos 229
Cosmos 230
Molniya 1
Cosmos 231
Cosmos 232
Cosmos 233
Cosmos 234
Cosmos 235
Cosmos 236)
Cosmos 237)
Cosmos 238
Cosmos 239
Cosmos 240
Zond 5
Cosmos 241
Cosmos 242
Cosmos 243
Cosmos 244
Cosmos 245
Molniya 1
Cosmos 246
Cosmos 247
Cosmos 248
Cosmos 249
Soyuz 2)
Soyuz 3)
Cosmos 250)
Cosmos 251)
Cosmos 252
Zond 6
Cosmos 253
Proton 4
Cosmos 254
Cosmos 255
Cosmos 256
Cosmos 257
Cosmos 258
Cosmos 259
Cosmos 260
Cosmos 261
Cosmos 262

1969

Venus 5
Venus 6
Cosmos 263
Soyuz 4)
Soyuz 5)
Cosmos 264
Cosmos 265
Cosmos 266
Cosmos 267
Cosmos 268)
Cosmos 269)
Cosmos 270
Cosmos 271
Cosmos 272
Cosmos 273
Cosmos 274
Meteor
Cosmos 275
Cosmos 276)
Cosmos 277)
Cosmos 278
Molniya 1
Cosmos 279
Cosmos 280
Cosmos 281
Cosmos 282
Cosmos 283
Cosmos 284
Cosmos 285
Cosmos 286
Cosmos 287
Cosmos 288
Cosmos 289
Luna 15
Cosmos 290
Molniya 1
Cosmos 291
Zond 7
Cosmos 292
Cosmos 293
Cosmos 294
Cosmos 295
Cosmos 296
Cosmos 297
Cosmos 298
Cosmos 299
Cosmos 300
Cosmos 301
Meteor
Soyuz 6
Soyuz 7
Soyuz 8
Intercosmos 1
Cosmos 302
Cosmos 303
Cosmos 304
Cosmos 305
Cosmos 306)
Cosmos 307)
Cosmos 308
Cosmos 309
Cosmos 310
Cosmos 311)
Cosmos 312)
Cosmos 313
Cosmos 314
Cosmos 315
Cosmos 316)
Cosmos 317)
Intercosmos 2

1970

Cosmos 318
Cosmos 319
Cosmos 320
Cosmos 321)
Cosmos 322)
Cosmos 323
Molniya 1
Cosmos 324
Cosmos 325
Cosmos 326
Meteor
Cosmos 327
Cosmos 328
Cosmos 329
Cosmos 330
Cosmos 331
Cosmos 332
Cosmos 333
Cosmos 334
Cosmos 335
Cosmos 336)
Cosmos 337 |
Cosmos 338 |
Cosmos 339 |
Cosmos 340 |
Cosmos 341 |
Cosmos 342 |
Cosmos 343)
Cosmos 344
Cosmos 345
Soyuz 9
Cosmos 346
Cosmos 347
Cosmos 348
Cosmos 349
Meteor
Cosmos 350
Cosmos 351
Molniya 1
Cosmos 352
Cosmos 353
Cosmos 354
Cosmos 355
Intercosmos 3
Cosmos 356
Venus 7
Cosmos 357
Cosmos 358
Cosmos 359
Cosmos 360
Cosmos 361
Luna 16
Cosmos 362
Cosmos 363
Cosmos 364
Cosmos 365
Molniya 1
Cosmos 366
Cosmos 367
Cosmos 368)
Cosmos 369)
Cosmos 370
Cosmos 371
Intercosmos 4
Meteor
Cosmos 372
Cosmos 373)
Zond 8)
Cosmos 374
Cosmos 375)
Cosmos 376)
Luna 17
Cosmos 377
Cosmos 378
Cosmos 379)
Cosmos 380)
Molniya 1
Vertikal 1
Cosmos 381)
Cosmos 382)
Cosmos 383
Cosmos 384
Cosmos 385
Cosmos 386
Cosmos 387
Cosmos 388)
Cosmos 389)
Molniya 1

1971

Cosmos 390
Cosmos 391
Cosmos 392
Cosmos 393
Cosmos 394
Cosmos 395)
Cosmos 396)
Cosmos 397
Cosmos 398
Cosmos 399
Cosmos 400
Cosmos 401
Cosmos 402
Cosmos 403
Cosmos 404
Cosmos 405
Cosmos 406
Salyut 1)
Soyuz 10)
Cosmos 407)
Cosmos 408
Cosmos 409
Cosmos 410
Cosmos 411)
Cosmos 412 |
Cosmos 413 |
Cosmos 414 |
Cosmos 415 |
Cosmos 416 |
Cosmos 417 |
Cosmos 418 |
Cosmos 419)
Meteor
Cosmos 420
Mars 2)
Cosmos 421)
Cosmos 422
Cosmos 423
Cosmos 424
Cosmos 425
Mars 3
Cosmos 426
Soyuz 11
Cosmos 427
Cosmos 428
Meteor
Cosmos 429
Cosmos 430
Molniya 1
Cosmos 431
Cosmos 432
Cosmos 433
Cosmos 434
Vertikal 2
Cosmos 435
Luna 18
Cosmos 436
Cosmos 437
Cosmos 438
Cosmos 439
Cosmos 440
Luna 19
Cosmos 441
Cosmos 442

To the end of September 1971

Luna 15 orbited the Moon 52 times before landing on the Moon. In contrast to Luna 9 and 13 it was able to manoeuvre and its orbit was altered twice.

Luna 16, launched in 1970, landed on the Moon, took samples of the soil and then brought them back to Earth. It was the first unmanned vehicle to do this. Luna 17 followed shortly after in November 1970, landing the first mobile laboratory, Lunokhod 1, on the Moon. This was controlled from Earth. It moved about the surface and carried out a number of experiments and measurements including soil analysis, and detection of extra-galactic radiation. In co-operation with the French, it carried a laser reflector for determining the exact distance of the Moon from the Earth.

(iii) Manned-flight programme. Sputnik 2 was the first indication of a manned spacecraft subsystem development. However, Sputniks 4, 5 and 6 in 1960 and Sputniks 9 and 10 in March 1961 were the real precursors of manned flight and were prototypes for the Vostok system that carried Gagarin into orbit. Sputnik 4 weighed over five tons and carried a dummy cosmonaut. It failed to return to Earth after 64 orbits, as planned, and instead entered a higher orbit to finally decay in 1962. Sputnik 5 carried the two dogs Belka and Strelka and was successfully recovered on the eighteenth orbit. This was the first time that living creatures returned from orbit. Sputnik 6 also carried two dogs but recovery failed and they thus became the first living creatures to die in return from an orbital flight. Sputniks 9 and 10 were launched in quick succession in March 1961. Each carried a dog and in both cases the animal was recovered successfully.

Vostok 1 was launched on 12th April 1961, with Cosmonaut Yuri Gagarin on board. He became the first man to orbit the Earth. The spacecraft was launched from Tyuratam and after one orbit it landed some 400 miles southeast of Moscow. Gagarin landed within his re-entry module. Vostok 2 followed a few months later in August with Herman Titov on board. He made 17 orbits and suffered from nausea after the fifth orbit. He used an ejection seat on landing, and left the re-entry module at 21,000 feet, descending by parachute. A year later the first paired flight took place with Vostok 3 and 4. They were launched within one day of each other on 11th and 12th August. Vostok 3 carried Andrian Mikolayev and Vostok 4, Pavel Popovich, and they stayed up four and three days respectively. Vostok 4 was launched so that it approached within four miles of Vostok 3. However, it seems that neither vehicle had manoeuvring capability and they slowly drifted 1,800 miles apart during the rest of the flight. Radio contact was maintained throughout the mission and both cosmonauts spent at least three hours floating freely about their respective space cabins.

June 1963 was another paired flight, this time with the first, and so far only, woman cosmonaut in the second spaceship Vostok 6. Valentina Tereshkova was launched two days after Vostok 5 which carried Valery Bykovsky on board. Some observers believe that Vostok 5 did not achieve its intended orbit thus preventing a repeat of the Vostok 3 and 4 mission. Vostok 6 was launched so that it passed within three miles of Vostok 5 twice in every orbit. It made 48 orbits before Tereshkova landed by parachute. Bykovsky stayed up for 81 orbits. Both vehicles carried an increased number of biomedical experiments, and Bykovsky also carried out an extensive series of scientific investigations.

In 1964 the Soviets started the Voskhod series of manned flights. The vehicle was slightly heavier than the Vostok, being some 11,731 pounds compared with the 10,428 pounds of the previous four Vostok flights. Photographs released indicated that the Voskhod was similar to the Vostok and probably an adaptation. Viskhod 1 was launched on 12th October 1964, with three men on board - Vladimir Komarov, the cosmonaut-pilot; Konstantin Feoktistov, scientists and spacecraft designer; and Boris Yegorov, a physician. Feoktistov and Yegorov were the first civilian cosmonauts in the Soviet space programme. The crew did not wear pressure suits during the flight which lasted for one day and involved 16 orbits. It was the first time that a physiologist was able to study the human being in a weightless state on the spot.

Voskhod 2 followed in March 1965 with two cosmonauts on board, Pavel Belyayev and Alexei Leonov. It was on this flight that Leonov left the spacecraft and spent 20 minutes outside the vehicle, and thus became the first man to walk in space.

This flight signalled the end of the Voskhod series, which clearly was an interim programme prior to the development of a new generation of spacecraft with much greater capabilities, namely, the Soyuz series. The Soyuz series began in April 1967 and is still going on. Soyuz 1 carried one cosmonaut, Vladimir Komarov, but he was killed upon re-entry when the lines of his parachute became entangled. His death must have put a brake on the Soviet space programme, and Soyuz 2 and 3 were not launched until October 1968. Soyuz 2 was unmanned and served as a rendezvous target for Soyuz 3. This carried one cosmonaut, Georgi Beregovoy, who manoeuvred to within 650 feet of Soyuz 2. He landed after 64 orbits. Soyuz 4 and 5 followed in January 1969. Soyuz 4 carried one cosmonaut, Vladimir Shatalov who carried out the first successful manual docking in space with Soyuz 5. Soyuz 5 carried a three-man crew of Yevgeny Khrunov, Alexei Yeliseyev and Boris Volynov. Khrunov and Yeliseyev transferred to Soyuz 4 and thus made the first exchange of crews between manned spacecraft. Volynov was left to land alone. These manoeuvres were testing techniques necessary for the development of manned space stations.

The world's first triple manned launch was the next step in the Soyuz programme. Soyuz 6, 7 and 8 were launched on consecutive days in October 1969. Soyuz 6 carried Georgi Shonin and Valeri Kubasov; Soyuz 7 carried Anatoly Filipchenko, Viktor Gorbatko and Vladislav Volkov; and Soyuz 8 carried Vladimir Shatalov and Alexei Yeliseyev from the earlier Soyuz 5 flight. The main aim of this record seven-man launching was to test common manual control from space and ground of three simultaneous flights. Shonin and Kubasov in Soyuz 6 carried out automatic welding experiments. The three-man crew of Soyuz 7 carried out independent navigation experiments, studies and photographs. Shatalov and Yeliseyev in Soyuz 8 made a number of manoeuvres with Soyuz 7, took pictures and communicated via the communications satellite Molniya 1.

Soyuz 9 came in 1970. It had two cosmonauts on board - Andrian Nikolayev and Vitalii Sevastianov - and they completed a record-breaking 18 days in space. They carried out an extensive scientific programme, but both men were adversely affected by prolonged weightlessness and took several weeks to recover upon their return. Soyuz 10 and 11 were launched in April and June 1971 as transport ships for the Salyut orbital space station, also launched in April. They both had modified docking systems and other instrumentation. After a record-breaking flight of 24 days in space, the three astronauts on Soyuz 11 were found dead on their return to Earth.

(iv) Planetary Probes. In 1960 the Soviet space programme widened to include a steady attack on the planets. Advantage has been taken of practically every so-called window (the period when minimum energy is needed to reach the planet) for Mars and Venus. However, the two probes launched to Mars in October 1960 both failed; this was revealed by NASA in 1962. In February 1961 there were two launches. Sputnik 7, weighing 14,292 pounds, went into a low Earth orbit. Then eight days later Sputnik 8, weighing the same amount, went into a similar orbit before launching the Venus 1 probe. This weighed 1,419 pounds and carried cosmic ray, magnetic field, charged particle, solar radiation and micro-meteoroid experiments. It missed Venus by 60,000 miles and radio contact was in any case lost some 4,700,000 miles from Earth.

In 1962 six planetary probes were attempted, five of them never left Earth orbit and were not announced by the Soviet Union. The sixth, Mars 1, launched in November, lost communication with the Earth 65,900,000 miles away. It carried similar experiments to the Venus probe and during its lifetime transmitted information on the Earth's magnetosphere, the interplanetary magnetic field and interplanetary cosmic ray flux.

There was a lull in the programme in 1963 followed by two successful deep space probes in 1964. These were given the general name of Zond. Zond 1 was placed on course for Venus but its communications failed before it reached its destination. Zond 2 was directed at Mars but again contact failed too soon. Zond 3, launched in 1965, die journey as far as the Mars orbit, but Mars was not there at the time. It took photographs of the back of the Moon as it flew past and filled in the pictures taken previously by Luna 3.

The same year saw three shots at Venus though one failed. Venus 2 passed close by the planet, and Venus 3 was assumed to have landed. However, communications failed from both probes as they approached. Two years later in June 1967 Venus 4 was launched and was presumed to have made impact on the planet. It sent back information for 94 minutes during its descent. When Venus 5 and 6, launched in 1969, reached the planet, they descended deeper into the Venusian atmosphere than Venus 4, making it clear that the latter's communications had failed before impact. It is likely that the same thing happened to Venus 5 and 6, because of the high temperature and pressure at the planet's surface.

Venus 7 was launched to the planet in 1970. First reports indicated that the communications had again failed during the descent; however, the Russians later announced that it had in fact landed on the surface of Venus and transmitted from there.

1971 saw the launching of two probes to Mars.

(v) Zond Programme. The first two Zond probes were directed at Venus and Mars respectively in 1964. Zond 3 was, a year later, a possible Mars probe, though it also made a successful lunar flyby taking pictures of the far side of the Moon at about 6,200 miles above the surface.

There was then a gap of three years in the programme followed by three launchings in 1968 which clearly marked a new phase, and a new satellite design. Little is known about the mission of Zond 4 which was launched into a near-Earth parking orbit. However, Zond 5 and 6 both flew around the Moon and then returned to Earth. Zond 5 landed in the Indian Ocean and Zond 6 soft landed in the Soviet Union. They both carried out photographic and biological experiments. Zond 7 followed in August 1968 with a similar circumlunar flight and, like Zond 6, it soft landed on return in the Soviet Union, making use of a skip entry involving aerodynamic lift to reduce the probe's speed.

(vi) Cosmos Programme. Cosmos is a blanket term covering a multitude of purposes and programmes. It includes a large number of purely scientific experiments which are usually discussed in the scientific literature, and a smaller number of military reconnaissance satellites. It is also the name used for rehearsals of manned probes, rehearsals or failures of lunar and planetary missions. Cosmos satellites are usually the precursors of later named programmes such as the Meteor, Molniya, Intercosmos, and so on. By the beginning of 1971 there had been some 400 Cosmos flights.

(vii) Meteor Programme. The Meteor programme is the Soviet meteorological programme. The first Meteor satellite described under that name was launched in 1969. Previously, observers believed that there had been 13 satellites of the Cosmos series used for meteorological purposes. Since then there have been four more Meteor satellites.

(viii) Molniya Programme. The Molniya 1 series of communications satellites started in 1965, though Cosmos 45 in 1964 is presumed to have been a test launch of a Molniya-type satellite. These communications satellites are launched into high elliptical and highly inclined orbits that have approximate 12-hour periods, which makes it possible for them to stay over the Soviet Union for some six to eight hours on each orbit. They are used for relaying black and white, and colour television, for relaying radio, telegraph, telephone, meteorological charts, and newspaper facsimiles. They serve the thirty ground stations of the 'Orbita' space communications network. The Molniya/Orvita system was devised to avoid the substantial engineering and logistics problems involved in providing a full communications network solely on the ground using micro-wave or cable systems, because of the vast distances involved and the difficult weather conditions in the north of the country. This space network operates for some 16 to 20 hours each day. By the beginning of 1971 there had been eleven launches of Molniya communications satellites.

(ix) Intercosmos Programme. The Intercosmos programme is a series carried out in co-operation with the other socialist countries in Eastern Europe - Bulgaria, Hungary, East Germany, Poland, Romania, and Czechoslovakia. Again initial flights were just designated as part of the Cosmos series, in this case Cosmos 261 launched in December 1968. The first official Intercosmos satellite came in October 1969 and was followed by Intercosmos 2 a couple of months later. The satellite carried out scientific experiments including investigation of solar flares, and the characteristics of the ionosphere. By the beginning of 1971 there had been four satellites altogether in this new series.

(x) Three other short-lived series remain to be mentioned:

    (a) *Polyot Vehicles.* There were two payloads in the Polyot series, the first launched in late 1963 and the second in 1964. Polyot 1 was the first space-craft to be able to manoeuvre extensively; Polyot 2 carried out similar movements on its first day in orbit.

(b) *Elektron Series*. Four vehicles were launched in pairs in the Elektron series during 1964. Elektron 1 flew to an apogee of 4,000 miles and studied the inner Van Allen radiation belt while Elektron 2 flew to an apogee of 40,000 miles and studied the outer belt. This was the first dual-payload Soviet launch. It was able to map the radiation belts and provide synoptic readings. Elektron 3 and 4 followed six months later in July with the same task of studying the radiation belts.

(c) *Proton series*. Three Proton flights took place during the period 1965 to 1966, and each was a highly impressive cosmic ray experiment. The experiment weight was set at 12.2 metric tons and was able to measure energies to a level of 100,000,000,000,000 electron volts in some of the highest fluxes ever detected by human instruments. Proton 4 launched in 1969 weighed 17 metric tons and was intended to continue the cosmic ray research.

(E) Launching Vehicles

Only since 1967 have any launch vehicles for orbital flight been put on public display, together with statistics on their performance. Little information was released in the earlier years, and even now the finer points of engineering have not been wholly revealed and some vehicles remain totally unveiled.

It is believed that there are six basic launch vehicles:

(1) Vostok is the standard Soviet launch vehicle. It is the original ICBM developed on the basis of work done at the Gas Dynamic Laboratory in Leningrad during 1954-57. It is made up of a central core liquid stage 91.8 feet long and 9.7 feet in diameter, which is the shape of an ordinary cylinder. There are four liquid stage strap-ons, each 62.3 feet in length and 9.8 feet maximum diameter, each one tapers toward the upper end as modified elongated cones. The four strap-ons carry the RD-107 engine which burns liquid oxygen and a hydrocarbon. The central core cylinder has a RD-108 engine which apparently only differs from the RD-107 in having four steering rockets instead of two, and a longer burning time. All four strap-ons and the core ignite before lift-off, and after the four strap-ons drop away, the core continues to burn. This gives a combined thrust (vacuum equivalent) of over 1,124,000 pounds. This basic launch vehicle has been used for the original Sputniks, and then with added stages it was used for the Vostok series, the Voskhods and Soyuz flights, and then, using the orbiting platform technique, for all the Luna flights, the Molniya 1 series, the Zond flights, and the Mars and Venus flights.

(2) Cosmos is the small launch vehicle. It is made up of the intercontinental ballistic missile (ICBM) plus an upper stage. The upper stage, designated RD-119 by the Soviets, was developed between 1958 and 1962 by the Leningrad Gas Dynamics Laboratory. It burns liquid oxygen and dimethyl hydrazine and has a thrust of 24,250 pounds. The total vehicle combination is 98 feet long and 5.4 feet in diameter, and most of the payloads it puts up are spin-stabilised. This launch vehicle has been used solely for the Cosmos series launched from Kapustin Yar.

(3) Skean is an intermediate vehicle. This launch vehicle has not been put on display, but it is presumed to be a medium-range ballistic missile given the code name Skean by the NATO countries. It has been used for payloads in the Cosmos series launched from Plesetsk.

(4) Proton is a non-missile space launch vehicle. There is little data on this vehicle. It has been announced to have a capacity three times that of the Vostok; this would mean it has a first stage thrust of 3,300,000 pounds. So far the largest payload it has put up is that of Proton 4 which weighed 37,000 pounds. This launch vehicle has been used for the Proton series, for the latest Zond flights, and for some of the Cosmos series.

(5) Scrag or Scarp is a military launch vehicle. Both these names are code names designated by NATO. Scrag is a three-stage liquid fuelled rocket, and Scarp is also liquid fuelled and probably has two stages. Both have been displayed at the military parades in Moscow and it is believed they have been used as part of the Cosmos series to test the fractional orbit bombardment system. They are launched from Tyuratam but at a different inclination from that of the Vostok vehicles.

(6) There is a good likelihood of a very heavy launch vehicle system, yet no name is known for it, nor whether it really exists. It was talked about as being in an advanced stage of development in 1967 by General Kaminin and Cosmonaut Popovich. It is believed to have a greater thrust than Saturn 5 and in its full development it would have a greater lift capacity. There may have been a setback in the development of this new launch vehicle as there were rumours that it had blown up on its pad in 1970.

(F) Research and Development

Space technology and rocketry are priority research fields in the Soviet Union. This means that they are top of the list in the allocation of resources; there is effective central co-ordination and control combined with flexibility in planning. In the 1950s and early 1960s, work on ICBMs and space rockets was carried out by various experimental bureaux, each responsible for a different part - the rocket itself, the rocket engines, the control systems, ground equipment, and so on. There were also a number of industrial and Academy Research Institutes involved in the design work. The work of all these separate teams was co-ordinated by the chief designer, the late Academician Sergei Korolev. Presumably the same type of organisation exists today.

Fields such as space and rocketry are at an advantage since the government uses its authority to enforce their priorities. Pressure from the centre has more effect than the formal requirements of the production plan, and difficulties and obstructions disappear. These priority fields also possess special production facilities for innovation. Experimental factories are attached to the experimental design bureaux and these are used for testing prototypes before serial production begins.

According to Professor Grigori Tokaty (in his piece on *Aeronautical Engineering Education and Research in the USSR*, Lawrence, Kansas, USA, 1960), the aeronautical research establishments come under the administration of four government bodies: the Ministry of Defence; the State Committee of the Aviation Industry; the State Committee of Armaments; and the Chief Managing Department of the Civil Air Fleet.

Tokaty names the *Central Aero- and Hydrodynamic Institute* (TsAGI) named after Professor N.E.Shukovsky as the equivalent of the American NASA. He lists three TsAGIs in all. Old TsAGI started in 1918-19 in Moscow. It is still in existence today, its facilities including a number of relatively old but still good wind tunnels, a first-class hydrodynamics laboratory, a bureau of information, a laboratory of airscrews and fans, and a precision workshop. New TsAGI came into existence in 1936 when a whole new research complex was built at Stakhanovo (now called Zhukovskaya), southeast of Moscow. The most important facility of the new TsAGI was a big wind tunnel officially known as T-101. It also has a number of other wind tunnels; a 'laboratory of spin' with a vertical wind tunnel, low turbulence wind tunnels, high speed wind tunnels, etc. Tokaty's 'newest new TsAGI' is the Institute of Hydrodynamics at Novosibirsk. The *Flight Research Institute* (LII) is situated close to new TsAGI. It has an aerodrome, experimental factory, ground laboratories and about 2,000 scientific personnel.

Tokaty lists three important military research establishments of the air force which come under the administration of the Ministry of Defence. These are the Scientific Testing Institute of the Air Forces (NIIVVS), the Scientific Testing Institute of Aviation Armaments (NIIAV) and the Scientific Testing Institute of Aviation Instruments (NIIAP). All three are situated to the northeast of Moscow at Chkalovskaya. They have first-class experimental facilities, including an important aerodynamical laboratory. The next best aerodynamical laboratory of the Soviet Air Force, according to Tokaty, is that attached to the Zhukovsky Aeronautical Engineering Academy in Moscow. This Academy is both a teaching and research establishment.

Other research institutes mentioned by Tokaty are the Central Institute of Aviation Engines (TsAM) in Moscow and the All-Union Institute of Aviation Materials (VIAN).

The Gas Dynamics Laboratory in Leningrad in another important research centre. Most of the work on the development of the ICBMs was carried out there.

Graduate and postgraduate training in aeronautics is carried out in the following higher educational establishments: the Zhukovsky Military Aeronautical Engineering Academy in Moscow; the Mozhaisky Military Aeronautical Engineering Academy in Leningrad; the Ordzhonikidze Moscow Aviation Institute; the Moscow Aviation Technological Institute; the Kharkov Aviation Institute; the Kuibyshev Aviation Institute; the Kazan Aviation Institute; the Ufa Aviation Institute; the Alma-Ata Aviation Institute; the Kiev Institute of the Civil Air Fleet; and the Leningrad Institute of the Civil Air Fleet. Tokaty names these the 'All-Union professional universities of aeronautical engineering and technology'. In addition he mentions the existence of two higher colleges of military aeronautical engineering; two chairs of aeronautical engineering in non-aeronautical military academies; four aeronautical engineering technikums (the Moscow Aviation Technikum is one example); and twelve departments and chairs of aeronautics in non-aviation universities (Moscow State University, the Moscow Higher Technical College, the Leningrad Polytechnical Institute, and Saratov State University are the best known).

Journals: Journals on space research include: *Aviatsiia i Kosmonavtika* (Aviation and Cosmonautics); *Kosmicheskaya Biologiia i Meditsina* (Space Biology and Medicine); and *Kosmicheskie Issledovaniia* (Space Research).

(G) Finance

No budget figures are published for the space programme as such. Using comparative techniques with US expenditure it is estimated that the Soviet share of GNP devoted to space must run at about 2 per cent (double the relative share of the United States, whose GNP is twice that of the Soviet Union). This would mean a figure of some 2,700,000,000 roubles in 1969 with the GNP of 138,500,000,000 roubles that year. The total defence budget was 17,700,000,000 roubles but again, not all the space research would be covered by this, a part of it is carried out in non-military research organisations.

(H) Plans and Policy

Soviet goals from the space programme are various, including the general support of science, technological insurance against being caught by some military surprise, and practical applications on Earth such as communications, weather reporting, navigation and military activities.

In general, the space programme has three basic trends: namely, exploration of near-Earth space, exploration of the Moon and circumlunar space, and interplanetary study (basically of Mars and Venus). One long-term professed aim of the Soviet programme is to put up long-operating manned orbital stations, and the first step towards this was the launching of a new series with Salyut 1 in April 1971.

7 Military Affairs

(A) Introduction

The "Workers' and Peasants' Red Army" was formed at the beginning of 1918. The "Workers' and Peasants' Red Navy" was organised at the same time. In 1925 the armed forces were divided into land troops (cavalry, artillery, armoured, etc.), air force and navy. These together with the OGPU (State Police) numbered 586,000 men in all.

Between 1929 and 1941 the strength of the armed forces in artillery, tanks, ships and the like was considerably increased. By May 1945 the number of men in the armed forces totalled 11,365,000 but by 1948 it had fallen to 2,874,000 because of demobilisation and age wastage.

(B) Organisation

According to the Constitution of the USSR, the Supreme Soviet has highest authority over the armed forces. It passes laws on questions of war and peace and organises the defences of the country.

The Council of Ministers and Central Committee of the Communist Party direct the general construction of the Armed Forces. Day-to-day authority lies in the hands of the *Ministry of Defence*. The General Staff of the Armed Forces, which is the central governing organ of the Armed Forces comes under this ministry.

The Armed Forces are divided for organisational purposes into the following sections: strategic rocket forces; the army; the air force; the navy; and the air defence command. To help in the running and organisation of the forces, the whole territory of the Soviet Union is divided up into military regions, with a military council in charge of each.

7(1) STRATEGIC ROCKET FORCES

The Strategic Rocket Forces possesses the country's basic fire power, and they are on a permanent battle-ready footing. Their task is to destroy the strategic and operational means of attack of the enemy, to wipe out his military and economic potential, and to break down his civil and military government.

The Strategic Rocket Forces have a strength of 350,000 men. They are armed with 1,300 operational inter-continental ballistic missiles (ICBM) and 700 operational intermediate-range and medium range ballistic missiles. The ICBMs include more than 1,000 liquid-fuelled and some 40 solid-fuelled rockets. Testing of multiple re-entry vehicles (MRBM) has been going on for more than two years and they may now be available for deployment. There have also been tests for the ICBM/fractional orbital bombardment system. The IRBMs and MRBMs are all liquid fuelled at the moment. Development continues of the solid fuelled missiles.

7(2) AIR-DEFENCE COMMAND (PVO)

The task of the Air-Defence Command is to protect the most important administrative and political centres, industrial regions, the transport and communications network and other objects of economic and strategic importance for the country.

PVO is a separate command of anti-aircraft artillery and surface-to-air-missile units. They use an early warning system based on radar and fighter interceptor squadrons. The total personnel of the PVO numbers about 500,000 of which about half are ground forces.

Their weapons include: anti-aircraft artillery, surface-to-air missiles, anti-ballistic missiles (ABM), fighters and early-warning aircraft.

7(3) THE ARMY

The total size of the Army, including the ground forces of the PVO, is estimated to be around 2,000,000 men. The Army is organised into a number of sections:

(i) The motorised infantry troops - this is the largest section, equipped with armoured transport, carbines, automatics, mine throwers, artillery, tanks and anti-tank guided missiles.

(ii) The tank troops - this section numbers around 400,000 men and has about 150,000 medium tanks at its disposal at full strength.

(iii) The rocket troops and artillery - this is a new section of the army combining the originally separate rocket and artillery troops.

(iv) The parachute troops.

(v) The engineers.

(vi) Chemical troops - this section has grown up because of the rapid development of nuclear and other forms of mass destruction. It includes radiation and chemical reconnaissance troops whose job is to de-gas and de-activate contaminated machines and areas.

(vii) The communication troops.

(viii) The transport troops.

7(4) THE NAVY (VMF)

The main task of the naval forces is to defend the interests of the Soviet Union at sea, to ensure the defence of her sea frontiers and, in a situation of war, to destroy the enemy's sea communications, naval bases, ports, and other strategic objects. VMF either carries these tasks out alone or in conjunction with the other armed forces.

The Navy's total strength is around 475,000 men, of which 75,000 belong to the naval air force. Other sections apart from the naval air force include the submarine fleet, the surface fleet and the shore rocket division. There are also specialised sections covering engineering, construction, aviation technology and so on. In total tonnage it is the second biggest navy in the world, its main strength lying in the submarine fleet.

The chief striking force of the Navy is its 80 nuclear-powered submarines armed with long range rockets (including those for submerged firing) and self-aiming torpedoes with nuclear warheads. It also has some 290 conventionally powered submarines. About 25 of these can fire ballistic missiles and 16 are equipped with anti-shipping cruise missiles with ranges of up to 300 miles. The remainder are attack submarines or training craft.

The naval air force possesses long-range jets carrying various types of rockets with both conventional and nuclear warheads. It also has bombers, helicopters, amphibious craft and transport aircraft.

The shore division is equipped with guided missiles.

7(5) THE AIR FORCE (VVS)

The Air Force is designed to operate either in conjunction with other divisions of the armed forces or on its own in independent operations. VVS covers three main categories: the long-range air force; the tactical or front-line air force and the air transport division. There are also the air divisions attached to PVO and the Navy.

The total strength of the Air Force is 480,000 men and 10,200 combat aircraft. Planes used include long and medium-range strategic bombers equipped with air-to-surface missiles (long-range air force) and fighters and light bombers (the tactical air force). The fighters are equipped with air-to-air and air-to-surface missiles. Interesting types include a variable geometry aircraft, similar to the American F-111, which was displayed at the Moscow Air Show in 1967, the supersonic strike version of the MIG-23 which may now be operational, and a vertical take off and landing variety (VTOL) which could come into service soon. The military transport planes have been developed to carry troops and heavy military equipment long distances and at high speed. Helicopters have also been widely developed.

In 1955 the armed forces numbered some 5,763,000 men. Because of scientific and technological developments, the numbers were reduced steadily during the period from 1956 to 1961. However, since then the numbers have remained fairly steady. The figure for 1970 was estimated to be 3,305,000 (Source: *The Military Balance 1970-1971*, P 6, (The Institute for Strategic Studies, London, 1970).

(C) Finance

Defence appropriations for 1970 were 17,900,000,000 roubles. This figure is the declared budget of the Ministry of Defence and does not include certain expenditures such as the cost of nuclear warheads, research and development expenditure on advanced weapons systems, and the military elements of the space programme, which are believed to be included in the budgets of other ministries. Total military expenditure could be around 23,000,000,000 roubles. During the 1960s, Soviet military research and development, which covers space and atomic energy efforts as well as military, increased about 60 per cent.

(D) Research and Development

"Statistics published in the Soviet Union concerning Higher Educational Establishments are incomplete. According to Korol, for instance, military schools are not included in these statistics. In his calculations Nicholas De Witt, includes 22 military (Academy type) Higher Educational Establishments, which he divided into eight Establishments for Command Staff and Political Officers, ten Engineering Establishments, and four Medical and Veterinary Establishments." (Source: Zaleski et al: *Science Policy in the USSR*, P 298-9, OECD, Paris, 1969).

"Through an extensive network of military and naval schools, the Soviet Defence Establishment possesses the means to increase substantially the military educational level of its personnel and to raise the battle readiness of its forces.

"In the Soviet military training system there are three general categories of training: intermediate, advanced and postgraduate. Intermediate and advanced schools lead to commissions, and have an upper age limit of 23 years for enrolment. Postgraduate education is for officers, and candidates must be under the age of 40. Almost all schools require candidates to take entrance examinations in mathematics, physics, Russian language and literature. Postgraduate students are selected from the results of competitive examination in their special discipline, history of the USSR Communist Party, and one foreign language (English, German or French). They cannot have less than two years practical work experience in their speciality immediately preceding application.

"Aviation schools include 12 advanced aviation schools, nine intermediate schools and nine advanced schools and two postgraduate aviation schools. There are no less than ten advanced engineering schools, seven intermediate and three postgraduate engineering schools. Courses in communication and radio engineering are provided in five advanced, eighteen intermediate and two postgraduate radio schools. The PVO forces sponsor most of the radio-engineering training. Political education is provided in eight advanced and two postgraduate schools. Training in western and eastern language is offered as postgraduate courses at the Military Institute of Foreign Languages in Moscow. Foreign language training is also included in the curricula of all military academies. Intermediate training in transportation is provided by the Leningrad School of Military Transportation im. M.V. Frunze. There are ten artillery schools. There are advanced anti-aircraft schools, and advanced tank command schools.

"The scope of equipment and manpower devoted to Soviet military education confirm the USSR's dedication of considerable financial support to military training. Education, an integral part of the military and naval establishment is recognised as the underlying foundation of the defence system". (Source: *US Naval Proceedings*, August 1969, P 134).

Since the Second World War the personnel of the armed forces has changed considerably. Of the officers 25 per cent have higher military or technical education. In PVO and VVS, 70 per cent of the officers are either engineers or technicians.

Journals: Journals on military questions include: *Aviatsiia i Kosmonautika* (Aviation and Cosmonautics); *Voenno-Meditsinkii Zhurnal* (Military Medicine Journal); *Voennye Znania* (Military Knowledge); *Voenny Vestnik* (Military Herald).

(E) Policy and Plans

Policy and plans are outlined in the programme of the Communist Party. This underlines the important role of the Communist Party within the armed forces and the need to strengthen Party organisation within the army and navy. The most important aims for the building up of the armed forces are as follows: strengthening the material and technological base of the forces; the development of its fire power; improving the organisation and discipline; preparation of political and technical cadres; strengthening educational work among the troops; improving relations between the army and civilian population and with the armies of other socialist countries.

8 Computer Industry*

(A) Introduction

Work on the first Soviet digital computer, the MESM, began in 1948 and was completed in 1951. The MESM, said to be the first sequence-controlled computer in continental Europe, was designed by Academician S.A. Lebedev at the Computer Centre (later reorganised into the Institute of Cybernetics) of the Ukrainian Academy of Sciences. The history of Soviet computers does not really begin, however, until 1953 - the year Stalin died. In addition to the increased educational facilities and greater emphasis on scientific studies which date from that time, the post-Stalin period saw a lessening of ideological opposition to the concept of cybernetics or automation. Admiral Axel Berg was an influential supporter of cybernetics and was one of the first to advocate the organisation of cybernetics research in scientific and industrial establishments. As a result, when the idea became ideologically acceptable, he was appointed chairman of the Soviet Academy's Scientific Council on Cybernetics.

Designs of the first member of the important Ural family was completed in 1954, and serial production began the following year. Between 1953 and 1954 the BESM-1, Strela, and Lem-1 computers became operational. Research initiated in the mid-1950s resulted in the appearance of more than a dozen new digital computers by 1960. The most advanced computer research, particularly in the earlier years, was conducted under the leadership of Academician S.A. Lebedev, designer of the MESM. After the completion of the MESM, he was appointed director of the *Institute of Precise Mechanics and Computer Engineering* (IPMCE) in Moscow. In 1952 he completed the design for the first BESM, the first large electronic computer in the country. In the same period, work on computers was also going on in the USSR Academy of Sciences' Institute of Electronic Control Computers under I.S. Bruk, and in the Institute of Mechanics and Instrument Design of the Ministry of Radio Industry under Yu. Ya. Basilevskii. The first Soviet computer with alphanumeric input/output (I/O) capabilities, the Setun, was developed in 1958 as part of a graduate student project at Moscow State University. Centres of computer research also developed in the 1950s in Armenia around the Institute of Computing Machines founded in 1956-57 under S.N. Mergelyan as director, and in Byelorussia around the Institute of Physics and Mathematics of the Byelorussian Academy of Sciences under the direction of Ivan V.Lebedev. (No relation to S.A. Lebedev).

In the early 1960s, the government recognised the benefits that could be realised from the introduction of computer methods - namely automation - to industry, and initiated a large-scale industrial automation effort, which led to the development of numerous

---

* This chapter draws heavily on material published in *Soviet Cybernetics Review*, Rand Corporation.

production control machines. Two examples are the Kiev-67 special-purpose computer and the Dnepr series of machines designed by Glushkov's team at the Ukrainian Academy's Institute of Cybernetics.

Although automation, or the science of control, is one of the USSR's newest and most rapidly developing branches of knowledge, the country still lags behind the United States in automation technology, by some five to ten years. (Based on *Soviet Cybernetics Review*, volume 4,1, P 6, January 1970).

(B) Organisation

There is no single ministry or committee responsible for overall research, development and production in the computer industry. Computer technology is apparently not regarded as a top priority and it therefore becomes the secondary responsibility of a number of ministries concerned with its applications. For example, in the Ukraine: "responsibility for the application of computers is with the heads of ministries, departments, enterprises and organisations, while planning for the efficient use of this equipment is under the Ukrainian Gosplan. Structural subdivisions are being developed within ministries and departments for problems of planning and introduction of automated control systems and computers. The cybernetics section of the 1970 Ukrainian State Plan for the development of the national economy includes problems of computer software, further development of the theory of digital automata, and development of scientific bases and methods of optimal control." (Source: *Soviet Cybernetics Review*, volume, 4, 7, P 37, July 1970).

The Ministry of the Machine Construction and Instrument Industry is one of the key ministries concerned with computer technology. It is developing automated systems for controlling entire branches of industry. Since 1966, when the order to establish these went through, a certain amount has been established. Teletype machines have been installed at 180 plants; branch information stations for collecting current information have been established in 43 regions of the country; a department for the introduction of computing equipment has been established under the ministry's technical administration; a communication centre and an information and computing centre connected via communication channels with the branch stations have been established in Moscow; and a department for the development and introduction of a branch automated control system has been established at the Experimental Scientific Research Institute of Metal-Cutting Lathes.

The development of an automated control system is being retarded, however, because there is no single policy on branch computing centres. Each branch devises its own policy, and in the same city, different ministries may organise their own branch information stations even though it is likely that one would be sufficient to process all information.

The *Central Statistical Administration* (CSA) is another key organisation in the computer industry. It establishes and adminsiters the large network of District (City) Machine Accounting Stations that have been set up in the past few years. As of January 1, 1969, 830 profit-based machine accounting stations and computer centres were in operation in districts and cities, and the organisation of 230 had been started. Since a government directive in 1966 on the need to set up a *State Network of Computer Centres (SNCC)*, the CSA has been regarded as a possible body for the implementation of this project. However, the whole subject is still under discussion and quite controversial.

Other organisations connected with the organisation of computer research, development and production include the Scientific Council on Cybernetics of the Academy of Sciences, the State Committee on Radioelectronics, the Ministry of the Radio Industry, and the Ministry of the Electronics Industry.

(C) Production

"In the current five-year plan, the growth of computer production in the USSR is proceeding at a high rate, and is significantly in the lead in instrument construction as a whole; the proportion of computers in total instrument production is also growing. Recently, the transition from first-generation to second-generation computers was completed. The production of series of small, medium, and large general-purpose and control computers has been established (mir. BESM-4, M-220, Minsk-22 and Minsk-32, BESM-6, the Dnepr series, IV-500, and so on). Inherently commendable centralised monitoring and regulating units are being serially produced. The production of punchcard equipment, including alphanumeric models, has sharply increased. More than five desk computer models have been developed, three of which have been put into serial production. The production of magnetic tape and drum external storage devices, of alphanumeric printers and of papertape perforators has increased.

"Important measures have been taken toward establishing assembly and installation organisations to handle the introduction of automated control systems and computers. In the past few years, important experience has been gained in the development and introduction of automated control systems based on computers.

.."However, in spite of the relatively high rate of computer production, the production level and the technical level of equipment being produced do not meet national economic requirements. Computers and especially input/output devices, produced by industry, are not always reliable or sufficiently productive. Inadequate software sharply lowers efficiency and scope of computer applications...." (Translation of excerpts from unsigned editorial in *Pribory i sistemy upravleniya*, No. 1 PP 1-2, 1970; translated by Irene Agnew in *Soviet Cybernetics Review*, volume 4, 7, P 3, July 1970).

The Ministry of Instrument Construction, Means of Automation and Control Systems is the leading producer of computers. Although production more than doubled during the 1965-70 five-year plan period, the demand for computers is still far from satisfied. Apart from the volume of production, the technical level is also not up to that required.

One of the reasons for poor production figures may be the lack of integrated organisation. In an article published in *Izvestiya*, in October 23, 1969, I. Manyushis, chairman of the Lithuanian Council of Ministers, described the situation in Lithuania, including the setting up of a special association to help boost production:

"In Lithuania there are a comparatively large number of enterprises under all-union ministries, which produce similar or related products. Therefore the establishment and operation of production associations is of utmost importance. The Sigma Association, a profit oriented territorial association of computing and office equipment manufacture, under the USSR Ministry of Instrument Construction, Means of Automation and Control Systems, has been operating in the Republic for some four years.

"The Sigma Association is comprised of seven plants and three design bureaux, and is managed by a special group which has been accorded the rights of an All-Union Main Administration. This type of organisation is capable of more rapidly organising production, more easily establishing economic co-operative relations, and more effectively utilising modernisation funds. It is more efficient and better qualified than individual enterprises to immediately solve day-to-day problems.

"From 1966 to 1968, the average annual increase in production of the Sigma enterprises was 25.7 per cent, compared with 10.6 per cent for the preceding three years. The average increase in profit was 29.5 and 8.7 per cent respectively..." (Source: *Soviet Cybernetics Review*, volume 4, 2, P 20, February 1970).

A similar dispersal of productive effort, as existed in Lithuania, may be the reason for poor production figures in other parts of the country.

At the present time transistorised, second-generation computers account for nearly all computer manufacture. These include the BESM-6, the largest and fastest of the known Soviet computing machines, which went into production at the Moscow SAM Plant in 1966; less than a dozen appear to have been built. There is the Kiev-67, announced in 1967 which was specifically designed for controlling cathode-ray technological processes in the production of electronic components. This machine was to help automate computer production, since most computer components are being assembled manually. The Nairi-1, completed in 1964, was a very successful production machine. More than 500 were manufactured, which is more than any other Soviet machine.

On January 20, 1970, *Pravda* announced that the M-1000, the first of its third generation machines, had been accepted for production.

A major handicap to Soviet production has been the poor quality of electronic components and the very high rejection rates of defective units. Only recently have manufacturers begun to assume any responsibility for software or for the installation, debugging, and maintenance of the machines they manufacture.

Computer production plants include the Moscow Calculating Machines Plant, the Scientific Research Institute of Calculating Machines, the Penza Plant, the Severodonetsk Instrument Construction Plant in the Ukraine, the Minsk Ordzhonikodze Plant, and the Leningrad Electrical Machines Plant.

(D) Computer Applications in Industry

Automation or the science of control is one of the USSR's newest and most rapidly developing branches of knowledge. It is, however, sometimes difficult to distinguish between interactive real-time, production control systems, and management information systems since the Russian word for control (upravlenie) can often be translated more accurately as management.

It is generally admitted by the Soviet Union that they lag behind in the development of production planning and control systems and they are very concerned about this. The lag is usually attributed to the fact that sufficient hardware for developing sophisticated control systems is not available, and this is further complicated by the diversity of incompatible computers produced in the country.

The control system installed in the Lvov TV Plant is frequently cited as an example of a very successful, sophisticated control system. The Lvov system was developed under the direction of Victor Glushkov at the Ukrainian Academy of Sciences' Institute of Cybernetics. It is basically a management control system with some production control functions, based on the Minsk-22 computer. The system was tailored to the Lvov TV Plant application and the experimental prototype first installed underwent extensive debugging and considerable on-site refinement. It has been such a success that the Institute of Cybernetics has been charged with generalising the system for application in other plants, particularly those engaged in mass production of a limited variety of products.

The first automated, multichannel monitoring system developed in the USSR was the Avtooperator System; it was installed in a chemical plant in 1961. This control system was followed by numerous others: for example, the Dnepr-22, the SOU-1, the Stal'-2, the Zenit-3, and so on. The Stal'-2 is a specialised computer intended for use in controlling the production and cutting of rolled metal, and is an integral part of the Stal'-2 automated control system. The Zenit-3 computer is intended for the centralised regulation and monitoring of technolgical processes. The Dnepr-22 appears to be an attempt to achieve dynamic control of entire plant complexes. For controlling large plants, the Dnepr-22 process control system is linked to the Dnepr-21 data processing system and is made to function in an interacting mode; both systems utilise the Dnepr-2 computer. This macrosystem is referred to as the Dnepr-2 Production Control System.

Work on process control automation is underway in several Soviet republics. In Byelorussia, for example, automated control systems are being introduced at the Minsk Tractor, Watch and Automatic Lines Plants. An experimental training programme has been established at the Byelorussian Academy of Sciences for providing practical experience and academic training for engineers specialising in automated machine design.

The major problem impeding the development of a truly automated industry is the inadequacy of existing hardware and software.

Traditionally, the Soviet computer designers and manufacturers have not provided the software for operating their machines; the customer is expected to design his own software. They have only recently begun to assume any responsibility for software or for the installation, debugging and maintenance of machines they manufacture. (Based on *Soviet Cybernetics Review*, volume 4,3, P 11, March 1970).

There are 39 automated control systems and subsystems operating in the Ukrainian Republic, including 31 systems for controlling technological processes, five systems for controlling machine building enterprises, and one system each for material-technical supply of automative transport and construction. Areas of industry where computers have been introduced include ship designing, ocean research vessels, instrument building, the space programme, etc. There are about 300 computerised monitoring installations operating in domestic enterprises; they act as 'advisers' to those carrying out the technological processes, and are not capable of fulfilling all the functions of automatic control.

The setting up of a state network of computer centres (SNCC) is a long-term objective as part of the development of automated systems for controlling entire branches of industry. SNCC is still in the discussion stages. Computers are used by the Gosplan, the State Planning Committees of the various republics and by many ministries. The purpose of the SNCC would be to produce an information system which could satisfy the diverse requirement of management at all levels, and which would at the same time be free from duplication and discrepancy.

(E) Finance

"Strictly speaking, there is no such official field as computer technology included in the national budget. A minister has the right to redistribute funds allocated to his organisation as he deems fit and, since each minister is judged by his success in his own particular field, he will frequently skimp on funds for the development of computer techniques in order to meet or surpass the quota imposed by the government on his organisations field of specialisation". (Source: *Soviet Cybernetics Review*, volume 4,1, P 41, January 1970).

According to an editorial in the journal *Mekhanizatsiya i automatizatsiya upravleniya* (Mechanisation and Automation of Control), computers pay for themselves in 0.5 to three years, and save 3 to 10 per cent of human and material resources when used efficiently. For example, the automated system for controlling thermal regimes of blast furnaces at the Krivorog Metallurgical Plant cost 150,000 roubles, but paid for itself in 0.6 years. The total annual economic benefits from the introduction of the Lvov automated system in 1968 were 464,300 roubles.

(F) Research and Development

There is no one body with overall responsibility for research and development in the computer industry. Some of the research institutes and organisations come under the USSR or republican academies of science, while others are administered by industrial ministries.

The main research centres are:

- Institute of Automation and Remote Control of the USSR Academy of Sciences.
- Institute of Cybernetics of the Ukrainian Academy of Sciences.
- Institute of Precise Mechanics and Computer Engineering in Moscow.
- Institute of Electronic Control Computers of the USSR Academy of Sciences.

Scientific Research Institute of Calculating Machines
of the USSR State Committee on Radioelectronics.

Institute of Mechanics and Instrument Design of the
Ministry of the Radio Industry.

Penza SAM Plant.

Severodonetsk Scientific Research Institute of Control
Computers.

Minsk Ordzhonikidze Plant.

Institute of Computing Machines of the Armenian
Academy.

Sigma Association of Lithuania.

All-Union Scientific Research Institute of
Electromechanics.

Computer Centre of the USSR Academy of Sciences
(space research).

"The Soviets long ago identified the research and development gap as a major shortcoming of their system; in some cases it was as long as 10 to 15 years between the time a scientific discovery was made and its subsequent application to computer technology. Scientists have made great progress in the field of solid-state physics; they have developed integrated circuit components and other advanced microminiaturised circuits - e.g. thin-film and hybrid circuits. However, they have yet to widely implement these circuits in their computer hardware; the first Soviet computer to utilise integrated circuitry was the Nairi-3 which was only announced during the past year (1968).

"Another problem contributing to the research and development gap is the large number of jurisdictions under which computers are designed and produced; one expert claimed that his projects in the computer field required the approval of as many as ten organisations. Lebedev, for example, before designing a new machine needed not only the approval of the USSR Academy of Sciences to which his Institute was subordinate, but also that of the State Committee on Radioelectronics which also had some control over his activities....".

Another example is the VNIIEM-3 computer. "It was developed at an institute of one of the ministries; the state committee that tested it found it to be more than satisfactory. But the Ministry of Instrument Construction, Means of Automation and Control Systems did not want to become involved with its production; it had its own machines of the same type. Production, as a result, has been delayed by several years, and as far as is known it still is not being serially produced. Neither USSR Gosplan nor the Committee on Science and Technology has been able to crack the stubborn opposition by the ministries on many such questions." (Source: Rudins: *Soviet Computers an Historical Survey, in Soviet Cybernetics Review*, volume 4, 1, PP 41-43, January 1970).

"The 1968 reorganisation of research and development should help generally to cut down the gap between development of an idea and its subsequent application.

"At the same time both the Ukrainian Institute of Cybernetics and the Minsk Ordzhonikidze Plant have had close links between research and production for some time. The Institute of Cybernetics has close links with other research and development organisations and industrial enterprises in the Ukraine. Between 1961 and 1965 the institute worked on 43 contracts with industrial enterprises, solved over 300 problems and the results were introduced into production. About 20 to 30 per cent of its budget comes from contracts with industry. It was this Institute which developed the very successful system for the Lvov TV factory.

Figure 6 The Development of Soviet Computers

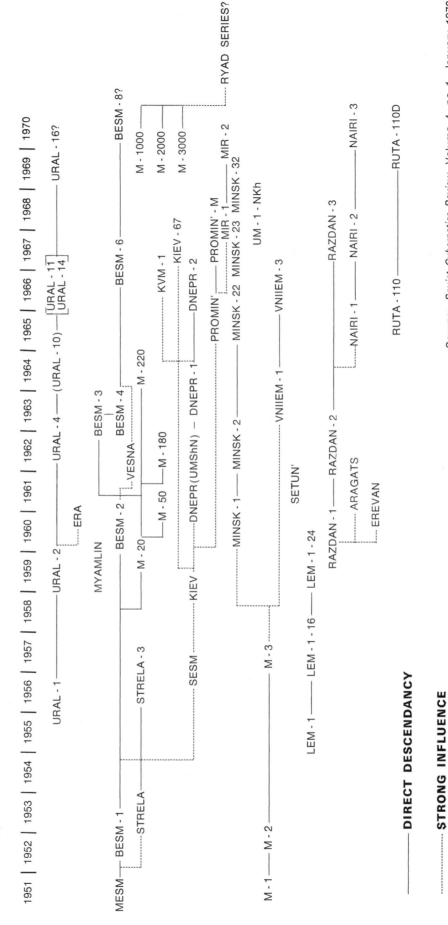

Source: *Soviet Cybernetics Review*, Volume 4, no 1, January 1970

"The development of Soviet computers is shown in the accompanying diagram (Figure 6). It is believed that the computer design effort is currently undergoing a large-scale orientation towards the development of third-generation equipment similar to the IBM system/360. The ASVT and Ryad projects are part of these developments.

"In addition, so far there has been no unanimity of opinion even on such a basic problem as the further development of computer technology. Some advocate computers of the general-purpose type. Others say they are expensive, awkward and unreliable, and that the main emphasis must be directed toward the construction of inexpensive, reliable, special purpose machines. Some like S.A. Lebedev, advocate the construction of high-speed, high-capacity machines on a par with the world's best. Others feel that the USSR should concentrate on improving already developed machines - primarily in terms of reliability and software - while still others advocate the development of multi-machine systems as a way of improving operational speed and capacity and, at the same time, avoiding the cost development and production of new, large computers."
(Source: Rudins: *Soviet Computers an Historical Survey, Soviet Cybernetics Review*, volume 4, 1, P 43, January 1970).

Glushkov has presented the Soviet situation and the thinking of the planners within a framework of five generations of computers:

First (1955-60): electron tube systems

Second (1960-65): solid-state systems

Third (1965-70): miniaturisation, low degree of integration, combined alphanumeric systems, parallel operation, machine language translators, input languages such as Simula-67, Algol-68, PL-1 standardised coupling of the central processor to peripheral equipment, time sharing.

Fourth (1970-75); Display screens, high degree of integration, extended miniaturisation, specialised processors.

Fifth (1975-): optoelectronics, laser elements, 10^{14} byte external disc memory, voice input and output.

Journals: Journals concerned with research into cybernetics include *Kibernetika* (Cybernetics); *Automatika i Vychislitel'naya Tekhnika* (Automation and Computer Technology); and *Automatika i Telemekhanika* (Automation and Telemechanics).

(G) Training

A number of articles have appeared recently in the popular press and in the technical literature complaining about the inadequacy of the training of specialists in various facets of computer technology. There are insufficient mathematicians, engineers, programmers, and economists acquainted with computational aspects, and the level of knowledge and experience of those being produced by the universities is insufficient for the problems facing the computer field. Qualified programmers are in particularly short supply. Part of the problem rests in the European tradition of the university, which is not oriented towards training people in applied technical skills. Nor are there higher educational institutions devoted to providing an education in these areas. Most training in programming is done either on the job or through specially organised extension-type courses.

(H) Plans and Policy

One of the main forward plans is the setting up of a State Network of Computing Centres, though the best way to organise this is still under discussion.

According to Glushkov research and development of computers will progress from third-generation instruments through fourth generation (1970-75) to fifth generation (1975-) based on the use of laser elements.

9 Power Industry

(I) ELECTRIC POWER PRODUCTION

(A) Introduction

Compared with most of the western world's power industry, its counterpart in prerevolutionary Russia was quite backward. The total output in 1913 was 2,000,000,000 kilowatt hours (kWh). The power of the biggest hydroelectric station was 16,000 kilowatt (kW).

In 1920 the Soviet Government elaborated the GOELRO plan for the electrification of Russia. According to this plan, which was to be implemented in 10 to 15 years, 30 district power stations with a total capacity of 1,500,000 kW would be built in various economic regions, and the capacity of existing stations would be raised to 250,000 kW, making is possible to generate 8,800,000,000 kWh of electricity annually, compared with the pre-war figure of 2,000,000,000 kWh for 1913.

The Soviet power industry grew rapidly before the Second World War. By 1935 it ranked second in Europe and third in the world for power generation. However, the power system suffered a severe setback during the war. More than 60 big stations with a total capacity of 5,800,000kW and nearly 10,000km of supply lines were destroyed. The pre-war 1940 capacity level had been regained, however, by the end of 1946. Since then development has continued to be rapid. Development is based on the building and expansion of big district power stations. In 1928 such stations accounted for only 39 per cent of the power generated in the country; in 1940 the proportion had risen to 81.2 per cent, and in 1968 it was 94 per cent.

Since the revolution power industries have been developed in the various republics of the Soviet Union, where previously there had been no such industries. The eastern regions of the country are also becoming important power producing areas - they now account for some 30 per cent of all electricity output - because of their rich fuel and water resources.

(B) Organisation

9(1) MINISTRY OF POWER AND ELECTRIFICATION

The Ministry of Power and Electrification is responsible for the electricity supply industry, which includes generation, transmission, distribution and sales. The minister and his board approve routine and repeat projects themselves. Prototype or unique projects have to be recommended to the Council of Ministers for their approval.

The organisation of the ministry can be summarised under the following headings (see also Figure 7.):

    Personnel

    Advanced condition of plant and maintenance

    System operation

    Power station operation, inspection and plant research institutes

    Planning

    Atomic energy

    Power station construction technology

    Construction of thermal power stations

    Construction of hydro power stations

    Financial

The Ukraine, Kazakhstan and Uzbekhistan republics have their own ministries of power and electrification, responsible for the administration of the electricity supply in the particular republic. These three republic ministries are subordinate to the All-Union Ministry of Power.

The other republics have central offices of power and electrification.

The three *Ministries of the Gas Industry, Oil-Extracting Industry and Coal Industry* are responsible for the supply of gas, oil and coal to the power industry. There is also a *State Committee for the Use of Atomic Energy*.

The *Ministry of Heavy Power and Transport Engineering* is responsible for the factories which produce boilers, turbines, and other equipment for power stations.

The *Ministry of the Electro-Technical Industry* controls the Electrosila Works at Leningrad and a branch factory at Novosibirsk, which produce most of the generators for the Soviet Union. It also controls the Zaporozhy transformer and switchgear factories.

Construction and Erection Trusts are the general contractors for the construction and erection of power plants. They are organised on a geographical basis (Mosenergo, Donbasenergo, Kalinienergo, etc.) and are responsible through regional offices to the Minister of Power. The trusts are awarded contracts by the operating organisations of the plant, and can subcontract specialist work to special erection trusts. Specialisation in the construction of power stations is developing all the time; there are now trusts dealing with such items as heat insulation work, application of anti-corrosive coatings, installation of instruments and automatic equipment and so on.

(C) Finance

When a project for a power station has been approved, either by the Council of Ministers of the Ministry of Power, the banks are authorised to make finance available. Money is released to the construction and erection trust against certificates signed on behalf of the appropriate operating organisation. Then payments are made in stages set out in the project report. The banks are responsible for investigating any deviations from the approved estimates and for obtaining explanations.

Figure 7 Structure of Ministry of Power and Electrification

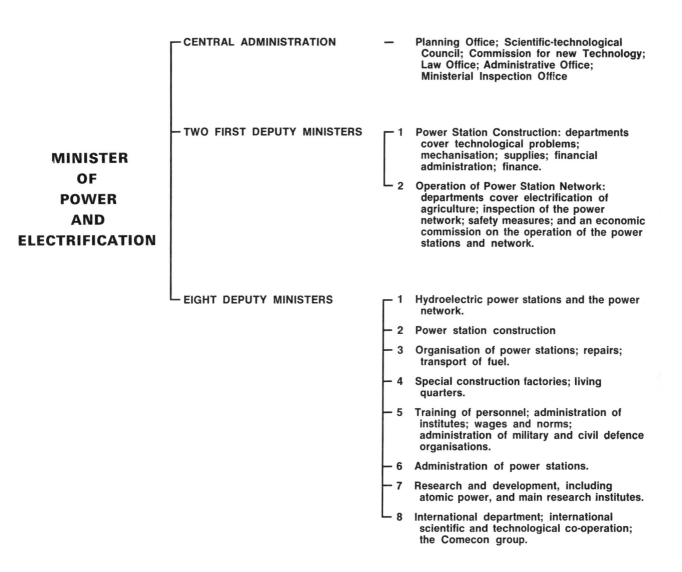

(D) Production

Electric power production in 1970 was 740,000,000,000 kWh. Some 94 per cent is generated by power stations. The capacity of these was estimated to be some 124,000,000 kW, at the beginning of 1967, and had risen to 133,000,000,000 kW by 1968. The expected figure for 1970 was 170,000,000 kW.

The bulk of electric power (some 80 per cent) is produced by thermal stations and among these is the world's biggest - the Pridneprovskaya station with a capacity of 2,400,000 kW. The main trend in power development is the construction of large thermal stations, the size of Pridneprovskaya or bigger, with turbines of 300,000, 500,000 and 800,000 kW capacity. Hydropower stations will develop mainly in Siberia, Central Asia, and Kazakhstan where 80 per cent of the hydropower resources are concentrated (see Figure 8.)

The growth of electric power production and the capacity of power stations is given in Table 13.

*Table 13**

	1913	1940	1958	1966	1969
Capacity of power stations (thousand kW)	1,141	11,193	53,641	123,000	153,790
Production of electricity (million kWh)	2,039	48,309	235,350	544,703	689,050
Of this, hydroelectric power stations accounted for: capacity	16	1,587	10,863	23,075	29,645
power production	35	5,113	46,478	91,809	115,181

*(Source: Narodnoe Khoziastvo *SSSR v 1969 g*, P 216, (Moscow, 1970)

(i) Thermal Power Stations. More than 80 per cent of Russia's electricity output is generated by thermal power stations. The main trend is towards the construction of very big stations with a capacity of 2,400,000 kW or more, and with turbines of 300,000, 500,000 and 800,000 kW capacity. In 1966 there were 12 thermal power stations with a capacity of 1,000,000 kW or more, and by 1970 this number was planned to be increased to 30. The 2,400,000 kW Pridneprovskaya Station was the first to have a 300,000 kW turbine. In 1968 a 500,000 kW turbine at the Nazarovo and an 800,000 kW one at the Slavianskaya power stations went into operation.

The fuels used by the power stations are usually the local onces, even if low grade, as this means reduced transportation costs. This situation does not apply, however, in the European part of the country.

Many thermal power stations are designed as both heat and power plants so that they provide heat in the form of steam or hot water to industrial and municipal customers. These district heating systems are well developed, particularly in the towns. In 1968 the district power stations supplied 433 million gigacalories of heat to customers.

(ii) Hydro Electric Power Stations. The significance of hydroelectric power for the USSR

Figure 8 Soviet Electric Power Stations

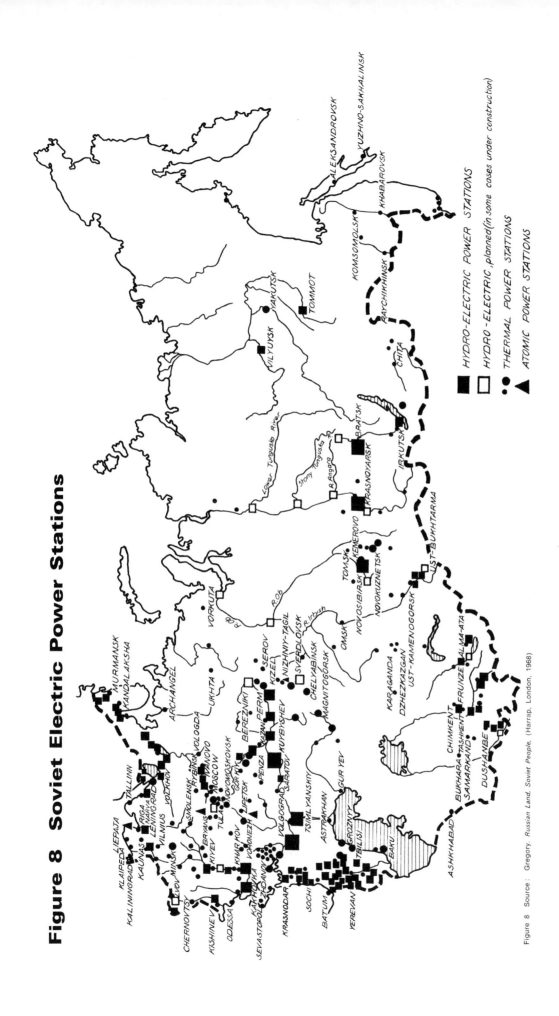

Figure 8   Source: Gregory. *Russian Land, Soviet People*. (Harrap, London, 1968)

is determined by the fact that the energy potential of the country's hydro-resources is approximately 4,000,000,000,000 kWh a year. Eighty per cent of the hydro-resources are situated in the east of the country, and it is there that hydropower construction will mainly develop.

On the Angara river, the Irkutsk and Bratsk power stations will be joined by the Ust-Ilim and Boguchan stations to complete the Angara chain of stations with a capacity of more than 14,000,000 kW. Four stations planned for the Yenisei River will have a total capacity of 27,000,000 kW, including the Krasnoyarsk which went into operation in 1970.

More stations are being added to the chains that exist on the European Russian rivers - the Volga, Dneiper, and Kama. In Central Asia and Kazakhstan some big power stations are being constructed, including the 2,800,000 kW capacity Nurek station on the Vaksh river. One experimental tidal station has been built at Kislaya Guba, a gulf of the Barents Sea. It has been operating since 1969, and has a power capacity of a few hundred kilowatts. Bigger tidal power stations are being planned and construction started in 1970. They are being built on the White Sea coast.

(iii) Atomic Power Stations. The Soviet Union put into operation a small 5,000 kW atomic power station in 1954. It was the first of its kind in the world. Since then several more atomic power stations of different designs and types have been built. Their overall capacity is around 1,800,000 kW. One is the Novovoronezh on the River Don which has slow neutron reactors using water as a neutron moderator and a heat removing agent; another, the Byeloyarsk Atomic Power Station, which went into operation in 1964, works on thermal neutrons and has a graphite moderator. A large breeder reactor (with 150,000 kW capacity) at Shchevchenko on the Caspian Sea is planned to go into operation at the end of 1970. As well as producing electricity, it will be used to desalinate sea water, producing some 120,000 cubic metres of fresh water a day.

Work has started on an atomic power station in the Ararat valley. The station will consist of two reactors with a total capacity of 880,000 kW. It will give an important boost to the power requirements of the Armenian Republic which needs power for the development of its resources of copper, molybdenum and gold. Other stations are being built on the Kola Peninsula, the Chukotka Peninsula in the far north east, and in the Urals. The Novovoronezh Station is being expanded to a capacity of 1,500,000 kW. A number of large graphite-water reactors are at present being built. Each unit will have a power of 1,000,000 kW and the first of these is being constructed near Leningrad.

According to Soviet calculation, the cost of electricity generated by nuclear power can compare favourably with that of electricity generated from coal-fired stations if the atomic power stations have a capacity of a million kilowatts or more. The main development is therefore to construct large atomic power stations and this will take place particularly in the European part of the country.

The Soviet Union has also developed small mobile power reactors for generating electricity in out-of-the-way places, for example the north of the country where there are rich deposits of minerals and other resources but little development of transport or power. They include a reactor on wheels and one which can be transported by plane and then reassembled. The largest of these reactors is the *Sever* with an output of 1,500 kW.

(E) Transmission

The setting up of a nationwide unified power grid is one of the main tasks of the power industry. The first step towards this - a unified power grid for the European part of the USSR - began to be developed after the Second World War, and by 1962 the power systems of the Centre, the Urals, the South (the Moldavian and Ukrainian Republics) and the Volga regions had been linked together. Since then the power systems of the North-West and of the Caucasus have been added making a grid with a total capacity of over 100,000,000 kW. The next stage is the unification of the European power grid with the Central Siberian and North Kazakhstan power systems. A 20,000,000 kW grid is now being set up in Central Siberia.

Standard lines for the long distance transmission of electricity are 600 mile 500kV lines. There were over 5,000 miles of such lines in 1969. They are now being overtaken in size by 1,250 or even 1,600 mile lines carrying 800 to 1,500 kV. These are direct current lines, though research is going into the possibility of alternating current transmission.

Of the electricity supply, 85 per cent is effected by power grids, which at the end of 1968 had 385,000 kilometres of 35 to 500 kV transmission lines.

The Soviet power systems are now being linked up to those of other socialist countries. The *Mir* power grid links power stations in the West Ukraine and North West of the country with ones in Hungary, the German Democratic Republic, Poland, Czechoslovakia and Rumania, and Bulgaria. The aggregate capacity of all the power stations involved is 47,000,000 kW, and the length of its 200 kV transmission lines is 13,000 kilometres. The *Mir* power stations are controlled from a central panel board in Prague.

(F) Applications in Industry

Electric power is an important base for the development of technology and industry. The point was made by Lenin when he saw electrification of the country together with power in the hands of the Soviets as the two requisites for building communism.

In industry electric motors have almost completely superceded steam and other engines. Compared with 1928 when they accounted for 64.9 per cent of all engines employed in industry, the figure for electric motors is now nearly 95 per cent. In coal mining, mechanical engineering and the chemical industry electrification is nearly 100 per cent. The electrical engineering industry produces a large variety of electric motors. In 1967 more than 5,000,000 were produced with a total capacity of over 29,000,000 kW.

Electricity is also used in industry for chemical processes and as a source of heat. Electrolysis is used to produce aluminium, sodium, magnesium, copper, nickel, lithium and other non-ferrous and rare metals. Electric furnaces are used to smelt super hard cemented alloys, high-grade steels (transformer steel, stainless steel, high speed steel and the like) and ferro alloys (ferro-silicon, ferro-chromium, ferro-manganese and ferro-tungsten). They are also used for hardening metals by heat treatment.

Electric power is being increasingly used for transport. In 1967 electric trains hauled nearly 45 per cent of the country's rail freight. The suburban railways of many major cities have been electrified; electric power is used for underground trains, trams and trolleybuses within the cities. In 1967 there were 7,778 km of tramway lines and 6,188 km of trolley bus lines. Moscow, Leningrad, Kiev, Tbilisi and Baku have underground trains.

More and more electricity is being used in agriculture. Rural consumers are being connected to the state power stations through 20 kV transmission lines, making it possible to dismantle small and uneconomic village power stations. The number of collective farms with electricity has grown from 15 per cent in 1950 to 99 per cent in 1967. The number of electric motors in use on collective farms has grown from 56,000 in 1950 to 1,500,000 in 1967. On the state farms in the same period their number has gone up from 49,000 to 1,430,000. The total capacity of the electric motors of both state and collective farms was 15,790,000 kW in 1967.

The domestic consumer is also using more electricity. In 1966-67 8,200,000 washing machines, 4,900,000 household refrigerators, nearly 2,000,000 vacuum cleaners, 9,400,000 television sets, 13,500,000 hot plates, 9,900,000 electric irons and 1,600,000 electric kettles and coffee pots were manufactured.

Electricity is also widely used in medicine.

(G) Exports

In the five-year period from 1971 to 1975 the Soviet Union will export 1,500,000,000 kWh to Poland. Over the previous three and a half years she supplied Poland with more than 800,000,000 kWh. Hungary is a major purchaser with a planned annual figure of 4,000,000,000 kWh. A new transmission line from the Ukraine to near Budapest has just been constructed to carry this power. Exports of electricity to Bulgaria are also increasing and will reach 3,000,000,000 kWh a year after the completion of the Moldavia-Dobrudja line. Estimates show that the export of Soviet electricity to the socialist countries in Europe will have increased some four times by the end of the next five-year period (1971-75) compared with the previous five-year period.

The Soviet Union is also going to supply electricity to Finland via the power grid of the Kola Peninsula in the Murmansk region, and this same grid is being used to supply electricity to some parts of Norway.

Power stations form another type of export. In 1968 about 20 per cent of the 152 power station projects being worked on in the Teploelectro Project Institute were for export.

(H) Research and Development

Research and development within the power industry is centered in scientific research institutes which come under the jurisdiction of the various ministries concerned, and in the appropriate theoretical institutes of the Academy system.

9(2) TEPLOELECTRO PROJECT INSTITUTE

The Teploelectro Project Institute is one of the main power research and development institutes. It was established in 1927 as the design division of the 'Energostroy Trust' for the construction of power stations and electrical networks. It was raised to the status of a separate Trust in 1932, and was later re-designated as the "All-Union State Design Institute Teploelectro-project". It has 12 regional offices, with a total of more than 12,000 employees. The institute is responsible for the design of nuclear and conventional power stations, for projects in the USSR and elsewhere in the world, and for the designing of district heating systems in the USSR. The design of nuclear stations is concentrated in the Moscow, Leningrad and Sverdlovsk offices. Between 14 and 45 project reports are produced annually, made up of roughly equal proportions of new stations and extensions to existing plants. In 1968 the institute worked on 152 power station projects, covering about 10,000,000 kW of capacity; 80 per cent of these were for installations in the USSR, and the rest for export. About 2,000,000 kW of nuclear capacity is designed each year.

There are individual trusts, responsible through regional offices to the Ministry of Power, which deal with specialist problems such as heat insulation work, application of anti-corrosive coatings, installation of instruments and automatic equipment and so on. Factory research and development is well developed in certain branches of the power industry. The Electrosila combine in Leningrad, which produces water turbines and generators, is a good example. It has substantial research and development facilities of its own including an affiliate of the All-Union Scientific Research Institute for Electro-Mechanics with a staff of 1,500. The affiliate controls all factory laboratories and shops within the combine unifying the development of new machines which are later produced in the combine's factories. Although the affiliate is technically a branch of the Institute of Electro-Mechanics, in practice it is independent and forms an inseparable part of the enterprise. Since 1963, when the combine first came into existence, the affiliate has developed 146 new machines, all of which have been introduced into production without difficulty. One problem of factory-based research and development is that it can lead to duplication of designs. The Electrosila, Electrotiazhmash and the Novosibirsk plants have all insisted on producing turbine and generator equipment to their own design, resulting in considerable duplication between them.

(i) *Thermo-electric power*. The basic trend in the development of thermal power is towards an increase in the power of the stations, transition to big units and high steam parameters, all of which helps to bring down the cost of the power generated. Table 14 shows the development of these trends.

*Table 14**

Power (1,000s kW)	Number and size of turbines (no. x 1000 kW)	Number and size of boilers (no. x tons hr)	Cost of 1kW (in %)
600	6 x 100	12 x 230	100
1,200	6 x 200	6 x 640	85
1,800	6 x 300	6 x 950	74
2,400	8 x 300	8 x 950	61

* Source: Vredenskii, ed: *SSSR 1917-1967*. P 329, Moscow 1967.

The increasing size and power of the thermal power stations means that less fuel is needed to generate 1 kWh of electricity. One of the chief means of reducing the amount of fuel needed is by raising the steam parameters. In the USSR the figures for steam pressure have risen successively from 29 atmospheres (atm) to 90 atm, 130 atm and 240 atm, at the same time the temperature of the steam has risen from 400°C, through 500°C, and 565°C. There is an experimental unit working at 300 atm and 650°C. Fuel economy in the power stations with these steam parameters depends to a big degree on the individual power of the unit plants. Raising these is another important trend in the technological development of the Soviet power industry (see Table 15).

*Table 15**

Power (1,000s kW)	Steam pressure (atm)	Steam temp.(°C)	Fuel used (Cal/kWh)
100	130	565	2,030
200	130	565	2,000
300	240	580	1,830
600	240	580	1,810

*Source: Vredenskii, ed: *SSSR 1917-1967*, P 329, Moscow 1967.

The Soviet power machine industry has consequently gone over completely to the production of large turbines. In 1970 the Electrotyazhmash works in Kharkov completed the blueprints for a 1,000,000 kW turbogenerator. Still bigger turbines of 1,200,000 kW capacity, are being designed at the Electrosila works in Leningrad. The latter is designed to be only 70 per cent heavier than the 300,000 kW model. In addition the productivity of the steam generators is being raised. At present boilers are being constructed which produce up to 1,000 tons of steam an hour at a pressure of 240 atmospheres and temperature of 540-565°C, with an intermediate heating of the steam at 565°C. There are plans for boilers to produce 1,950 tons an hour. Increasing the steam producing capacity of the boilers is one of the most important tasks within the power industry as the boilers form part of the important energy chain: boiler - turbine - electric generator.

Another important trend of development is the combined heat-power station. The dual role of generating electricity and producing heat raises the efficiency of the fuel. The use of natural gas in thermal power stations is increasing. If the gas enters the turbine at a temperature of 1,200°C, then gas turbines are more economic than anyoothers in thermal power generation. However, it is difficult to construct materials that can stand up to such high temperatures and this is one problem which the power engineers are working on.

Attempts to raise the coefficiency of productivity led to the idea of combining the steam and gas turbine cycles creating a "steam-gas apparatus". The first such station went into operation in 1963 in Leningrad. In 1970 the Kharkov turbine plant produced the first steam-gas turbine in the USSR with a power capacity of 200,000 kW. It will be installed in the Novinomyssk thermal power station in the south of the country. The plant is also planning to manufacture a 400,000 kW steam-gas turbine.

(ii) Hydroelectric power. An important characteristic of Soviet hydroelectric power engineering is the complexity of the construction problems - energy, transport and irrigation problems have to be worked out simultaneously with the construction of the power station dams and cascades. Usually more than one dam is necessary to control the river flow satisfactorily. For example there are seven hydroelectric power stations on the Volga, and similar complex systems have been built on other rivers.

Despite their high capital costs and lengthy construction times, hydroelectric power stations have a number of advantages. They can solve several economic problems at the same time; the cost of their power is low; they are easily mechanised and need less personnel to run them than thermal stations. They are also important for helping to even out the peak load in power grids. Increasing the power of individual plants (from a few hundred kW to 500,000 kW) lowers the cost and periods of construction of the hydroelectric power stations.

Growth in the power of hydropower/units is being accompanies by a rise in their qualitative indicators. The efficiency of the rotary blade turbines is around 94 per cent. The weight of rotary blade turbines has been reduced from 29 to 15 kg/kW, and that of radial axial from 20 to 7 kg/kW.

An area of experimentation is in tidal power stations. The first one went into operation in a gulf of the Barents Sea in 1969. It has an output capacity of 400 kW and is serving as an experimental tryout before constructing larger stations in the same area.

Experiments are being carried out at the USSR Academy of Sciences' High Temperatures Institute into MHD generation. They have had an experimental 40 kW generator working there for several years. It has served as a prototype for a 25,000 kW MHD generator to be installed in a semicommercial power station in Moscow. This was under construction in 1970. Both these generators are of the open cycle type, but the staff of the Institute are also working on an explosion impulse MHD generator with a super-conducting magnetic system.

(iii) Atomic power. Reactors of different designs and types have been built in the USSR - mainly pressurising water reactors and light water-cooled graphite-moderated reactors. Research is being carried out to determine the best reactors for the nuclear programme. Widespread use of reactors working exclusively with thermal neutrons would rapidly exhaust the country's uranium resources. Thus quite a lot of development work

has been done on fast breeder reactors (FBR). Three experimental FBRs have been built at Obninsk, and on the experience of these, a fast breeder reactor, with 350,000 kW capacity, at Shchevchenko on the Caspian Sea which comes into operation some time in the early 1970s. Fast breeders can make use of virtually all natural uranium and thorium resources, not just uranium-235.

Work is also being done on the direct conversion of atomic energy into electric power. In 1964 the Kurchatov Institute of Atomic Energy in Moscow set up the 'Romashka' experimental unit. Its operating system is based on a high temperature nuclear reactor and converter. The converter has thermoelectric elements made of silicon-germanium alloy. Heat produced in the reactor by fast neutron fission of uranium-235 heats one side of each element while the other side is kept cool, so that an electric current flows in the element. The elements are coupled together to produce a current of 88 amps. The 'Romashka' has an output of 500 watts.

Another trend in atomic energy technology was the building of small mobile stations for use in remote areas. Pure gas oil is used instead of water to convey heat from the reactor. Since this does not become radioactive, concrete shielding walls are only required for the reactor itself, which makes the units lighter and more compact. However, interest in this development has lessened recently.

For long-term prospects, considerable work is being done on thermonuclear fusion. A number of different reactors are being studied. The best known and most promising at the moment is the *Tokamak* model of a fusion reactor. This is based on magnetic containment within a toroidal scheme and the parameters achieved by 1969 were a temperature of 30,000,000°C, in a plasma of maximum density 3×10^{13} particles per ar.cm contained for a period of about 20 milliseconds. The figures necessary for thermonuclear fusion are temperatures of over 100,000,000°C, plasma densities of 10^{14} per cu.cm and a containment period of one second or more.

(iv) Power transmission. The main direction in research and technological development of power transmission is concerned with the tremendous distances that the electricity has to be carried over. Long distance transmission lines have now reached up to 1,600 miles in length and they carry up to 1,500 kV. These are DC lines; methods of AC transmission are being investigated intensively. Superconductive power transmission, where electricity can be transmitted virtually without any power losses, is another important research topic.

Journals: Research journals concerned with problems of the power industry include *Teploenergetika* (Thermal Power Engineering); *Elektricheskii Stantsii* (Electricity Power Plants); *Elektrichestvo* (Electricity); *Elektrotekhnika* (Electrotechnology); *Energetik* (Power Engineering); *Atomnaya Energiia* (Atomic Energy); and *Magnitnaya Gidrodinamika* (Magnetic Hydrodynamics).

(I) Plans and Policy

A 20-year plan covers the years 1961 to 1980 and gives general perspectives for the development of energy production. The main targets are that production of electrical energy should be raised to between 2,700 and 3,000 MWh by 1980 and that installed capacity should reach to between 540,000,000 and 600,000,000 kW. This means a yearly increase of 45,000,000 to 50,000,000 kW in the period covering 1970-80.

The bulk of this power production will be provided by thermal power stations; at the same time the development of hydropower resources continues particularly in the eastern regions of the country. Since oil, gas and coal are also important raw materials for the expanding chemical industry, research and development will continue into other sources of cheap power, in particular nuclear power.

The setting up of a unified grid system for the whole country is another major target. The 1966-70 plan provided for the building of a 2,500 km DC line from Siberia to the centre of the country. By 1980 it is hoped to be able to transmit 225,000 million kWh of electricity a year from the East to the West.

(II) FUEL PRODUCTION AND BALANCE

The perspective in Russia is towards the greater use of gas and oil as fuel for the thermoelectric power stations. This is because of the changing fuel balance within the country. The production figures for the different fuels, taken as a percentage of the whole, are given in Table 16.

Table 16*

	Coal (%)	Oil (%)	Natural gas (%)	Peat (%)	Shale (%)	Wood (%)
1913	48.0	30.5	–	1.4	–	20.1
1940	59.1	18.7	1.9	5.7	0.3	14.3
1958	58.8	26.3	5.5	3.4	0.7	5.3
1966	40.7	36.7	16.5	2.3	0.7	3.1
1969	37.3	39.9	18.3	1.4	0.7	2.4

*Source: Narodnoe Khozaistvo *SSSR v 1969 g*, P 196, Moscow 1970.

(i) Coal Production. Figures for the development of coal mining are given in Table 17.

Table 17*

	1913	1940	1958	1966	1969
Total (1,000s tons)	29,153	165,923	493,236	585,604	607,802
Coal (" ")	27,987	139,974	353,030	439,170	467,316
Brown Coal(" ")	1,166	25,949	140,206	146,434	140,486

*Source: Narodnoe Khozaistvo *SSSR v 1969 g*, P 201, Moscow 1970.

Areas of coal mining are shown in Figure 9. Two new centres which are being particularly developed in the next 15 year programme are those in northern Kazakhstan and in the upper basin of the Yenisei river in Siberia. These two basins will produce the cheapest coal in the country and it will be used mainly to fuel giant power stations.

Mechanisation of coal mining is quite well developed in Russia. In 1970 coal production topped 600,000,000 tons and the long-term plan is to increase this by some 230,000,000 tons. Of the increase, 150,000,000 tons will come from new pits, and the other

Figure 9 Soviet Fuel Resources

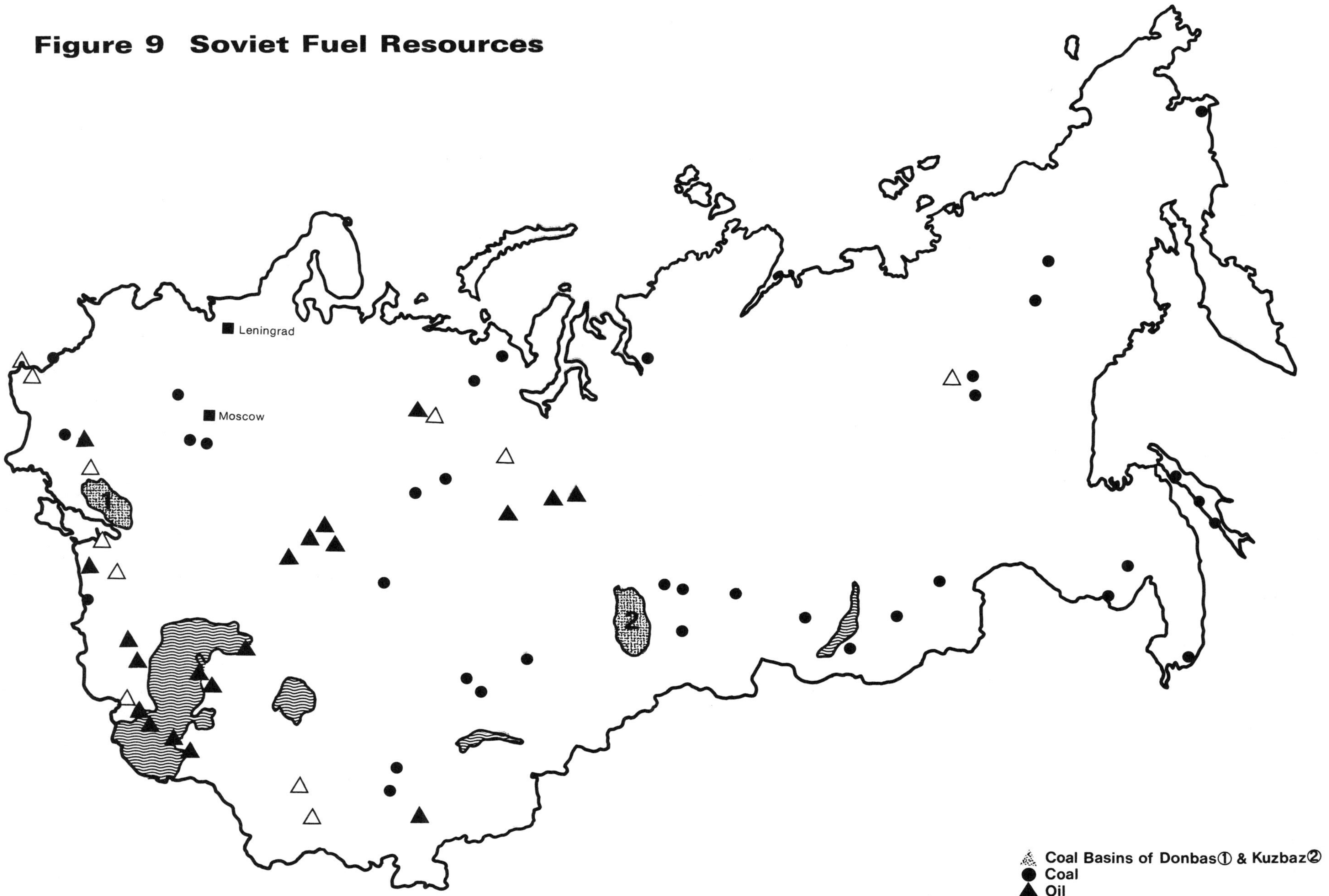

Coal Basins of Donbas① & Kuzbaz②
● Coal
▲ Oil
△ Gas

Figure 9 Source: Vvedenskii, *SSSR 1917-1967*, (Moscow, 1967)

80,000,000, from expanded and modernised existing pits.

(ii) Oil Production. Table 18 gives figures for the production of oil.

*Table 18**

	1913	1940	1966	1970 (planned)
Total (million tons)	10.3	31.1	265	345-355

*Source: Vredenskii, ed: *SSSR 1917-1967*, P 217, Moscow 1967.

Areas of oil fields are shown in Figure 9. Whereas before the revolution the main centre was round Baku, now oil is also obtained in the Volga basin, in the Tatar and Bashkir Autonomous Republics. Two new large centres of oil production are Tyumen, in Western Siberia, and the Mangyshlak Area, in Kazakhstan, though these fields have been discovered only recently and production started in them in the late 1960s. Another area where large deposits of oil have been found is in the arctic regions of the country.

The bulk of oil is obtained from gushers.

In 1964 the total amount of oil refined was of the order of 180,000,000 tons (oil production that year was 223,600,000 tons). The plan for 1970 was for almost 70 per cent more oil to be primarily refined than in 1963, and 50-70 per cent more gasoline, kerosine, diesel fuel and lubricating oil produced. The octane ratings of automobile gasolines were to be raised to 95 and more, and the sulphur content reduced, that of diesel fuels to 0.2 per cent or less. In order to achieve these goals new refineries with annual production capacities of 12 to 18 million tons of crude oil were to be built, and existing refineries were to be modernised. In 1966 there were 52 oil refineries in operation and 8 under construction.

(iii) Gas Production. Table 19 gives figures for the production of natural gas.

*Table 19**

	1940	1958	1966	1970 (planned)
Total (million cu.m)	3,400	29,900	145,000	225-240,000
Natural Gas	3,200	28,100	143,000	196,000

*Source: Vredenskii, ed: *SSSR 1917-1967*, P 217, Moscow 1967.

Natural gas is mostly obtained in the Volga region, the Ukraine, the Caucasus and in Bukharskii (see Figure 9). There are 500 deposits of natural gas in the country, many of them discovered only over the past 10 years. The most important of the recently discovered fields are in Siberia and Central Asia. It is estimated that the Urengoi deposit in the basin of the River Ob is the world's biggest with reserves equal to

Figure 10 Soviet Oil Pipelines (1967)

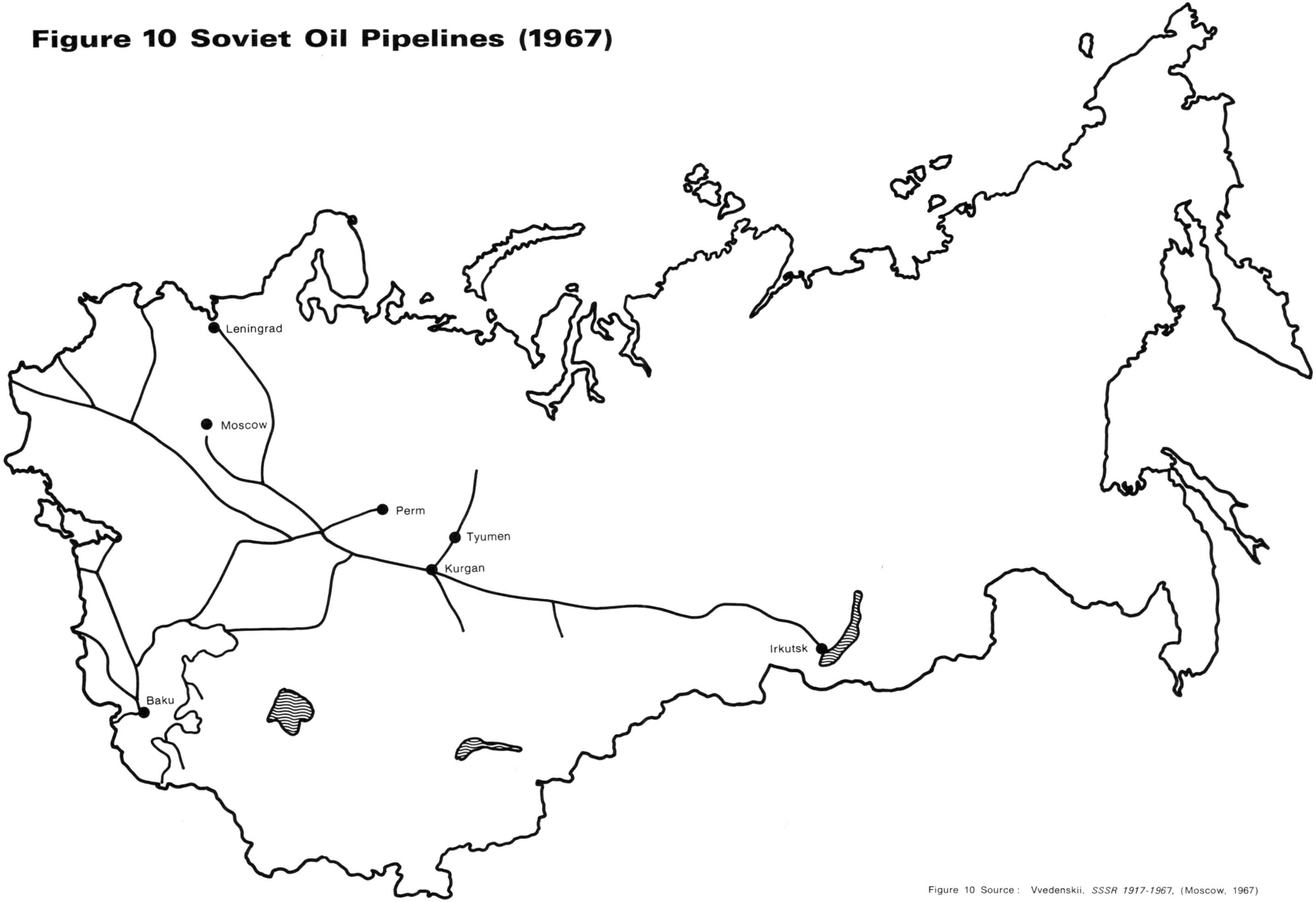

Figure 10 Source: Vvedenskii, *SSSR 1917-1967*, (Moscow, 1967)

2,600,000,000,000 cu.m. Another big deposit, in Uzbekistan lying between the Caspian and Aral Seas, has estimated reserves of 46,500,000,000 cu.m. The estimated stocks of natural gas for the whole country are 12,000,000,000,000 cu.m, and the plan is eventually to achieve a production of 600,000,000,000 cu.m a year.

There are technical difficulties in the development of oil and gas extraction. In 1968 a number of deposits of high pressure gas had to be closed temporarily because there was insufficient production of casing pipes and other fixtures needed for high-pressure wells. This was mentioned in an interview in *Pravda* with the Minister of the Gas Industry, A.K. Kortunov (March 15, 1968). In the same interview the minister made the point that they were hoping to increase the number of wells where gas was extracted simultaneously from two levels in the same well. This would both economise capital expenditure and increase production. In an interview given in April, 1968, the deputy minister of the oil industry, R. Sh. Mingareyev, mentioned the unsatisfactory organisation of work in prospecting and commercial drilling (Bakinsky Rabochii, April 20, 1968). This is particularly true of test drilling where a number of completely different systems are in use. In addition, there is no one body with overall control - various geological departments, trusts and prospecting groups all had an interest in the work. One exception is the Tatar Autonomous Republic, where in 1957 test drilling was placed exclusively into the hands of the drilling organisations. However, it has apparently been difficult for this more simplified structure to be extended.

(iv) Transport of Oil and Gas. Figure 10 shows the extent of oil pipe lines in 1967. An extensive oil and gas pipeline network is being built, primarily to bring the two fuels from Siberian fields to the western, more industrialised, areas of the country. Large pipes are being planned with diameters of 2.5 metres, though at present 4-foot gas pipes are the maximum size in use.

The total length of gas pipelines in Russia is more than 35,000 miles, and another 10,000 miles should be added by the end of 1971 (see Figure 11). There is also a pipe network linking the USSR with the Comecon countries and beyond. The oil *Druzhba* system has a total length of 3,000 miles of pipelines linking five countries (see Figure 12). It was opened in 1962.

Natural gas is also supplied to these countries.

(v) Labour. Fifty thousand oil extracting and refining workers are employed in Siberia and their numbers are increasing. The bulk of the enterprises under the Ministry of Oil operate within the new planning and incentives system.

(vi) Exports. The Soviet Union exports much of its oil and petroleum products. The figures for some recent years are as follows:

	1962	1964	1967 estimated
Oil (tons)	26,300,000	36,700,000	80,000,000
Oil products (tons)	19,100,000	19,900,000	

Source: "The Soviet Oil Refining Industry", *Bulletin of the Institute for Studies of the USSR,* P 49, (July, 1966).

Oil and petroleum products occupy an important place among traditional Soviet export goods as earners of foreign currency. In 1967, 45,000,000 tons of the 80,000,000 tons of oil and oil products exported were sold to capitalist, chiefly Western European, countries. These exports accounted for some 10 per cent of the oil needs of those countries. In 1964 the USSR sold its oil to the capitalist countries for 14 dollars a ton and to the socialist countries for 22 dollars a ton. As the socialist countries develop their petrochemical and automobile industries their imports of Soviet oil and

Figure 11 Diagram of Soviet Gas Flow

The total length of Soviet gas pipelines is over 35,000 miles. Nearly 10,000 miles will be added by the end of 1971

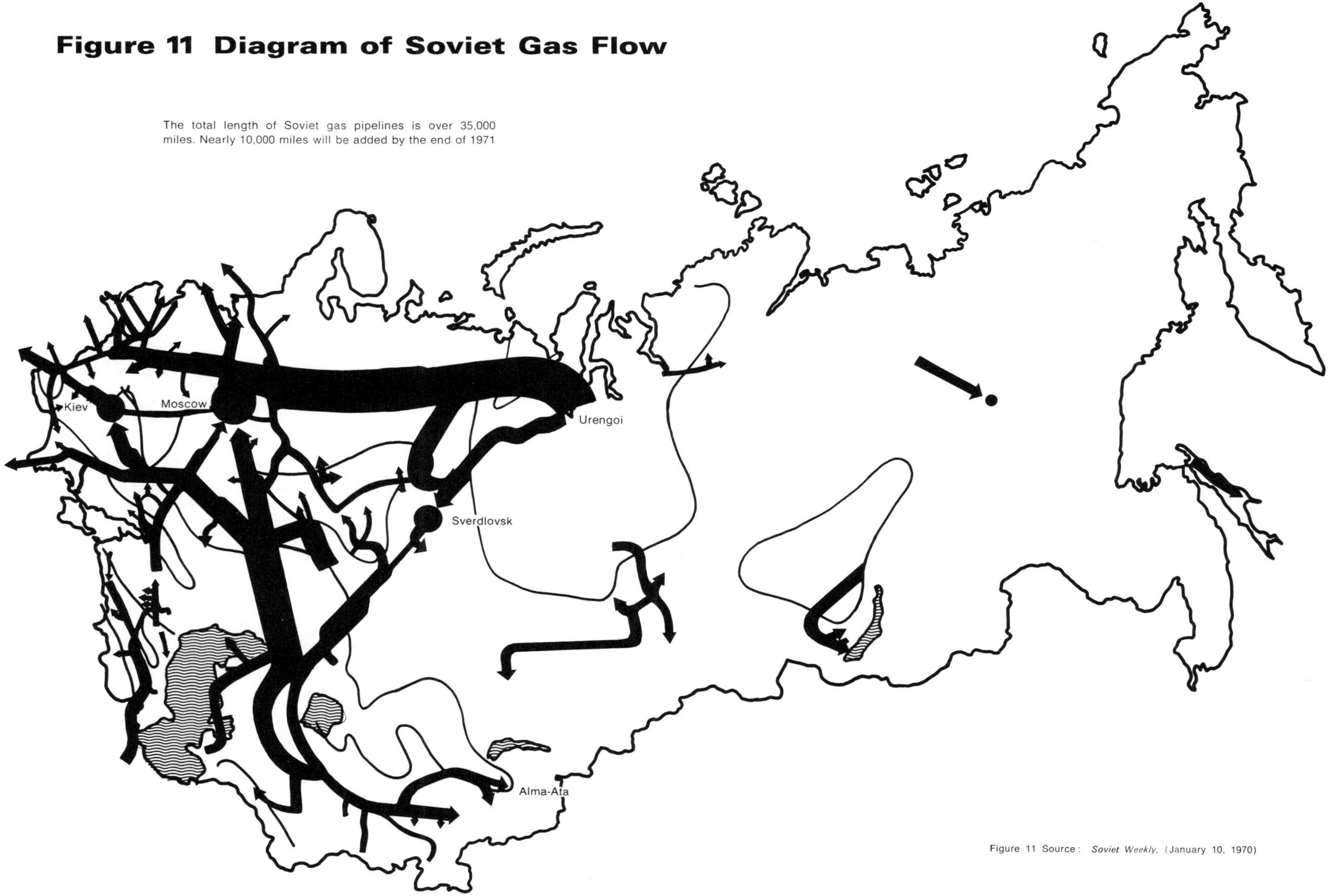

Figure 11 Source: *Soviet Weekly*, (January 10, 1970)

Figure 12 The route of the Druzhba pipeline

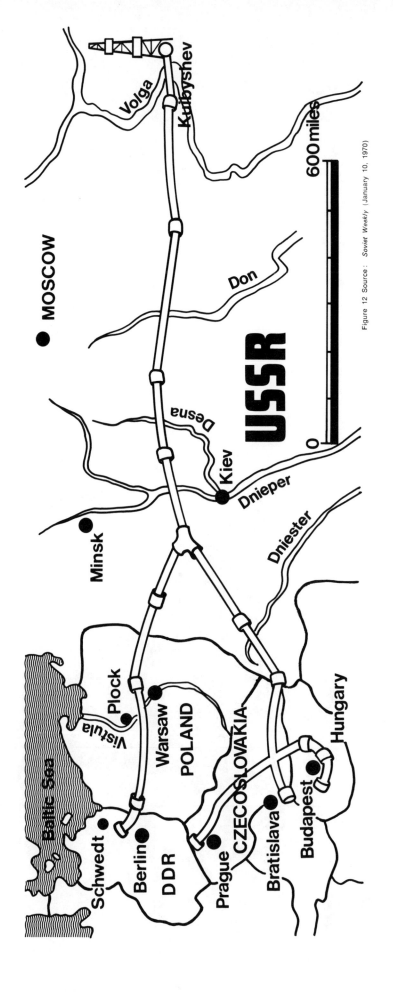

Figure 12 Source: *Soviet Weekly* (January 10, 1970)

oil products will rise and it is likely that they will become the main buyers of Soviet oil and petroleum products.

Soviet natural gas, which in the past had been sold only to socialist countries, was sold to Austria in commercial quantities in 1968. The pipelines are now being extended to Italy and West Germany, and negotiations are in hand with France.

A noticeable event on the oil and gas market is the trend taking shape in the USSR not only to export but also to import a certain quantity of oil and gas. Soviet foreign trade organisations have signed contracts with the state oil enterprises in Iran, Iraq and Algeria. They will supply the USSR with oil or gas in payment for Soviet machines and equipment.

(vii) Technological Developments. The discovery of enormous oil and gas fields in such inhospitable areas as Western and Northern Siberia has called for new developments in Soviet oil and gas technology. Helicopters have been used for laying pipe lines, and the hovercraft principle was adapted for moving derricks into position in the swampy taiga. The permafrost in these regions poses a problem and its characteristics are being studied at Moscow University, and the Irkutsk Permafrost Research Institute. It is necessary not only to develop metals for pipes able to withstand the very cold temperatures, but also to increase the size of the pipes (ones with diameters of 2.5m are planned) to cope with the enormous flow of gas.

The oils from the eastern deposits contain quite high percentages of sulphur. Research is therefore directed to finding new and improved methods for winning high grade products from these oils. Hydrocracking is one promising method; however, it is not widely used industrially since it needs large quantities of expensive hydrogen.

10 Metallurgy and Engineering

(I) FERROUS AND NON-FERROUS INDUSTRIES

(A) Introduction

Before the revolution the production of ferrous and non-ferrous metals was concentrated in the south of the country. Since 1917 large metallurgical centres have been built in the central, western and eastern regions with the aim of making better use of the abundant raw materials and fuel resources, and of bringing production closer to the populated areas.

(B) Organisation

The organisation of the ferrous and non-ferrous industries comes under the *Ministry of Ferrous Metallurgy* and the *Ministry of Non-Ferrous Metallurgy*, headed by a Minister of Ferrous Metallurgy and a Minister of Non-Ferrous Metallurgy respectively. These ministries have overall administrative responsibility for production, research and development in their respective fields. They work within the directives laid down by the Council of Ministers and Gosplan.

(C) Production

Figure 13 shows the main deposits of ferrous and non-ferrous metals and rare elements, and the main centres of their production.

The Soviet Union has vast iron ore reserves (exceeding 58,000,000,000 tons according to the Minister of Ferrous Metallurgy). The main deposits are at Krivoi Rog in the Ukraine, in the Urals, around Magnetogorsk, and in the Kursk Magnetic Anomaly area. A large metallurgical centre is being constructed around the Kursk deposit, rather similar to Magnetogorsk in the Urals, but on a larger scale. A special scientific council of the Academy of Sciences of the USSR has been set up to decide on the best way of developing the various resources of the anomaly, which include bauxites, marl, writing chalk and fine sand in addition to high grade iron ores.

The Soviet Union occupies first place in both reserves and the mining of manganese ores. The known reserves exceed 2,500,000,000 tons, the bulk of these being in the Ukraine and Georgia.

The bauxite reserves in the northern urals, the north-west region and Kazakhstan have formed the basis of the aluminium industry which has grown up since the revolution.

Table 20 gives the production figures for pig iron, steel, ferrous rolled stock and steel pipes:

*Table 20**

(in million tons)	1913	1940	1958	1968	1970
Pig Iron	4.2	14.9	39.6	78.8	85.9
Steel	4.3	18.3	54.9	107	116
Rolled stock of ferrous metals	3.6	13.1	43.1	85.2	92.6
Steel pipes	0.1	1.0	4.6	9.9(1966)	12.4

*Source: Vredenskii, ed: *SSSR 1917-1967*, P 218, (Moscow 1967) and *USSR Industrialisation*, P 20, Novosti Press Agency.

In world output of iron and steel the Soviet Union comes second after the United States. However, she still does not produce enough to satisfy her needs. Expansion of production is being achieved in a number of ways. First, new plant is being built. The emphasis here is on constructing large capacity units - blast furnaces (with volumes of 2,700 cu.m or more), open hearth furnaces (producing from 500 to 1,200 tons of steel a day, high-speed rolling mills, and continuous steel pouring facilities. Mechanisation and automation within metallurgical plants is being increased with the installation of equipment for the continuous rolling, pipe-rolling and finishing of rolled goods.

Another development is the use of oxygen and natural gas in metal smelting which can double the productivity of metallurgical plants.

Measures are being taken to improve the quality of metal, increase the output of alloy steel, thin-walled steel pipes and bent shapes from sheet, strip and ribbon steel. These developments would mean a better quality and wider assortment of metallurgical products. All these developments should help cut down the consumption of metal in the engineering and other industries.

Improvement in productivity is another important factor in increasing production. An article in *Pravda* in 1968 pointed out that 77 per cent of the blast furnace capacity, 40 per cent of the open hearth steel furnace capacity and one third of the rolling mills in the Ukraine were not yielding the quantity of metal they were designed to give. The main reasons for this, given by the Minister of Ferrous Metallurgy, Mr. Kazanets, was lack of co-ordination, between factors such as supply, maintenance and marketing. By 1968, 67 metallurgical plants were working according to the new profit incentive system, introduced by the 1965 economic reforms, and their productivity record was better than that of plants operating under the old system. By 1970 all iron and steel plants were due to go over to the new system.

The Soviet Union has a number of export markets for her iron and steel production equipment. She is presently giving technical assistance in the building of 26 metallurgical plants in 14 countries. Soviet engineers are to participate in building a blast furnace in Sweden and a metallurgical complex in France. Italian, Japanese and West German companies have already bought Soviet equipment.

The production of the non-ferrous industry has grown from practically nothing before the revolution to a well established industry today. Three hundred enterprises have gone into operation since 1917. The aluminium and titanium-magnesium industries, the

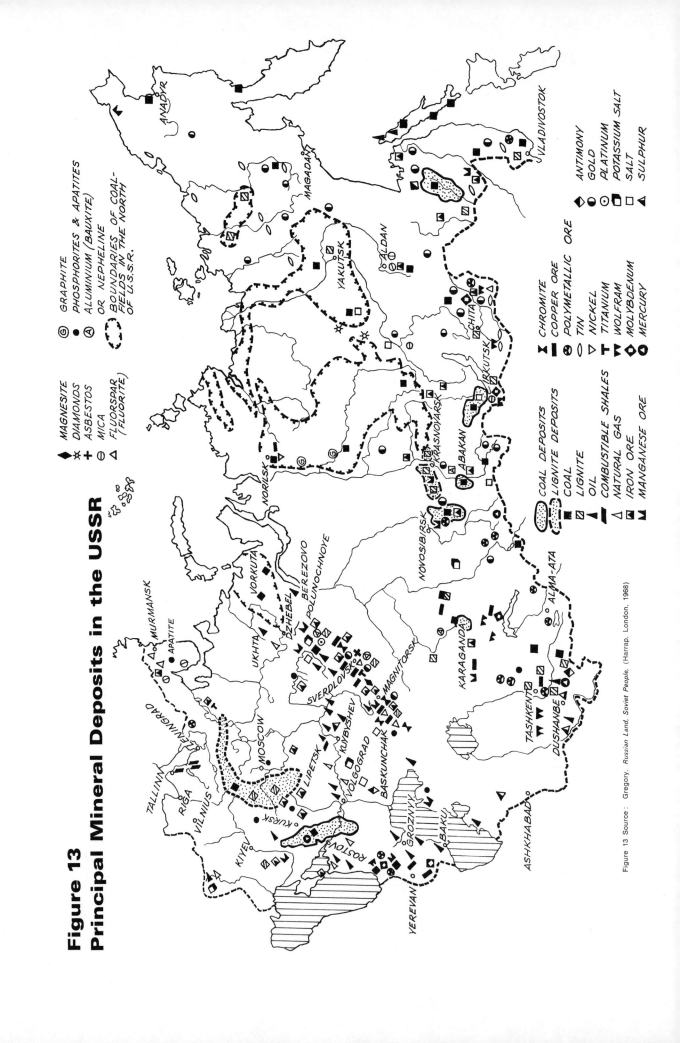

Figure 13
Principal Mineral Deposits in the USSR

Figure 13 Source: Gregory, *Russian Land, Soviet People*, (Harrap, London, 1968)

production of rare metals and diamonds have been the fastest growing. The next five-year plan (1971-75) emphasises the production of aluminium, copper, titanium-magnesium and nickel-cobalt.

The products of the non-ferrous metallurgical industry are exported to 45 countries. Copper and zinc from Kazakhstan and antimony are important exports.

(D) Research and development

In 1964 the then State Committee for Ferrous and Non-Ferrous Metallurgy (the forerunner of the present two separate ministries) had 100 research and development establishments under its administration. Fifty of these were research institutes and the other 50 were concerned with project and design work. There are also research institutes concerned with ferrous and non-ferrous metals attached to the Academy and other industrial ministries.

Two key institutes for the iron and steel industry come under the Ministry of Ferrous Industry.

10(1) STATE INSTITUTE FOR THE DESIGN OF IRON AND STEEL PLANTS (Gipromez)

The State Institute for the Design of Iron and Steel Plants was founded more than 40 years ago. It is responsible for the design and performance of specific plants and works, from the blast furnaces onwards. It does not concern itself with coke ovens or sinoter plants which are the responsibility of other bodies (e.g. Giprokoks). Gipromez not only designs the plant but also carried out the detailed engineering. It does not actually manufacture, but the Gipromez designer is responsible for commissioning and for the satisfactory performance in collaboration with the manufacturer. It has no responsibility for training operators. This is done by the Customer. Other responsibilities of Gipromez include long-term development of the regions, economic research, co-operation with other institutes in the USSR, and plant design for overseas clients and customers; the organisation employs some 10,000 people in the Moscow central office and its branch offices.

10(2) BARDIN CENTRAL SCIENTIFIC RESEARCH INSTITUTE FOR FERROUS METALLURGY

The Bardin Central Scientific Research Institute for Ferrous Metallurgy is the largest institute for ferrous metallurgy in the USSR. It co-ordinates the work of all the research institutions working in this field. Its research programme is settled by twice-yearly meetings of representatives from all these institutes; the ministry has to approve the programmes which are planned for two years ahead, though money is voted each year. Seventy per cent of finance comes from government sources and 30 per cent from research contracts with industry. The institute employs 3,000 people of whom 2,000 are engaged in the research laboratories and 1,000 in the pilot plant.

The institute's research field ranges from blast furnaces to high alloy steels. Four of its departments have the status of institutes: these are new technology, quality steels, precision alloys and physical metallurgy. The other departments, without institute status, are technical economics, central laboratories, continuous casting, the technical department, project design and pilot plant laboratory. The institute possesses experimental factory facilities and it pioneered work on the continuous casting of steel, a field in which the Soviet Union is well advanced.

10(3) BAIKOV INSTITUTE OF METALLURGY

The Baikov Institute of Metallurgy comes under the administration of the Academy of

Sciences of the USSR, and its interests tend to more fundamental research than the Bardin Institute, including research into non-ferrous metals. It is smaller than the Bardin Institute, and had a staff of 1,500 in 1964. Its major activities include vacuum treatment, and laboratory work on deoxidation equilibrium in iron and in nickel solutions.

Other important research institutes include the *USSR Institute of Aviation Materials*, the *Paton Electric Welding Institute* in the Ukraine, the *Institute of Steel and Alloys*, the *All-Union Research Institute of Non-Ferrous Metals* in Ust-Kamenogorsk, and the *Central Research Institute of Non-Ferrous Metallurgy*.

Although the pattern of research within the iron and steel industry is for the research institutes to be situated away from the centres of production, there is a very close liaison between the two sides. Staff from the Bardin Central Scientific Institute into Ferrous Metallurgy visit the factories, and representatives from the factories in their turn visit the institute. This co-operation contact, helps to make for a smooth introduction of the results of research and development into production. A number of large plants carry out their own research work into improving metallurgical processes. Examples are the *Dneprospetsstal*, the *Chelyabinsk*, the *Krasny Oktyabr*, the *Electrostal* and the *Zlatoust*.

The iron and steel industry is a priority industry within the economy, (similar to defence and aerospace) and as such it has good developmental facilities.

Journals: The main research journals of the ferrous and non-ferrous industries include: *Automaticheskaya Svarka* (Automatic Welding); *IAN SSSR Seriia Metally* (Proceedings of the USSR Academy of Sciences. Metal Series); *Kuznechno-Shtampovochnoe Proizvodstvo* (Forging and Stamping Production); *Liteinoe Proizvodstvo* (Foundry); *Metallovedenie i Termicheskaya Obrabotka Metallov* (Metallography and Heat Treatment of Metals); *Metallurg* (Metallurgist); *Ogneupory* (Refractories); *Svarochnoe Proizvodstvo* (Welding) *Stal'* (Steel); and *Tsvetnye Metally* (Non-Ferrous Metals).

(E) Plans

Writing in Pravda on July 20, 1968, the Minister of the Ferrous Metallurgy Industry, Mr. Kazanets, said: "With the object of further raising the technical level of blast furnace production, we plan to build only 2,700 cu.m or bigger furnaces and gradually to dismantle obsolete furnaces, while further increasing blast temperature.. In the coming five-year plan period metallurgists are going to treble the share of capital investments, earmarked for measures to widen the assortment and improve the quality of production. Emphasis will be on increasing the output of rolled sheets, which will account for between 43 and 44 per cent of the total by 1975. There will be a substantial increase in the output of cold-rolled electro-technical steel, sheet products with anti-corrosion coating, tin, various types of sheets with plastic plating. About 500 new hot-rolled and 360 cold-rolled steel shapes will be produced to widen the assortment".

(II) ENGINEERING AND MACHINE TOOLS INDUSTRIES

(A) Organisation

The machine tools industry comes under the administration of the *Ministry of General Machine Building*, the *Ministry of Medium Machine Building*, and the *Ministry of the Machine and Hand Tools Industry*. However, specialised branches of the machine tools and engineering industries have their own ministries. For example, the Ministry for Machine Building for the Light, Food and Household Equipment Industries; the Ministry of Instrument Making, Automatic Devices and Control Systems; the Ministry of Communal Building and Road Building Machinery; the Ministry of Tractor and of Agricultural

*Table 21**

	unit of measurement.	1913	1940	1958	1966	1970
Turbines	1,000 kW	6	1,179	6,647	15,172	16,200
Turbine generators	1,000 kW	468	468	5,186	13,447	10,600
AC electric motors over 100kW power	1,000 kW	-	527	3,328	5,511)	
AC electric motors under 100kW power	1,000 kW	300	1,300	10,400	22,600)	36,300
Metal cutting machines	1,000	1.8	58.4	138	192	
Automatic and semi-automatic lines for machine building and working	sets	-	1	88	219	
Forging-pressing machines	1,000	-	4.7	26.2	38.4	
Metallurgical equipment	1,000 tons	-	24	176	252	317
Oil apparatus	1,000 tons	-	15.5	70.9	148	126
Chemical equipment	million roubles	-	-	112	417	463
Textile machines	1,000	4.6	1.8	14.4	23.9	19.8
Main line diesel engines	sections	-	5	712	1,529	1,785
Main line electric engines	units	-	9	344	600	2,427
Main line freight transport wagons	1,000	12.9	30.9	40.3	40.2	58.3
Main line passenger carriages	1,000	1.5	1.1	1.8	2.0	
Automobiles (total of which	1,000	-	145.4	511.1	675.2	916
lorries and buses		-	139.9	388.9	445.0	572
cars		-	5.5	122.2	230.2	344
Tractors	1,000	-	32	220	382	459
Combine harvesters	1,000	-	12.8	65	92	99.2

* Source: Vredenskii, ed: *SSSR 1917-1967*, (Moscow, 1967)

Machinery Construction; the Ministry of Heavy Power and Transport Machine Building; the Ministry of Chemical and Oil Equipment Production; the Ministry of the Electronics Industry and the Ministry of the Electrical Engineering Industry.

These ministries administer their branch of industry according to the directives laid down by the Council of Ministers and Gosplan.

(B) Production

The production figures for the main products of the machine building and engineering industry are given in Table 21.

Since 1958 the Soviet Union has produced more metal-cutting machine tools each year than the United States. Their quality is generally high, and because of the large annual production the age structure of machine tools is young relative to the United States. However, as with other equipment, there is a shortage of spare parts, which means that a high proportion of the total stock is employed in repair shops. According to Soviet sources, in 1964-65 more than 1,000,000 machine tools, out of the total stock of 1,600,000 employed in industry were being used in repair shops, leaving some 600,000 in the main production shops of engineering factories. (Source: Zaleski et al: *Science Policy in the USSR*, P 493, OECD, Paris 1969)

Machine tool factories in Russia tend to specialise in their main line production, but their auxiliary production is less specialised. Almost all factories have their own foundries and forging and tooling shops, and only a very small proportion of these items is produced at specialised plants (0.6 per cent of the total engineering and metal working output in 1962).

Important machine building and engineering plants are the Moscow and Gorky automobile plants, the Volgograd, Kharkov and Chalyabinsk tractor plants, the machine building giants of the heavy, power and electrical-engineering industries (Uralmash, the Kramatorskii machine-building plant, the Leningrad metallurgical plant, Electrosila, the Luganskii diesel engine plant, the Rostselmach agricultural machinery plant), and the machine tool plants - the Minsk, Vitebsk and Krasny Borets are all in the Byelorussian Republic, which is an important centre of machine tool production.

Soviet machine tools are exported to more than 60 foreign countries, but she also imports some.

(C) Labour

The machine building and engineering industry employs more than 9,000,000 workers, engineers, technicians and office staff.

Table 22 gives a breakdown of the engineering graduates from university and higher educational establishments, including part-time courses, according to field in 1965.

During the 1960s there was a large increase in the numbers of engineers for the electrical engineering and electro-instrument making industry. In 1961 they made up 7.7 per cent of the total engineering compared with 15.6 per cent by 1965. The production of engineers for the machine building and instrument making industry remained steady at around 30 per cent of the total.

(D) Research and Development

In the engineering field as a whole, the pattern of research and development is traditional. There are specialist research institutes for developing new products of significance for just one branch of engineering; while major research institutes are responsible for the development of products suitable for a number of engineering fields and which will eventually be mass produced. The main idea here is that the wider the application of a new product, the more centralised its development should be, so as to eliminate unnecessary duplication. However, the situation does vary from field to field. In a number of industries - tractor, automobile and radio and television, for example - there is considerable dispersal of research and development facilities with a resulting duplication in products.

Research and development within the machine tool industry differs from that in the rest of the engineering field in the close liaison that exists between centralised and factory research and development, and in the major part played by factories in development work. Most of the numerous important research institutes come under the administration of the Ministry of the Machine and Hand Tools Industry.

*Table 22**

	Figures in 1,000s
Total graduations	157,400
Geology and survey of mineral resource deposits	3,200
Exploitation of mineral resources deposits	4,000
Power engineering	7,000
Metallurgy	4,800
Machine building and instrument making	46,000
Electrical Engineering and electroinstrument making	24,600
Radio Technology and communications	14,000
Chemical Technology	10,100
Forestry engineering and technology of woods, cellulose and paper	2,900
Technology of food products	4,800
Technology of consumer goods	3,200
Construction	21,300
Geodesy and Cartography	900
Hydrology and meteorology	1,000
Transport (operations)	9,600

*Source: Zaleski et al: *Science Policy in the USSR*, P 138-9, (OECD, Paris 1969)

10(4) EXPERIMENTAL RESEARCH INSTITUTE OF METAL CUTTING MACHINE TOOLS (ENIMS)

The Experimental Research Institute of Metal Cutting Machine Tools is the most important. Its research covers a wide range of fields and it also co-ordinates the research of other institutes working in the machine tool and metal cutting fields. It prepares state standards for machine tools. The institute has a staff of 1,500. A large experimental factory (*Stanko-konstruktsiya*) is attached to the institute.

A number of special design bureaux (SKBs) are responsible for the design of particular products. Many of these work in close liaison with a factory, depending on the factory to apply their ideas, particularly at the batch production stage. Then there are factories responsible for the production of a particular type of machine tool which have their own research and development facilities. Examples are the milling machine factory in Gorky, and the Krasny Proletarii plant in Moscow for turning lathes.

Other leading research institutes include the *All-Union Research Institute for Metallurgical Machine Building*, and the *Institute for Machine Science*.

Journals: The main research journals include *Vestnik Mashinostroeniia* (Machine building Journal); *Stanki i Instrument* (Machines and Tools); *Mashinovedenie* (Machines); *Mashinostroitel'* (Machine builder); and *Stroitelnye i Dorozhnye Mashiny* (Construction and Road-Building Machines).

(E) Plans

Writing in the magazine *Ogonyok* in 1970, the Minister of the Machine and Hand Tools Institute, Mr. Kostousov, said of future plans: "One of our main tasks in the next few years is to develop new types of standardised lathes and milling and grinding machines. That will make it possible to start large-scale manufacture of a range of interchangeable parts and units. We also want to raise the productivity of such machines ... To achieve this we shall use diamonds and abrasives, mechanise and automate auxiliary operations, introduce new machining methods and so on. We shall also step up the output of precision machine tools, many of which will be programme controlled .... We are also working on new types of machine tools based on the use of ultrasound, electroerosion, electro-chemistry and lasers which work parts from refractory steels, glass, porcelain and various alloys with incredible precision." (Source: *Soviet Weekly*, P 11, June 6, 1970).

11 Chemical Industry*

(A) Introduction

There was practically no chemical industry in Russia before the revolution, except for a few small factories. Since 1917, the Soviet government has built a large number of chemical factories and enterprises. However, despite continuous efforts to expand the industry, its level of development does not yet satisfy all the needs of the economy. In 1958 the Soviet Union had a relatively small chemical industry which produced a restricted range of traditional chemicals. That year the government decided on a rapid expansion of the industry, and since then chemical production has in fact grown very rapidly.

(B) Organisation

The Ministry of the Chemical Industry, the Ministry of Oil-Refining and Petrochemical Industry and the Ministry of Chemical and Oil Equipment Production are all concerned with the chemical industry. However, a high percentage of chemicals are produced outside the chemical industry proper (and this percentage is increasing), so a number of other ministries are also involved in the production of chemicals and petrochemicals. The ministries work within the plans laid down by the overall planning authorities.

(C) Production

The share of the chemical and petrochemical industries in the overall volume of industrial production came to 4.4 per cent in 1960, 5.8 per cent in 1965 and by 1970 it is planned to have grown to around 7.2 per cent.

In 1964, 7,000 different reagents were produced by Soviet industry. Together with the output of socialist countries and elsewhere the range available was 10,000 to 12,000 grades. This was three to four times the range available in 1958, but still some way behind other countries. Soviet production of a number of chemicals was clearly inadequate. Packaging is another problem. Few chemicals are produced in small packages, and the quality tends to be poor.

Between 1958 and 1965 the production of chemicals increased almost two and a half times, and between 1966 and 1970, the rate of growth of the chemical industry was 12 per cent, compared to a rate of 8 per cent for industry as a whole. Despite these

---

* Much of the information in this chapter is based on G.W. Hemy's article "The Soviet Chemical Industry" published in *Chemistry and Industry*, PP 207-215 February 17, 1968.

tremendous increases, production still lags behind the plan. Delays in building adequate new plant have been some of the main reasons for this.

Table 23 shows the production of the main branches of the chemical industry, expressed in millions of roubles.

*Table 23**
(*in million roubles*)

	1958	1965	1970 (planned)
Chemical mining	100	320	480
Basic Chemicals	680	1,900	4,750
Synthetic dyes	100	180	270
Resins and plastics	290	850	2,500
Chemical fibres	280	900	1,800
Organic chemicals	680	1,750	3,500
Paints and lacquers	530	1,000	1,500
Rubber and asbestos	1,060	2,100	3,200
Other	530	1,500	3,000
TOTAL	4,250	10,500	21,000

*Source: G.W. Hemy, "The Soviet Chemical Industry", *Chemistry and Industry*, P 207, (17 February 1968).

The accompanying map (Figure 14) shows the distribution of the chemical industry throughout the country.

(i) Chemical mining. Chemical mining includes the mining and preparation of apatite, other phosphate ores, potash salts and other chemical minerals. It accounts for some 3 per cent of the total chemicals production.

(ii) Basic chemicals. Basic chemicals cover the production of inorganic acids, salts and alkalis, chlorine and ammonia, and the production of fertilisers. The branch grew from 16 per cent of the total in 1958 to more than 22 per cent in 1970. The low 1958 figure reflected the shortage of fertilisers particularly nitrogenous ones.

There has been a continuous shortage of fertilisers. The targets set by the plans have not been fulfilled despite effort to boost fertiliser production since 1953. Krushchev declared that between 16,500,000 and 17,500,000 tons of artificial fertiliser would have to be produced in 1959, and between 28,000,000 and 30,000,000 tons in 1964. In fact under 13,000,000 tons were produced in 1959, and 31,253,000 tons in 1965. The main reason for the failure to meet output targets would seem to have been the delay in building adequate plant, but 55,400,000 tons were produced in 1970. By 1975 output of mineral fertilisers is to reach 90,000,000 tons.

In 1963 nitrogenous fertilisers accounted for 43.2 per cent, phosphate fertilisers for 29.2 per cent, potassium fertilisers for 16.8 per cent and phosphorite meal for 10.8

Figure 14 Main centres of production in the chemical industry

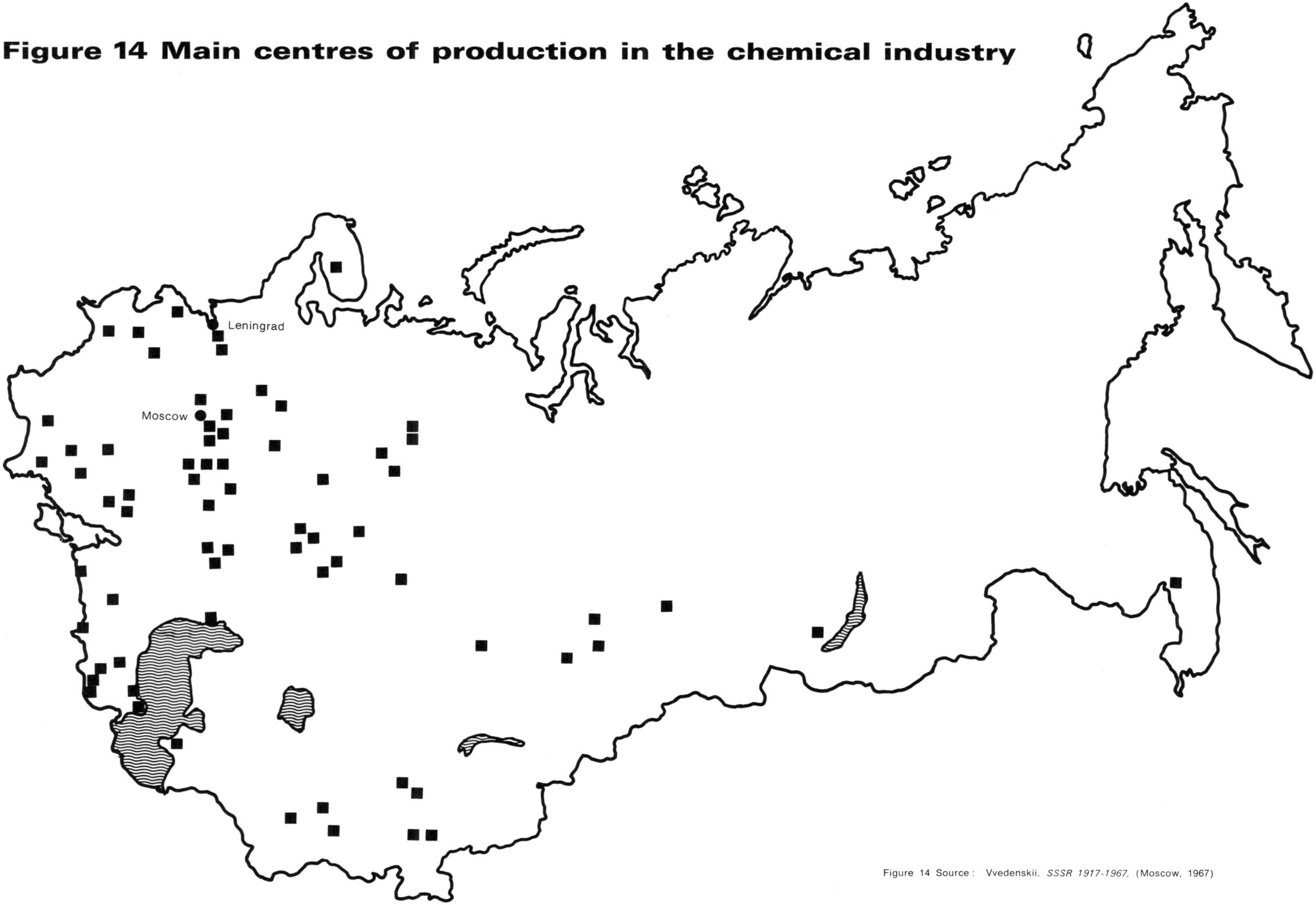

Figure 14 Source: Vvedenskii. *SSSR 1917-1967.* (Moscow, 1967)

per cent of total fertiliser output. The output targets for 1970 were 37.3 per cent for nitrogeneous fertilisers, 33.4 per cent for phosphate, 19.4 per cent for potassium, with phosphorite meal output remaining at the same level of roughly 9.3 per cent. Production of boron fertiliser was to rise to 200,000 tons in 1964 and to remain at this level until 1971.

The bulk of Soviet exports consists of phosphate fertilisers (65.3 per cent), followed by potassium (22.4 per cent) and nitrogenous (12.3 per cent) fertilisers. The amount exported was planned to be 4,000,000 tons in 1965 or some 11.4 per cent of total production - the rest being available for domestic use. This means more for the home market than in previous years. In 1958, out of a total output of 12,400,000 tons, 10,626,000 tons (82.6 per cent) was retained for domestic use, while 2,154,700 tons was exported. In 1962, production had risen to 17,300,000 tons, of which 13,645,000 tons (77 per cent) was allocated to Soviet farming, while 3,923,700 tons (22.7 per cent) was exported.

(iii) Dyestuffs. The absolute production of dyestuffs is increasing but the percentage that they represent of the total production of chemicals is falling. The reason is that the Soviet Union was relatively well provided with dyestuffs in 1958.

(iv) Resins and plastics. Resins and plastics formed 6½ per cent of the total production of chemicals in 1958. It is intended that this percentage shall increase to some 12 per cent by 1970. The products include pitch and bitumen plastics. The actual tons produced rose from 237,000 tons in 1958, to 801,000 tons in 1965, and were planned to rise to 2,100,000 to 2,300,000 tons in 1970.

(v) Chemical Fibres. Chemical fibres formed 6½ per cent of the total chemicals production in 1958 and their percentage was planned to increase to 9 per cent of the total by 1970. Production figures are as follows:

	1958	1965	1970
Artificial fibres (1,000s of tons)	153	330	520
Synthetic fibres (" " ")	13	78	280

In 1958 a resolution was adopted which concerned speeding up the production of polymer materials, including chemical fibres. For this reason the production of artificial and synthetic fibres was particularly well developed during the period of the seven-year plan 1959-65. Production was more than doubled compared with the previous seven-year period. During this time the industry acquired 12 new factories and six other enterprises. In 1966 new enterprises accounted for 53 per cent of the total production of the industry. Over the 20-year period, 1946-65, the capital expenditure for the development of the industry totalled 1,580,000,000 roubles, including 1,100,000,000 of roubles spent over the period of the seven-year plan.

The development of the production of chemical fibres has been marked not only by quantitative increases but also by improved quality. As well as viscous and cupro-ammonium fibres, synthetic and acetate fibres are now produced. Table 24 shows the structure of the manufacture of chemical fibres according to types (as percentage of the total).

Chemical fibres are now being produced for technical purposes as well as for consumer goods (see Table 25).

(vi) Organic chemicals. Apart from the traditional organics, organic chemicals include synthetic rubber, carbon black and all petrochemicals. They have accounted for 16 per cent of the total chemicals production since 1958.

Petrochemicals are an important growth industry. Much of the development planned for the next five-year plan (1971-75) is due to take place in Siberia using the oil and natural gas from the vast fields that have recently been prospected there.

Table 24*

Type of Fibre	1940	1948	1955	1958	1965	1966
Artificial	100	99.4	92	92.3	81	79
Viscous	90.2	87.6	83.8	83.9	72.3	70.3
Cupro-ammonium	9.8	11.6	6.8	6.9	4.1	4.3
Acetate	-	0.2	1.4	1.5	4.3	4.4
Synthetic	-	0.6	8	7.7	19	21
Polyamide	-	0.6	7	7.2	16.1	18.1
Polyester	-	-	-	-	1.8	1.6
Polyacrilonitrite	-	-	-	-	0.7	0.9
Perchlorovinyl	-	-	1	0.5	0.4	0.4

*Source: Ivanova and Mogilevskii: "The Development of The Chemical Fibres Industry in the USSR", *Khimicheske volokna*, 1967 (5); translated in *NLL Translations Bulletin*, P 369, vol 10, no 4, April 1968.

Table 25*

Designation of fibre	1940	1948	1955	1958	1965	1966
For general consumer goods	100	100	84.9	74.9	70.6	68.1
Textile fibres	77.5	62.2	30.3	28.5	24.3	22.9
Staple fibres	22.5	37.8	54.6	46.4	46.3	45.2
For technical purposes	-	-	15.1	25.1	29.4	31.9
Viscous	-	-	13.2	22.7	21.1	21.6
Synthetic	-	-	1.9	2.4	8.3	10.3

Source: Ivanova and Mogilevskii: "The Development of The Chemical Fibres Industry in the USSR", *Khimicheske volokna*, 1967 (5); translated in *NLL Translations Bulletin*, P 370, vol 10, no 4, April 1968.

(vii) Paints and lacquers. The percentage share of paints and lacquers in the total production of chemicals has decreased from 12½ per cent in 1958 to some 7 per cent. The total amount produced has increased from about 696,000 tons in 1958 to 1,066,000 tons in 1965.

(viii) Rubber and asbestos. The share of rubber and asbestos in the total production

of chemicals has fallen from 25 per cent in 1958 to 20 per cent in 1965, and is planned to reach 15 per cent in 1970. Improvement in the service life of lorry types is one reason for the decline as these have always constituted an important part of this category. The figures for the production of motor tyres are 14,400,000 in 1958, 26,400,000 in 1965 and 34,600,000 in 1970.

(D) Labour

In 1965 the chemical industry employed some 935,000 workers. Of these, almost 75 per cent were employed in production, just under 20 per cent were engineers or technicians and the remainder were employed in clerical jobs or were apprentices and juniors.

Hemy describes the labour force as being relatively highly qualified; more than 55 per cent had a higher qualification, about 43 per cent a secondary qualification, and only 2 per cent a lower qualification or no qualification at all. In some branches, namely the nitrogen industry, synthetic rubber and resins and plastics, 65 per cent of the workers had higher qualifications.

The labour force is also relatively stable. Hemy mentions that 45 per cent of all workers in 1958, and 53 per cent in 1964, had more than 10 years' total length of service. In addition, in 1963, almost 15 per cent had 15 years unbroken service with the same enterprise.

Most production workers work on a three shift system (42 per cent of the total and as high as 58 per cent in rubber and asbestos) though in chemical fibres a four shift system is almost as frequent, accounting for 31 per cent on a three shift system. Day work, that is a single shift system, is common in basic chemicals where about a third of the total force is employed.

The average hours worked per day were 6.81 in 1965.

Over the period 1958-65 it is claimed that labour productivity increased by 51 per cent. Dyestuffs (25 per cent increase) and rubber and asbestos (20 per cent) had the lowest increases while resins and plastics (90 per cent increase) and chemical mining (75 per cent) had the highest. (Source: Hemy G.W. "The Soviet Chemical Industry", *Chemistry and Industry,* February 1968).

(E) Exports

Soviet trade in chemicals is increasing. It multiplied some four times between 1958 and 1965, exports increasing from 60,000,000 roubles to 246,500,000 roubles, and imports from 112,000,000 roubles to 450,000,000 roubles. The share of chemicals in the USSR's total trade, however, has remained approximately at the same level - around 3.9 per cent of her exports and 6.2 per cent of her imports.

The two largest export groups are general chemicals and fertilisers and crop chemicals. The three biggest import groups are general chemicals, dyes and paints, and rubber products.

(F) Finance

Table 26 gives details of capital investment in the chemical industry in Russia.

Between 1958 and 1965 the total amount spent increased four times, and the share of the chemical industry in total industrial investment more than doubled, accounting for about one-ninth of total industrial investment in 1965. About half the total investment was spent in rebuilding, reconstructing or enlarging existing plants. This is the preferred method wherever possible because it is cheaper per unit of capacity and generally quicker as well. The backlog of capital construction at the end of each year has risen almost in step with capital spending, plans being on average one year behind. The new capital coming onto the books of the chemical industry each year is below the total spending because of depreciation and the retirement of some capital during renovation, rebuilding or enlargement.

The 1966-70 five-year plan envisaged a stepping up of investment to 14,500,000,000 roubles for the period, compared with 9,100,000,000 roubles in the previous seven-year period.

Table 26*

	1958	1960	1962	1964	1965
Total capital investment (millions roubles)	455	890	1,164	1,984	1,924
Percentage of all industry	4.6	7.0	8.2	11.7	10.7
Spent on reconstruction renewal or expansion of existing plants		463	628	1,012	885
Unfinished at end of year	490	968	1,546	2,056	2,276
New capital in use during year	396	585		1,561	1,620

*Source: Hemy, G.W. "The Soviet Chemical Industry", *Chemistry and Industry*, P 210, (February, 1968).

According to Mr. Kostandov, the Soviet Minister for the chemical industry, writing in the Soviet Journal *Chemical Industry* in 1966: "Our immediate task is the proper allocation and the economic use of capital funds allocated to the chemical industry ... growth of production often lags behind the increase in fixed capital...due to numerous errors in planning, about 10 per cent of all projects recently coming on stream have not attained the designed capacity. At this time the inistry of Chemical Industry must give great attention to improving the efficiency of factories, to maximising use of capital funds by improving production planning, to the elimination of bottlenecks and, above all, to attaining maximum output at minimum capital cost. There is plenty of room for lowering consumption of materials and energy, which account for 65 to 70 per cent of total factory outlays...the low effectiveness of investment and the low yield on capital is mainly due to faults in planning and design of capital projects and in factory building..."

By the beginning of 1971 the majority of Soviet industrial enterprises including those belonging to the chemical industry, had switched over to the new system of organisation first introduced in 1965 with the economic reforms. According to Academician Fedorenko writing in the *Ekonomicheskaya Gazeta* the economic reform means that the chemical enterprises have an interest in the proper pricing of chemical materials so that prices reflect the effectiveness of their use as far as possible. The profit to the producer will increase according to the effectiveness of his products, and the consumer will select the most effective products, thus promoting technical progress. Fedorenko writes that the next step is to apply the new cost accounting methods to the management organisations in the chemical industry.

(G) Research and Development

Responsibility for research and development in the Soviet chemical industry is divided between ministerial establishments, academy institutes (both All-Union and Republican) and the VUZy. According to Amann, Bury and Davies (*Science Policy in the USSR*, P 418, OECD, Paris, 1969) theoretical and exploratory research is mainly carried out within the academy system and by VUZy. The academy institutes specialise in various fields; some deal with a particular branch of chemistry (for example, the Institute of General and Organic Chemistry and the Institute of Inorganic Chemistry) and others concentrate their activities on more narrowly defined fields (for example, the Institute of Catalysis of the Siberian Division of the USSR Academy of Sciences). Although the USSR Academy of Sciences' main interest is in fundamental research, much of the work appears to be directed towards practical ends. It has been noted that, for example, the work of the Institute of High Molecular Weight Compounds (under the USSR Academy) differs little in its work or experimentation methods from a large industrial research institute such as the Karpov Institute.

Amann, Bury and Davies list three categories into which the research and development establishments under the industrial ministries tend to fall: capital projects institutes without research laboratories (for example, Giproplast), research institutes with experimental facilities (for example, VNIINefthim and NIIPlastmass) and institutes concerned with capital projects and research(for example, the Institute of the Nitrogen Industry and Institute of Chemical Pigments). The project institutes are used to develop the designs supplied by the research institutes up to industrial process size. Where the research and project-making functions are not combined in one institute such as NIIPlastmass and a project organisation (Giproplast) working together almost exclusively.

Many prominent research institutes in the broad field of chemistry possess pilot-plant facilities. The Institute of Synthetic Rubber (Leningrad), for example, has its own pilot-plant and access to several factories outside Leningrad for proving work; the Institute for Synthetic Fibres (Kalinin), the Institute for Fertilisers and Insecto-Fungicides (Moscow) and the Institute for Polymerised Plastics (Leningrad), are also well provided for in this respect. However, these examples are almost certainly untypical. According to Soviet evidence, one of the main problems facing research and development in the chemical industry is the inadequate provision of pilot-plant facilities. New technology and capital equipment is often introduced into production directly from the research institute, capital-projects institute or factory without a period of trial and refinement. According to Amann, Bury and Davies there is no Soviet equivalent of such western firms such as Lurgi and Imhausen, which are responsible for research, the development of new processes, and projecting and building chemical plant. One solution to the problem of "scaling-up" a project from an experimental prototype to full-scale production specifications, which has been explored widely in recent years, is the mathematical simulation of chemical processes. However, although such models may be a useful guide in building new plant, they are said to be no real substitute for pilot-plant proving.

In industry, established chemical enterprises have also tended to lack either pilot-plant or factory laboratories; the contribution which industrial enterprises make to the research and development effort is small compared with that of the specialist institute network. However, there are significant exceptions to this pattern. Amann, Bury and Davies cite petrochemicals as an example of a field of the chemical industry where large experimental establishments for the proving of new products and processes are a characteristic feature. In fact "leading base factories" are organised for some important chemical products. These are responsible together with branch research institutes, for carrying out research projects and improving technology in their given field.

Journals: The main journals concerned with the chemical industry include: *Khimicheskaya Promyshlennost'* (Chemical Industry); *Khimicheskie Volokna* (Chemical Fibres); *Khimiia v Sel'skom Khozaistve* (Chemistry in Agriculture); *Khimiz Tverdogo Topliva* (Chemistry of Solid Fuel); *Neftekhimiia* (Petrochemistry); *Lakokrasochnye Materialy i ikh Primenenie* (Paints and Varnishes and their Applications); *Kauchuk i Rezina* (Rubber) and *Zavodskaya Laboratoriia* (Factory Laboratory).

(H) Plans and Policy

The chemicalisation of the national economy is regarded as one of the most important problems facing the country. "Chemicalisation" is a portmanteau word used by the Soviets to mean the widespread development of the chemical industry, by increasing its production and the number of its products, and by widening their use in other fields of the national economy, particularly in agriculture.

The main direction of development of the chemical industry is that of introducing new technical equipment and intensifying the basic methods of production. Basic aims are an increase in output volume by intensifying the use of equipment and by changing the parameters governing different chemical processes. At the same time as improving the technology the aim is to improve the quality of goods produced.

There are a number of additional problems. It is necessary to work out a general scheme of relationship between the sources of raw materials and the siting of the chemical industry enterprises. At the moment European Russia produces the overwhelming bulk of chemicals but in the long run Eastern Siberia could be the preferred chemical base because of the quantity of raw materials available there. It will be a question of balancing costs of labour and development in the eastern underdeveloped areas against the costs of transport to the industrialised western regions. The growth of the consumer branches of the chemical industry and the increased market for chemical goods, makes it necessary to determine the capacity and structure of this market, in particular it calls for the elaboration of pricing problems with respect to the demand for polymer materials, which will undoubtedly become the leading product of the industry.

12 Timber, Cellulose and Woodworking Industry

(A) Introduction

The timber, cellulose and woodworking industry is one of the country's oldest industries and makes up an important part of her exports. The traditional wood felling areas were in the centre and western regions of the country. However, in recent years the Soviet government has been trying to shift the industry more to the Eastern regions where there are the greatest supplies of raw materials.

(B) Organisation

The Ministries of the Timber Industry and of the Pulp and Paper Industry are responsible for the administration of the timber, cellulose and woodworking industry.

(C) Production

In 1961 forests covered an area of some 738 million hectares, or 32.9 per cent, of the total area of the country. The total reserves of timber amounted to some 80,000,000,000 cubic metres. Of the forests, 72.7 per cent are coniferous, and these provide more than two-thirds of the world's resources of coniferous timber. All in all the USSR accounts for more than half the world's timber resources. Siberia, the Far East and the Northern Regions have the richest resources, and a large number of pulp and paper making plants have been built there recently or are at present under construction. The European part of the country, which has been the main centre of the timber felling industry until quite recently, contains 17,000,000,000 cu.m of the country's timber reserves.

During the past 10 years or so, the Bratsk Timber Conversion Plant (which includes two heat and power stations, a timber port and eight factories and plants using 6,500,000 cu.m of raw material a year), the Baikal Pulp Mill and the Krasnoyarsk Mill in East Siberia, the Komsomolsk Mill in the Far East, the Kotlas Mill and the Syktyvkar Timber Conversion Plant in the North have been built. At the same time, older paper mills, at Arkhangelsk, Kondopozhsky, Solombalsk, Segezh and elsewhere, have been reconstructed and modernised. These measures have doubled pulp production and increased paper production by 70 per cent over the past 10 years. In the same period the production of cardboard has nearly tripled, and the output of cellulose for the production of synthetic fibre has more than doubled. The planned production figures for 1970 were 5,200,000 tons of pulp (12.7 per cent up on 1969 level), more than 4,000,000 tons of paper including 1,108,000 tons of newsprint (3.5 per cent increase on 1969), nearly 2,100,000 tons of cardboard (15.6 per cent rise on 1969).

Wood, cellulose and paper account for some 6.5 per cent of the USSR's exports, and some 2 per cent of its imports.

The aim of all areas of the industry is to increase production through introducing modern machinery and equipment at all plants and so increase the productivity of the labour force.

Table 27 gives figures for timber and paper production for the years 1958, 1964 and 1969.

*Table 27**

	1958	1964	1969	1970
Rough timber (million cu.m)[1]	250.9	276.9	-	289
Sawn timber (million cu.m)	93.7	110.9	-	
Cellulose (million tons)	2.1	2.9	4.6	
Paper (million tons)	2.2	2.9	4.0	4.2
Cardboard (million tons)	0.8	1.2	2.2	

[1] Excludes timber sawn by collective farms

*Source: Gregory, *Russian Land Soviet People*, P 346 (Harrap, London, 1968) and *Narodnoe Khozaistvo SSSR v 1969 g.* P 237, (Moscow, 1970).

(D) Research and Development

In 1962 there were 100 research, project and project design institutes directly involved in the timber, cellulose and woodworking industry. They included the All-Union Research Institute of the Cellulose and paper making industry, and the Central Paper Research Institute. There is also an Institute of Forestry attached to the Academy of Sciences. According to Academician Zhukov, the Institute's Director (writing in 1970 in *Soviet Weekly*), there is need for a national body to co-ordinate research work within the field of forestry. There are numerous research problems to be tackled including: environmental problems; guarding against forest fires, pest and diseases; study of forest land reclamations, the movements of forest animals; determining the most suitable distribution of trees, bushes, plants and the effects of sunlight on growth and the accumulation of mineral elements; the development of valuable hybrids; and the research into the use of chemicals.

Journals: The research journals of the timber, cellulose and woodworking industry include: *Bumazhnaya Promyshlennost'* (Paper Industry); *Gidroliznaya i Lesokhimicheskaya Promyshlennost'* (Wood Pulp and Sylvochemicals Industry); *Derevoobrabatyvayushchaya Promyshlennost'* (Woodworking Industry); *Iz. VUZ Lesnoi Zhurnal* (Proceedings of Higher Educational Establishments. Forestry Journal); *Lesnaya Nov* (Forestry News); and *Lesnaya Promyshlennost'* (Forestry Industry).

13 Food and Consumer Industries

(A) Introduction

Before 1917 the light and food industries were mainly of the character of cottage industries. Industrial processing of agricultural raw material was poorly developed. Since 1917 most of the enterprises of the light and food industries have been reconstructed and equipped with up-to-date technology, and many new ones have been built. The number of products has increased and the distribution of factories throughout the country is now more even.

The consumer industries, hwoever, have always had to take second place in capital investment to the heavy industries. This has meant that the supply of consumer goods has not met the demand of the population, with resulting shortages of many goods. In recent years this situation has improved considerably with increased output of consumer goods. Though at the same time the share of consumer goods in total industrial output takes a second place to that of heavy industrial producer goods (the proportion was approximately one to four at the beginning of the 1960s)

(B) Organisation

The main industries involved with the food industry are the *Ministry of the Food Industry*, the *Ministry of the Meat and Milk Industry*, and the *Ministry of Health*. The first two are primarily concerned with food production and distribution, working within the directives laid down by the overall state plan. The Ministry of Health is mainly concerned with quality control, particularly where food additives are concerned. The *Ministry of Light Industry* is responsible for light industry (primarily textiles and related products), and the *Ministry of Culture* covers the so-called cultural goods, or household and personal articles. The *Ministry of Machine Building for the Light Food and Household Equipment Industries* touches on all three areas.

Distribution is carried out directly from the factory to the trading organisation which retails the goods. Until the middle of the 1960s the latter had to take what was delivered to it and could not make its own arrangements with the manufacturing enterprises. Despite quality control inspectors in the factories and a governmental inspection service, the proportion of unusable goods on sale sometimes reached as high as 20 per cent. Consequently in 1964 two enterprises in the clothing industry were permitted to enter into direct contact with the trading organisations, both sides negotiating over what was needed and producing accordingly. The experiment was a success and since then has been extended to the whole of the country's trade and light industry.

(C) Production

Table 28 gives production figures for selected products of light industry.

*Table 28**

	1913	1940	1958	1966	1969	1970
Cotton material (mln square metres)	1,800	2,700	4,300	5,700	6,208	6,150
Wool " "	138	152	385	510	618	644
Linen " "	121	268	440	592	674	707
Silk " "	35	64	690	869	1,026	1,146
Leather shoes (million pairs)	68	211	356	522	636	676

*Source: Vredenskii, ed: *SSSR 1917-1967*, P 221 (Moscow, 1967) and *Narodnoe Khozaistvo, SSSR v 1969g*, P 249, 250 (Moscow, 1970).

Table 29 gives production figures for various household, personal and cultural goods.

*Table 29**

	1913	1940	1958	1966	1969	1970
Watch and clocks (1000s)	700	2,800	24,800	32,400	38,000	40,200
Radios	-	160	3,902	5,842	7,266	7,800
Televisions	-	0.3	979	4,415	6,595	6,700
Refrigerators	-	3.5	360	2,205	3,701	4,100
Washing machines	-	-	464	3,869	5,153	5,200
Vacuum cleaners	-	-	245	899	1,359	
Sewing machines	272	175	2,686	1,025	1,324	
Cameras	-	355	1,473	1,420	1,961	
Motorbikes	0.1	6.8	400	753	827	843
Bicycles, motorbicycles and mopeds	11.2	255	3,651	4,048	4,372	4,400
Pianos	-	10.1	66.6	168	170	

*Source: Vredenskii, ed: *SSSR 1917-1967*, P 221 (Moscow, 1967) and *Narodnoe Khozaistvo, SSSR v 1969 g*, P 263 (Moscow, 1970).

The production figures for a number of food products are shown in Table 30:

*Table 30**

Type of Product (1000s tons)	1913	1940	1958	1966	1969	1970
Meat [1]	1,273	1,501	3,372	5,724	6,483	7,100
Sausage-ham products		391	1,049	1,760	2,140	2,300
Fish, sea animals and whales - the catch	1,051	1,404	2,936	6,090	7,091	
Animal fat [1] (butter)	129	226	659	1,042	954	963
Whole milk products (evaluating counting the milk) [1]		1,300	6,000	12,900	18,300	19,500
Cheese and brynza	7.9 [2]	51	169	353	431	478
Granulated sugar	1,363	2,165	5,433	9,740	10,347	10,200
Vegetable oil	538	798	1,465	2,730	2,979	2,800
Confectionary	125	790	1,676	2,242	2,765	2,900
Conserves (mlns jars)	116	1,113	4,073	7,505	9,660	10,600

[1] The figures relate to industrial production and do not include production on the individual plots of the farmers, and for meat and whole milk products the figures do not include collective farm production either. The general/overall production of meat in 1966 was 10,800,000 tons, of animal fat was 1,157,000 tons and of milk was 75,800,000 tons.

[2] Within the 1939 USSR territorial boundaries

*Source: Vredenskii, ed: *SSSR 1917-1967*, P 222, (Moscow, 1967) and Narodnoe Khozaistvo *SSSR v 1969 g*, P 280 (Moscow, 1970).

Increased food production is changing the feeding patterns of the people. They are eating more meat, butter, eggs and fruits, while purchases of bread and potatoes are declining. In 1968 purchases per head of population showed the following increases on the 1950 figures: meat from 26kg to 51kg; milk products from 172kg to 290kg, sugar from 11.6kg to 38.8kg, and fish from 7kg to 15.9kg. Over the same period, purchases of potatoes dropped from 241kg to 125kg per head and bread products from 172kg to 142kg.

(D) Labour

In 1963 the number of people employed in the light and food industries constituted 26.3 per cent of the total industrial labour force. Labour productivity rose more slowly in these industries in the beginning of the 1960s than in the heavy industries. In 1963 it rose 3 per cent and 14 per cent in the light and food industries respectively compared to an increase of 16 per cent in industry as a whole. The economic reforms and direct trading links established in the second half of the 1960s may lead to improvement here. In addition, with increased emphasis on consumer goods production there is a corresponding increase in mechanisation. The shoemaking industry is largely automated with production in many factories based on conveyors. More than 90 per cent of the footwear produced is manufactured by the straight flow method. Another highly mechanised mass-production branch is the fur industry, where hat-making and fur-dressing operations are 85 per cent conveyorised. Bread bakeries are equipped with rigid annular conveyors and 86.5 per cent are mechanised. The textile, knitwear, garment, tanning, meat, fish and poultry processing, and other branches of the light and food industries have been similarly modernised, with increased hydro-acoustic installations on fishing and fish-surveying vessels, straight-flow lines for making butter, automatic bottle-washing, milk-pouring and corking lines. The share of automatic looms grew from 45 per cent in 1958 to 65 per cent in 1965.

(E) Finance

In 1970 retail trade turnover in the state and co-operative systems totalled 153,600,000,000 roubles (this is an increase of 7.4 per cent over 1969). While the amount of capital invested in the consumer goods industry rose, its share of the total invested capital dropped, between 1960 and 1963, from 14.2 per cent to 13.4 per cent. At the end of 1960 the total value of capital invested in all industry and in the consumer goods sector respectively was 90,100,000,000 and 12,800,000,000 roubles respectively. At the end of 1963 the corresponding figures were 123,900,000,000 and 16,600,000,000 roubles.

(F) Research and Development

There are more than 90 research establishments and another 40 project organisations working within the food industry. There are also 12 full-time higher educational establishments attached to the food and fishing industries. Within the light industries the figures are eight full-time higher educational establishments. The textile engineering industry has four research institutes, 26 special design bureaux and 48 factory design bureaux; the radio and television industry has 17 design bureaux; and the furniture industry has 40 project organisations (not counting factory design bureaux).

Journals: The main journals of the food and consumer industries are: *Vinodelie i Vinogradarstvo* (Viticulture and Wine-Making in the USSR); *Iz. VUZ Pishchevaya Tekhnologiia* (Proceedings of the Higher Educational Establishments. Food Technology); *Iz. VUZ Tekhnologiia Tekstilnoi Promyshlennosti* (Proceedings of the Higher Educational Establishments. Technology of the Textile Industry); *Kozhevenno-Obuvnaya Promyshlennost'* (Canned and Dried Vegetables Industry); *Maslozhirovaya Promyshlennost'* (Butter and Fats Industry); *Molochnaya Promyshlennost'* (Milk Industry); *Mukomolno-Elevatornaya Promyshlennost'* (Flour Mills and Elevators); and *Miasnaya Industriia SSSR* (Soviet Meat Industry).

(G) Food Additives

The addition of substances to foods is regulated in Russia by public health provisions and by technical instructions and directives issued by the Main Medical-Epidemiological Division of the Ministry of Health. The use, in food production, of substances which have not received the approval of the public health authorities is illegal. Permitted food additives must conform strictly to governmental food standards, technical specification, or special technological directives. Contaminants which may enter foods during treatment or processing are also controlled. Before any chemical substance is permitted for use as a food additive, it must be subjected to research and investigation to determine its possible toxicity, and, if necessary, its carcinogenicity. Such investigations are undertaken in the research institutes of the Ministry of Health and of the Academy of Medical Sciences, as well as by the general hygiene and food hygiene departments of state medical institutes and institutes for advanced medical studies.

The authorities recommend that additives be kept to a minimum in food production. For example, the USSR allows only three synthetic colouring matters to be used in the food industry (compared to 15 in Canada and 19 in West Germany). Various branches of the food industry (canning, vegetable and fruit drying, confectionary etc) are always experimenting in their laboratories and institutes to find new, more effective, more easily obtainable and more economical chemical substances to use as food additives. Every firm proposal for the introduction of a new additive presented by a trade research institute must be approved by the public health bodies. The basic criterion for a new food additive is the safety of both the additive and the foodstuffs treated with the substance. Authorisation to use a food additive is denied if the intended effect can be obtained by a safe and economically suitable technological process. Additives are also prohibited if their presence is likely to conceal technological deficiencies, spoilage, or to lower the nutritive value of a food. Permitted additives are also subject to constant observation and re-examination. Source: Stenberg, A.I. et al: *Food Additive Control in the USSR*, FAO Food Additive Control Series No. 8 (1969).

(H) Plans and Policy

The directives for the new 1971-1975 plan have, for the first time, envisaged a greater increase in production of consumer articles than in the production of heavy industry. The figures show an increase in production of consumer goods of 44-48 per cent, compared with 41-45 per cent increase for the production of the "means of production". At the same time the government maintains its belief that a high level of development of heavy industry, particularly of its key branches together with an effective rise in agricultural production are the prerequisites for the expansion of the consumer industries.

14 Construction Industry

(A) Introduction

One of the key aims of the Soviet government has been to establish a strong construction industry. Before the revolution the industry was technologically backward relying almost entirely on manual labour.

In 1918 a committee was set up to work out plans for state construction, to overlook all state construction projects coming under the aegis of the Supreme Council and Peoples' Commissariats. A specially military section was included within this Committee of State Construction to deal with military construction work. The large building programmes called for by the Eighth Party Congress in 1920 and later by the five-year plans meant a tremendous development of the construction industry. The production of building materials, especially cement and bricks, had to be increased. From 1929 to 1932, during the first five-year plan, 1,500 large enterprises were built. By 1932 35 per cent of all workers engaged in heavy industry belonged to the construction industry.

In 1936 all-union and regional construction trusts were set up to deal with specific kinds of work - construction of industrial enterprises, ventilation, heating, etc. During the fifth five-year plan (1951-55) a single construction organisation for Moscow (Glavmosstroi) was created. Later similar organisations were set up for other cities such as Leningrad, Kiev and so on.

(B) Organisation

14(1) STATE COMMITTEE FOR CONSTRUCTION

The State Committee for Construction (Gosstroi) is the chief controlling body within the construction industry. Its main responsibility is to formulate the national construction policy, including norms and standards. It is generally responsible for all civil construction. The chairman of Gosstroi is automatically a member of the Council of Ministers and is, in fact, a deputy prime minister of the country. Gosstroi works very closely with the economic planning authority Gosplan. It also directly controls a number of research, design and information institutes.

14(2) STATE COMMITTEE FOR CIVIL CONSTRUCTION AND ARCHITECTURE

The State Committee for Civil Construction and Architecture (Gosgrazhdanstroi) guides regional development and urban planning, administers building regulations, and supervises architectural design. The master plan for each town and all major architectural schemes are submitted to it for approval. The committee is under Gosstroi. Its chairman is a member of Gosstroi.

In addition to these two State committees, there are six Ministries with a direct interest in and responsibility for the construction industry. These are the *Ministry for Construction*, the *Ministry for Heavy Industrial Construction*, the *Ministry of Agricultural Construction*, the *Ministry of the Building Materials Industry*, and the *Ministry of Transport Construction*. The ministries work in liaison with the State committees according to the directives of the overall state plan. Both the State committees and ministries have their Republican counterparts.

At the city council level of large cities, such as Moscow, Leningrad and Kiev, there is a *Chief Administration for Architecture and Planning* which operates functionally under Gosgrazhdanstroi. The plans drawn up by these chief administrations are carried out by the respective city administration for construction.

Thus the execution task of designing and erecting buildings (including the operation of house factories and certain building materials plants) is decentralised to the republics, to certain regions and to local authorities. In a typical situation, such as that at Kiev, the architects and building staff have a dual responsibility. They answer to the Republican, in this case Ukrainian Council of Ministers, or the City Soviet for the formulation of the master plan and the completion of the building programme on time; but they must meet the requirements of the State Committee for Civil Construction with regard to building and planning standards, architectural merit and similar matters.

(C) Production

(i) Housing. Industrial methods of construction are being increasingly used, in particular for housing in the cities. Standard plans were apparently used for 95 per cent of all dwellings in 1965 and for 82 per cent of public buildings. In Leningrad the figure is 70 per cent. Overall large panel construction has grown from 1.3 per cent of the total housing in 1959 to 30 per cent in 1965, and it is expected to reach a limiting value of around 40 per cent. The large panel units are produced by *building "combinats."* These were formed over the past 10 years or so, following the decision to concentrate house building on large panel construction. There are about five or six combinats in both Leningrad and Moscow. Each one has a number of factories which produce the major concrete components for internal and external walls, floors, staircases, service core units and so on. Other components are brought in from the materials industry or are provided by specialised "trusts" dealing with such things as engineering services, foundations and so on. The scale of production of the combinats is very high - Combinat no. 1 in Moscow, for example, produces 25,000 dwellings a year.

Despite the increasing use of industrial methods, brick construction, using precast concrete floors and often covered with ceramic tiles, still forms a significant proportion of total housing.

The present housing norm for the period up to 1970 is 9sq. metres per person. This is to be raised to 12sq. metres per person. The stock of housing in 1967 provided just under 6sq. metres per person though the norm was exceeded in certain cities, for example, Tallin and Kiev.

(ii) Non-domestic. Non-domestic construction covers industrial buildings, offices, factories, power stations, gas lines and so on. Between 1956 and 1960, 4,900 large state enterprises went into operation. For the period 1961 to 1965 the figure was 3,290. These included a number of large thermal and electric power stations. Among the latter was the Bratsk power station. Nearly 400 chemical enterprises were constructed. There were a number of oil refining factories and oil pipelines, including the 'Druzhba' line which carries oil from the Soviet Union to Poland, Hungary, Czechoslovakia and the German Democratic Republic (see Chapter 9). A large volume of construction work took place in the eastern region.

(iii) Building materials. Table 31 gives figures for the production of building materials.

*Table 31**

	1928	1940	1950	1960	1965	1970
Cement (million tons)	1.8	5.7	10.2	45.5	72.4	95.2
Prefabricated ferro-concrete elements (million cu.metre)			1.3	32.4	63.5	83.0
including wall panels (million sq.metre)				5.5	33.8	
Bricks (1,000 million pieces)	2.8	7.5	10.2	35.5	36.9	43.0
Window glass (million sq.metre)		34.2	44.7	76.9	147.2	190.3	231.0

*Source: Vredenskii, et al: *SSSR 1917-1967*, P 228 (Moscow, 1967)

The plan for capital construction of enterprises of the building materials industry has not been fulfilled in recent years, falling some 4 to 7 per cent behind the average national level of plan fulfillment. The figures for new projects are lower still. Only 37 projects were commissioned against a planned figure of 69.

As Table 32 shows, the stock of basic building machinery has grown steadily in recent years.

*Table 32**

	1950	1955	1960	1965	1969
Excavators	5,900	17,500	36,800	70,000	92,000
Scrapers	3,000	9,300	12,200	21,000	26,000
Bulldozers	3,000	16,100	40,500	67,000	94,000
Travelling cranes	5,600	28,900	55,000	84,000	111,000

(The figures indicate the number of machines in stock at the end of the year)

*Source: Shchepetev, A.I. et al: "Fifty Years of Mechanisation in Building, *National Lending Library Translations Bulletin*, P 561, vol. 9, June 1967 (translation of article in *Mekhanizatsiya Stroitel'stra*, PP 10-14, 1967,81)

(iv) Rural construction. Special importance is being attached to rural construction.

The decision of the July (1970) Plenary Meeting of the Central Committee of the Communist Party provided for a big increase in this field, in response to the Party's call for an increased growth of agricultural production.

In the next five-year period (1971-75) the Ministry of Rural Construction is expected to build and commission a large number of animal-breeding complexes, poultry-raising establishments and stud farms. At present the ministry is carrying out preparatory work in the Ukraine, Byelorussia, Kazakhstan, the Northern Caucasus, and Western Siberia, for building industrial combines that will turn out the structures for animal-breeding complexes.

The plans all provide for the erection of industrial establishments to manufacture precast reinforced concrete structures for production premises, and building plants for the manufacture of large-panel, block and preassembly units for housebuilding. The level of fully-prefabricated construction of production premises will rise to 35 and 40 per cent, of housing to 30 to 35 per cent, and of cultural-domestic establishments to 20 per cent. The figure for elevator capacities erected from prefabricated elements will rise from 75 per cent to nearly 90 per cent. Greater emphasis is going to be laid on light-weight structures in future prefabricated building.

(v) Heating, mechanical and electrical work, auxiliary services. Large cities and newly developed urban areas associated with them are nearly always connected to a district heating network. The source of heat is an electricity generating power station. The power stations are designed on the basis of providing both electrical power and heat involving the use of "pass-out" or back pressure turbines. In certain places where the progress in building very large housing estates has outstripped the capacity of the combined heat and power stations to provide the necessary heat, temporary "pure" heating stations to deal with these areas alone have been provided, although it is intended that in due course the areas will be connected to a combined heat and power station. Natural gas is widely used for generating power but in some cases the direct heating stations use heavy fuel oil, either as a means of coping with peak demands, or as a standby in the event of a disturbance of the natural gas supply.

As a general rule, mechanical and electrical works outside the buildings are carried out by engineering designers and constructors from organisations separate from those which design and execute the internal engineering services. The bulk of the internal electric wiring is carried out on site after the buildings have been erected and during finishing processes.

(D) Labour

Labour productivity has been increasing steadily. This is partly due to the increased industrialisation of construction and the introduction of large-panel work. It is also a result of rises in the qualifications of the building workers.

In 1965 7,200,000 men were employed in the construction industry. Of these 84.4 per cent were workers, 9 per cent engineering-technical workers and 6.6 per cent professional workers. While the number of men employed had grown by 45.7 per cent compared with 1955, the number of engineering technical workers had doubled. The numbers of machine operators, fitters, welders, and concreters is also growing because of industrialisation. The building materials industry employs some 3,000,000 men. There is quite a high labour turnover at some construction sites. Attempts are being made to improve the conditions of the construction workers, by increasing wages, providing more flats and better facilities on the site itself.

The work of restoration and preservation of buildings of historic and architectural importance is in the hands of specially skilled architects, artists and technicians. In Leningrad some 5,000 workers are apparently employed on such work.

(E) Finance

Capital investment in the construction industry is given in Table 33.

Table 33*

	(in 1,000 million roubles) Total capital investment	State and co-operative enterprises and organisations	Collective farms	Individuals building own flats and houses
Total for 1918-67	647.3	549.6	56.6	41.1
1918-28	4.1	1.7	0.03	2.4
1st five-year plan	7.3	6.6	0.3	0.4
2nd five-year plan	16.6	15.0	1.0	0.6
3½ years of 3rd five-year plan	17.3	14.8	1.3	1.2
1st July 1941 - 1st January, 1946	17.5	14.3	1.5	1.7
4th five-year plan	41.2	34.1	3.2	3.9
5th five-year plan	77.7	65.7	6.7	5.3
1956	22.4	18.6	2.3	1.5
1957	25.3	21.1	2.2	2.0
1958	29.4	23.9	2.8	2.7
1959-65	281.0	240.8	24.8	15.4
1966	51.7	45.1	4.9	1.7
1967 (plan)	55.8	47.9	5.6	2.3

*Source: Vredenskii, ed: *SSSR 1917-1967*, P 230 (Moscow, 1967)

The *All-Union Bank for Capital Investment Finance* (*Stroibank*) finances state and co-operative investment through local branches. Many construction jobs are entirely financed by loans from Stroibank.

A number of organisations within the construction industry have gone over to the new cost accounting system of management introduced in the 1965 economic reforms. Some 40 building and assembly trusts and similar organisations were working under the new system by 1970, and the whole industry is planned to go over to the new system. This gives the construction organisations greater economic independence and interest in the job. It is hoped that the new organisation will improve the effectiveness of capital expenditure. In future the plan is for more of the finance for capital construction to come from the construction enterprises themselves together with state credit.

In housing there is a standard price per dwelling which varies to take account of local factors. In Kiev the price is 122 roubles per square metre compared with a building cost of 109 roubles; this yields a profit of 12 roubles per square metre which is allocated in investment in new machinery and land for the factory, and for the provision of amenities for the staff.

(F) Research and Development

The State Committee for Construction, the State Committee for Civil Construction and Architecture and the various Ministries connected with the construction industry all have a number of research institutes and scientific stations under their control. The total number is in the region of 270 organisations.

The *Academy of Construction and Architecture of the USSR* is a specialised organisation which comes under the control of Gosstroi. It has a counterpart the *Academy of Construction of the Ukrainian Republic*, which comes under the State Committee for Construction of the Ukrainian Republic. The Academy of Construction is concerned with improving construction materials, standardisation of construction and the like. The research and design institutes under Gosstroi are also mainly concerned with these questions - for example: *Central Scientific Research Institute for Organisation, Mechanisation, and Technical Assistance,* the *Scientific Research Institute for Reinforced Concrete,* the *Central Scientific Research Institute for Design of Steel Construction,* the *Scientific Research Institute for the Economics of Construction,* the *State Institute for Prototype Design and Technical Research.* Gosstroi also runs an information institute - the *Central Institute for Scientific Information on Building and Architecture.*

The institutes coming under Gosgrazhdanstroi are more concerned with planning: for example, the *Central Scientific Research Institute for Experimental Planning of Housing,* the *Central Scientific Research Institute for City Construction and Regional Planning,* the *Scientific Research Institute for Architectural Theory and History.*

Thirty two research institutes employing some 2,347 scientists, come under the aegis of the various Ministries connected with the construction industry. The institutes are concerned with the problems of their particular ministry. For example, there is the *Scientific Research Institute for Rural Construction* and the *State Institute for Prototype Design of Rural Farms* under the Ministry of Rural Construction. The Building Materials Ministry controls scientific research institutes for cement, ceramics, sanitation technology, automation of building materials production, and so on.

On a more regional level there are research institutes to deal with the local conditions, climatic and otherwise.

There is a special school in Leningrad to train workers for the restoration of buildings of historic and architectural importance.

There are a number of problems that the construction industry research organisations are at present concentrating on, aimed at improving the productivity and technological level of the industry. According to Mr. Kachalov, deputy chairman of Gosstroi, (writing in *Stroitel'naya Gazeta,*, August 8, 1969) research in the field of carrying structures, foundations and bases of structures should be directed to developing lighter types, with greater strength, longer service, better anti-corrosion properties, and greater soundproof capacity. The design of prefabricated panels and blocks should allow for greater flexibility and variation in the finished dwelling. Greater attention should be paid to wooden structures and the work done by the research organisations in this field. Development of the construction industry must take place in the most suitable areas from a geographical and technical point of view.

And finally, according to Mr. Kachalov, reducing the weight of the structures of buildings may be considered one of the major problems in the further scientific and technological progress of construction. Mr. I. Dmitriev, head of the construction department of the central committee of the party, has made the same point emphasising that increased usage of light concrete structures, of extra-strong steels, aluminium alloys combined with light warm linings would cut down the consumption of materials. Other problems of research would include the mechanisation of underground work, building dams on permeable soils, building in permafrost areas, and methods for building in winter.

Journals: The main journals of the construction industry include *Arkhitektura SSSR,*

(Soviet Architecture); *Byulleten Stroitelnoi Tekhniki* (Bulletin of Building Technology); *Gidrotekhnicheskoe Stroitelstvo* (Hydraulic Engineering); *Iz. VUZ Stroitelstvo i Arkhitektura* (Proceedings of the Higher Educational Establishments. Construction and Architecture); *Mekhanizatsiia Stroitelstva* (Mechanisation of Construction); *Promyshlennoe Stroitelstvo* (Industrial Construction); *Selskoe Stroitelstvo* (Farm Construction); *Steklo i Keramika* (Glass and Ceramics); *Transportnoe Stroitelstvo* (Transport Construction); and *Tsement* (Cement).

(G) Plans and Policy

The main aim of the current five-year plan appears to be to improve the productivity and organisation of the construction industry so that the money invested and the materials used go further than at present. Organisations within the industry are going over to the new cost-accounting methods introduced in the 1965 economic reforms. The aim here is to improve the organisation of production, improve planning and increase economic incentives. The construction organisations will get greater independence, and the role of the profit motive will increase thus encouraging the workers and organisations to take a greater interest in the organisation and speed of work. More of the capital for financing construction work will come from the construction enterprises themselves, together with State credit. Another important trend in the construction industry is that of industrialisation. With the whole economy of the country developing at its present rate, it is not possible to draw more men into the industry. This means that productivity has to be raised to deal with the growing volume of construction work. The 1971-75 plan has set a target of a 35 per cent increase for the building materials industry. The number of prefabricated buildings and structures made of large-sized parts, units, panels and blocks in 1975 are planned to be almost double that of 1970. Fully pre-fabricated building techniques should account for about 30 per cent of the volume of construction and assembly work.

Proposals have been made to develop the building industry in the Far East, with an investment of 1,000 million roubles. Fifty five pre-fabricated ferroconcrete works will be built in the coastal regions and the Amur Valley. There will be a 200 per cent increase in the capacity of the Vladivostok House-Building Combine.

15 Agriculture

(A) Introduction

Before the revolution Russian agriculture retained influences of feudalism and serfdom. Large areas were owned by landlords and the peasants who worked it had to take their small farms on lease. There were no plants in Russia for producing tractors, combines or mineral fertilisers. Even simple machines were imported from abroad. Tsarist Russia had one of the lowest crop yields of cereals. However, exports were high, and the peasants had to sell grain in order to pay taxes and debts.

Two of the slogans of the 1917 revolution directly concerned the peasants: "bread" and "Land". (The third slogan was "peace"). On November 8, 1917, immediately after the revolution, the Second Congress of Soviets passed a Land Decree which confiscated all the land belonging to the landlords and church without compensation, giving ownership to the State and people. Sale and purchase of land was prohibited, and by 1918 more than 150,000,000 hectares (some 375,000,000 acres) of confiscated land had been divided up among the peasantry. This created an enormous number of small peasant holdings: some 23,700,000 in 1927. The Soviet government was dependent on agriculture to provide the surplus capital needed to build up an industrial base in the country. In fact, with no investment coming from abroad, agriculture had to support the cities, provide a small export surplus to pay for essential machinery and technical goods, as well as providing a living for the peasants themselves. At the end of the 1920s and beginning of the 1930s, large co-operative farms (kolkhozy) and State farms (sovkhozy) were created to improve agricultural productivity and organisation. They remain the two basic farming units to the present day, apart from the small individual plots that the collective farmers have been allowed to retain.

The comparative failure of Soviet agriculture has been evident for some time, accepted both by outside commentators and also, by implication, by the Soviet authorities themselves. Low productivity per head of the population, high costs of production, and periodic food difficulties are the main symptoms. In addition, it has only been able to provide limited supplies of raw materials for important processing industries such as food, textiles and leather goods. The Soviet government is continuously preoccupied with problems of agricultural policy and administration, an indication of the persistance of the unsolved problems and their vital political importance. (Source: Strauss, Erich, *Soviet Agriculture in Perspective*, P 21 Allen and Unwin, London, 1969).

(B) Organisation

Many bodies are concerned with the organisation of Soviet agriculture, and there is no one institution with overall control. Development plans for agriculture are drawn up by the Communist Party and the state planning bodies. The State fixes the price of farm goods and guarantees to buy a fixed quantity over a period of years. Within this context the collective and state farms and other agricultural organisations operate with a certain amount of autonomy.

15(1) MINISTRY OF AGRICULTURE

In 1953 local agriculture administration was transferred from the Ministry of Agriculture to Communist Party and State authorities at the district level. During the period of Krushchev's government the ministry lost all its executive powers and was largely confined to the organisation of agricultural research. However, in 1964, after Krushchev's resignation, its executive functions were restores. It operates on an all-union and at Republican levels. In 1966-67 it had 98 higher educational establishments subordinate to it.

There is also the Ministry of Land Reclamation and Water Conservancy responsible for irrigation and other similar works.

15(2) STATE FARMS (SOVKHOZY)

The State farm is a government owned and operated enterprise, in which the workers are paid wages as in a factory. Many have been established on formerly un-cultivated land, such as the virgin lands in Western Siberia and Northern Kazakhstan. They usually specialise in the production of grain, industrial crops, seeds, grapes, pedigree cattle or poultry.

In 1966 there were 12,200 State farms in the Soviet Union. Each one employed on average some 650 workers and covered an average of some 7,300 hectares of cultivated land; they possessed an average of 2,075 head of cattle and 4,073 head of sheep per farm. In total there were 7,900,000 workers employed on State farms and the sown acreages they covered was 88,500 hectares.

The sovkhozy account for 54 per cent of state buying of grain. Their share in the production of other crops is as follows; raw cotton - 20 per cent; potatoes - 35 per cent; sheep - 56 per cent; cattle and poultry - 41 per cent; and milk - 42 per cent. State farm workers and other employees get regular wages and also bonuses for surpassing the plans. They are entitled to the same rights as industrial workers, getting paid holidays, grants and pensions. The state farm is run by a manager, who is appointed by the appropriate state organisation, usually the all-union or Republican ministry of Agriculture. With the recent economic reforms, the government has decided to introduce cost accounting on all state farms. This is intended to give them greater independence in their economic activities, raising the significance of profits and profitability. State farms (as collective farms) receive from the government plans of their sales to the State of their agricultural products covering six years ahead.

The State farm workers, like the collective farm workers, are allowed their own individual plot of land. (See Section 15(4)).

15(3) COLLECTIVE FARMS (KOLKHOZY)

A collective farm is a collectively owned enterprise uniting the farmers of one or several villages. The land belongs to the State but is granted to the peasants who cultivate it for their permanent use.

There are more than 37,100 collective farms, including 600 collective fisheries. The average number of people working on the farms was 18,400,000 in 1966, and the land under cultivation then was 103,200,000 hectares. Each collective farm consisted on average of 418 farming households, had an average of 6,000 hectares of arable land including 3,000 hectares under crops. Each possessed an average of 1,072 head of cattle, 667 pigs, 1,490 sheep and 38 tractors.

The collective farmers are the owners and managers of the collective farm and all its wealth. The general farmers' meeting is the supreme management body of the collective farm's affairs. It elects a chairman, management board and auditing commission. The peasants work in brigades (usually numbering around 100 to a brigade) on assigned tasks. Until recently the measure for work and payment on a collective farm was a work-day unit. The incomes according to work-day units were distributed at the end of the year, and, before that, advance payments were made both in cash and kind. The work-day unit is a relic from the first days of collective farms, when many of them were quite small. The trend in recent years has been towards amalgamation (since 1947 the number of farms has fallen from some 220,000 to the present 37,100). Those that have consequently consolidated their economic position have switched over to monthly wages according to the established rates paid to state farm workers.

The collective farms sell the bulk of their produce to the state through state procurement agencies. Since 1965, fixed procurement plans covering a number of years have been established. These allow for the regular annual purchase of 55,700,000 tons of grain for the period 1966-70. Similar stable yearly plans have been fixed for the same period for other farm and animal produce. At the same time, purchase prices were raised for wheat, rye, barley, millet, rice and other cereal, and for certain kinds of animal produce, taking into account regional factors.

The Third Congress of Collective Farmers, held in November 1969, drew up a set of new model rules for collective farms. The new rules record the right of collective farmers to guaranteed remuneration for their labour and to social insurance and pensions. Each collective farm is recommended to draft its own rules on the basis of the model rules and to approve them at a general meeting of the collective farm. The congress decided to introduce a single system of social insurance for collective farmers in 1970. A centralised fund for the whole of the USSR has been created for this. Collective farms will pay 2.4 per cent of their wages fund into it. Members of collective farms are able to get temporary disability allowances, vouchers to sanitoria and rest homes, and are granted other types of social insurance in accordance with uniform standards for all collective farms, and similar to those used in the social insurance of state farm workers, including old age pensions. The new model rules extend the economic independence and initiative of the collective farms. The collective farmers themselves can work out and approve a plan for agricultural work, the starting point being a fixed plan-quota for a number of years for the sale of products to the state. When they have fulfilled the planned quota, the collective farms can sell grain and other products to the state at higher prices. They also retain their right to sell their products in the collective farm market.

15(4) INDIVIDUAL PLOTS

Each family on a collective farm is allowed a plot of land for their private use, and may own the necessary tools and as much livestock as can be raised on it. In 1966 the personally-owned husbandry of collective farmers and state farm workers and employees was 29,300,000 head of cattle, 16,500,000 pigs, 33,300,000 sheep and goats, and millions of poultry. These private plots play an important part in agricultural production. Their size was increased in 1965 to help stimulate production.

The relative position of state and collective farms and of individual plots for the year 1961 is given in Table 34.

15(5) COLLECTIVE FARMERS CONGRESS

The First All-Union Congress of Collective Farmers was held in February 1933. The second took place two years later. The third Collective Farmers' Congress was held in November 1969. There were 4,500 delegates representing 18,000,000 workers on collective farms. The work of the Congress was to discuss the running and organisation of the collective farms.

The Congress elected the *National Collective Farms Council*, consisting of 125 members.

*Table 34**

	Percentage of arable land in USSR under				Percentage of all		
	all crops	grain	industrial crops	potatoes and veg.	cattle	pigs	sheep
Collective farms	54	53.5	77.9	35.2	44.9	45.2	48.8
State farms and other government enterprises	42.7	45.6	21.3	18.5	26.1	28.9	29.7
Individual plots and allotments	3.3	0.9	0.8	46.3	29	25.9	21.5

*Source: Gregory: *Russian Land-Soviet People*, P 301, Harrap (London, 1968).

15(6) COLLECTIVE FARMERS' COUNCIL

The Third Collective Farmers' Congress, held in November 1969, adopted a resolution on the forming of *Councils of Collective Farms*. They are being set up with the aim of furthering the development of collective farm democracy. It is hoped they will stimulate discussion on the most important questions concerning the life and work of collective farms, helping to sum up experience gained in the organisation of production, and to draft recommendations on the fullest utilisation of production reserves.

In local districts, the councils will be elected at meetings of collective farm representatives. In the regions, republics and at the centre, the councils will be elected at meetings of representatives of the Collective Farm Councils. The Third Congress elected an All-Union Council of Collective Farms with 125 members. The Minister of Agriculture was elected the chairman.

15(7) STATE PROCUREMENT COMMITTEES

The State Procurement Committees are marketing organisations, set up in 1961. They are intermediate bodies between the State and Farms, catering for the greater economic autonomy brought about by the economic reforms. They are multi-tier organisations operating on different geographical levels.

Previously the job of procurements had been covered by a Ministry of Procurements. This no longer exists.

15(8) SUPPLY ASSOCIATIONS

The Soyuzselkhoztekhnika, or Agricultural Technical Association, took over from the Machine and Tractor Stations which were disbanded in 1958, the job of supplying and servicing both the state and collective farms with plant and spare parts. Set up in 1961, it is a multi-tier body operating centrally and regionally through special councils. Leading members of state and collective farms and agricultural machinery plants are represented on the councils.

Another supply organisation, the *Soyuzselkhozkhimia*, has been set up to supply fertilisers and other agricultural chemicals.

15(9) AGRARIAN INDUSTRIAL ASSOCIATIONS

The Agrarian Industrial Associations were referred to by Leonid Brezhnev in his speech to the Third Congress of Collective Farms, held in November 1969. These associations will gradually emerge, where, given the appropriate specialisation and co-operation of agricultural and industrial enterprise, agriculture will combine organically with the industrial processing of its produce. A number of specialised State farms and collective farms in the Krasnodar Territory, in the Ukraine and in Moldavia are starting to operate along these lines.

(C) Climate, Land and Crops

The USSR stretches from the subtropical zones of the South to the Arctic Ocean, and so naturally has very varied climates. The lowest temperatures are registered in the vicinity of the town of Oimyakon, in north eastern Yakutia. Here the mean temperature in January is minus 50C and it can drop to minus $70^{o}C$. The hottest part of the country is Central Asia where the mean temperature of the warmest month is more than 30^{c} and the maximum registered in the town of Termex is about $50^{o}C$.

There are four climatic zones in the Soviet Union: the arctic, sub-arctic, temperate and subtropical - which take in some 13 different natural regions (see Figure 15).

(i) Arctic Zone. The Arctic Zone covers the islands of the Arctic Ocean and the Taimir peninsula. The climate is very severe. The average temperature in August is around $0^{o}C$. There are many icebergs and much of the land is covered with snow all the year round. Between 70 and 80 per cent of the surface has no soil or vegetation, while the remainder supports mosses, small plants and lichens. A few birds, insects and lemmings survive there.

(ii) Sub-Arctic Zone. The Sub-Arctic Zone covers the southern part of Novaya Zemlya, and stretches from the edge of the Arctic Ocean down to the Arctic Circle in the western part of the USSR and less far south in the more easterly regions of the country. It takes in two natural regions: the tundra and forest-tundra.

> *(a) Tundra.* The region is characterised by an abundance of moisture with a lack of warmth and by permafrost. There are no trees but much of the surface is covered with mosses and lichens. It is warmer than the arctic region but the average temperature in July does not exceed 10 to $11^{o}C$. The surface of the region is excessively damp and there is a large flow of rivers during the summer. In winter most of the rivers freeze to the bottom. The soils are acid and poor in humus. The main animal inhabitants are the northern caribou, lemming, polar fox, wolf, snowy owl, and tundra partridge. In summer the lakes and bogs are filled with water birds. The tundra is used for pasturing the domesticated caribou, partly for the fur trade. Some vegetables can be grown in the southern regions.
>
> *(b) Forest-Tundra.* This region stretches as a narrow band from west to east varying from some 20km to 200km wide to the south of the tundra region. Its summer is a little warmer than than in the Tundra proper. It is an extremely marshy

and boggy region. Forest, podsolised and bog soils predominate. There are bushes and a sparse growth of trees. The animals are a mixture of tundra and taiga types, the latter including elk, brown bear, squirrel, hare and hazel-grouse. The polar fox has the most commercial importance. Winter and spring pasturing of the caribou is concentrated in the forest tundra. A number of fairly large cities have been built in this area since the revolution (for example, Norilsk, Murmansk) and a number of industrial centres have grown up (for example, the Vorkuta coal mines, iron and salt workings in the Khatange valley, nickel and copper at Monchetundra and Norilsk). Around these areas vegetable growing is being developed.

(iii) Temperate Zone. The Temperate Zone is a large area stretching from the arctic circle down to the southern borders of the country. It takes in the taiga, areas of mixed forests, broadleaved forests, the forest-steppe, steppe, semi-desert, and desert.

(a) Taiga. The taiga has a cold, comparatively moist climate. The vegetation consists mainly of fir forests and the soils are podsolised, boggy and frozen. The taiga is the largest natural region of the USSR. Its southern boundary runs close to a line stretching from Leningrad through Ivanovo, Gorky, Kazan, and Tomsk to Krasnoyarsk. The taiga region is warmer in summer and colder in winter than the tundra and forest tundra. The average July temperature ranges from some $12^{o}C$ in the north to some $19^{o}C$ in the south. Coniferous forests predominate in the taiga, containing fir, larch, pine and silver fir. In the western and eastern regions where the climate is temperate the forests are mainly fir. Apart from the coniferous forests there are also woods of birch and asp. The taiga forests have a comparatively high productivity. The taiga of western Siberia is boggy and contains a lot of high quality peat. The animal world is quite rich. There are brown bear, elk, squirrel, lynx, hare, grouse, various species of woodpecker, wood grouse and field vole.

Timber is the chief natural resource of the taiga. The fur industry (sable, squirrel, hare, kolinsky, etc) is also important. The taiga has good potential for the development of agriculture (grain, flax, potatoes, vegetables, and fodder grass). The meadow and wood pasture provide a base for expanding milk and meat production.

(b) Mixed Forests. The mixed forests region can be further divided into the west-European and Far-Eastern regions.

The *West-European Region* has a comparatively moist climate. There are mixed conifer and broadleaved forests. The summers are cool with an average July temperature of around $18^{o}C$, but the winters are comparatively mild with an average January temperature of $-3^{o}C$. It has a rich network of rivers. The fauna and flora of the region are enriched with western species. Besides the brown bear, fox, wolf and elk, there are also European deer and mink, wood marten, dormouse and others. A considerable part of the forests has been cut down, leaving only 30 per cent of the area wooded. The amount of fir and oak is decreasing giving way to birch and larch. Because of the reduced forested area a number of animals, including the sable, have become very rare. About a third of the region is cultivated. The main crops are potatoes, vegetables, flax, grain and sugar beet in places.

The *Far-East Region* of mixed forests lies along the Amur basin, and is characterised by its monsoon climate. During winter, when the main winds blow from off the continent, the weather is cold with an average January temperature ranging from $-28^{o}C$ in the north to $-18^{o}C$ in the south. In summer, the average temperature is around $22^{o}C$, the monsoons blow in from the sea and there is widespread flooding. The fauna is a mixture

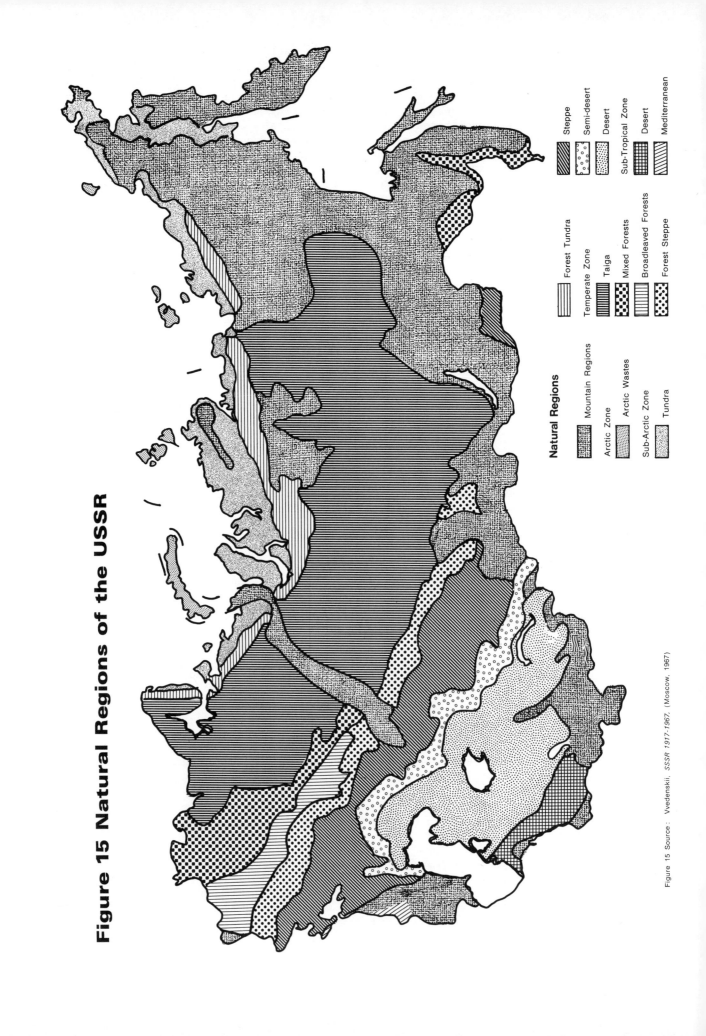

Figure 15 Natural Regions of the USSR

Figure 15 Source: Vvedenskii, *SSSR 1917-1967*, (Moscow, 1967)

of northern taiga species and southern species of west Asian and Indo-Malaysian origin. The area is the most cultivated region of the Far East of the country. Crops include wheat, oats, rye, soya, rice potatoes and vegetables.

(c) *Broadleaved Forests*. The region of broadleaved forests is situated in the western-European plain, forming a small belt between the mixed forest region to the north and the forest-steppe to the south. The main forests are oak but few of them are left. Most of the area is cultivated, arable land predominating.

(d) *Forest-Steppe*. The forest-steppe region has areas of forests growing on grey forest soils alternating with grassland areas on the black soil. It stretches from the Carpathians to the Altai. The average July temperature is around 22°C. There are periodic droughts. The surface river network is less than in the forested zones. The forests are mainly oak in the western-European region and silver-birch and larch in Western Siberia. The fauna is a complex mixture of forest and steppe species. The forest-steppe is one of the most cultivated regions in the USSR. Between 70 and 80 per cent of it is ploughed up. The main crops are wheat, sugar beet, maize, sunflower and hemp.

(e) *Steppe*. The steppe region stretches from the western boundaries of the USSR to the Altai, with islands of steppe occuring further east. It it warmer and drier than the forest-steppe. The average July temperature is 23°C. The river network is small. There are dangers of wind erosion. The western regions are cultivated to the same extent as the forest-steppe, but less so further east. The steppe and forest-steppe are the main granary of the country. Crops include wheat, maize, sunflower millet, melons, and in the southern regions there is a lot of horticulture and vineyards. Fertilisers, wind breaks and snow retention are the main methods of land improvement.

(f) *Semi-desert*. The climate of the semi-desert region is dry and sharply continental - the average July temperature being 24°C. Most of the rivers are transitory. Soil waters lie at a great depth and are often salty. Salt domes are widespread. The fauna includes steppe and desert species. Suslik or gopher, field vole and jerboa are prevalent. The semi-desert is a region of selective cultivation using artificial irrigation. There is abundant pasture.

(g) *Desert*. The desert region is situated in Astrakhan, Kazakhstan and Central Asia and stretches down roughly to the 40th parallel. It is characterised by an extremely dry and continental climate, a lack of fresh surface water, the salinity of the soils and the sparsity of the vegetation. The average July temperature can go up to 29°C. In the oases and irrigated areas rice, cotton and other crops, requiring high temperatures, are grown. The rest of the land is available for pasture. The future for agricultural development of the region lies in the use of the reserves of underground water.

(iv) *Sub-Tropical Zone*. The sub-tropical zone is a small area to the south of the country taking in a desert area of Central Asia to the south of the 40th parallel and a mediterranean area on the eastern edge of the Black Sea.

(a) *Desert*. The winter of the sub-tropical zone's deserts is warmer than in the temperate desert region with above-zero temperatures and no persistent snow cover. The average July temperature is around 32°C. In the oases the more heat-loving species of fine-fibre cotton and sesame are grown. The weather is such that there can be successive harvests of different crops. The basic problem, as in the temperate desert zone, is that of adequate water supplies.

(b) Mediterranean. The mediterranean region has a hot dry summer and comparatively warm moist winter. Near Yalta, the average January temperature is 3.5°C and the average July temperature 24°C. There are mixed forests. The fauna includes wild species such as leopard and wild cat. Orchards and vineyards are the two main forms of cultivation. In some areas tea and citrus fruits are grown.

(v) Mountain Regions. Mountains cover large areas of the USSR: from the Carpathian in the far west through the Urals and Central Asian mountain ranges to the mountainous country of eastern Siberia, and the Kamchatka.

(D) Production and Labour

The Soviet Union holds second place in the world for overall agricultural production and leads the world in the production of wheat, rye, potatoes, sugar beet and flax fibre.

(i) Field crops. One of the main reasons for the increase in agricultural production in the USSR has been the expansion of the area of land under cultivation (see Table 35).

*Table 35**
The Sowing Area of the USSR (million hectares)

	1913	1940	1945	1960	1966	1969
Total sowing area	118.2	150.6	113.8	203.0	206.8	208.6
of which:						
grain	104.6	110.7	85.3	115.6	124.8	122.7
industrial	4.9	11.8	7.7	13.1	15.1	14.4
potato and vegetables	5.1	10.0	10.6	11.2	10.3	10.0
fodder	3.3	18.1	10.2	63.1	56.6	61.5
Fallow areas	-	28.9	23.2	17.4	16.8	16.9

*Source: Vredenskii, ed: *SSSR 1917-1967*, P 234 (Moscow, 1967) and *Narodnoe Khozaistvo SSSR v 1969 g*, P 308 (Moscow, 1970).

Expansion of the sowing area has been mainly achieved by development of the virgin and unused land in the eastern regions of the country. Over the short period from 1954 to 1960 more than 40,000,000 hectares of new lands were sown.

Fertilisers have also played a part in increasing production. The use of fertilisers has grown from 3,200,000 tons in 1940 to 30,500,000 tons in 1966. This represents a growth from 16.5kg for one hectare of land in 1940 to 138.2kg in 1966. By 1970 the use of mineral fertilisers was planned to have increased to 55,000,000 tons.

The growth in the use of fertilisers, the increased level of mechanisation, better seeds, etc., have ensured the increased productivity of the basic agricultural crops. The figures are given in Table 36.

The expansion of the sown area and the growth of crop yields has ensured a considerable growth in production, as can be seen in Table 37.

*Table 36**
Crop yields of the most important crops
(centner per hectare)*

	1913	1940	1945	1960	1966	1969
Grain	8.2	8.6	5.6	10.9	13.7	13.2
Raw cotton	10.8	10.8	9.6	19.6	24.3	22.5
Sugar beet	168	146	66	191	195	211
Flax fibre	3.2	1.7	1.5	2.6	3.3	3.7
Sunflower	7.6	7.4	2.9	9.4	12.2	13.3
Potatoes	76	99	70	92	104	113
Vegetables	84	91	58	111	125	126

(* 10 centners - 1 ton)

*Source: Narodnoe Khozaistvo, SSSR v 1969 g, P 228 (Moscow, 1970)

*Table 37**
Production of Basic Crops
(million tons)

	1913	1940	1945	1960	1966	1969	1970
Grain	86.0	95.6	47.3	125.5	171.2	162.4	167.5
Raw cotton	0.74	2.24	1.16	4.29	5.98	5.71	6.1
Sugar beet (factory)	11.3	18.0	5.5	57.7	74.0	71.2	78.3
Sunflower	0.75	2.64	0.84	3.97	6.15	6.36	6.1
Flax fibre (thousand tons)	401	349	150	425	461	487	
Potatoes	31.9	76.1	58.3	84.4	87.9	91.8	96.6
Vegetables	5.5	13.7	10.3	16.6	17.8	18.7	20.3

*Source: Narodnoe Khozaistvo, SSSR v 1969 g, PP 286-7 (Moscow, 1970).

In order to fully satisfy the country's food requirements it is necessary to increase the total average annual wheat harvest to some 190 to 200 million tons, raw cotton to 7,000,000 tons, sugar beet to 90,000,000 tons, sunflower seed to 7,000,000 tons and potatoes to 115,000,000 tons. It will be necessary also to increase the production of vegetables and fruit and to provide greater variety.

(ii) Animal farming. The development of animal farming is above all characterised by the increase in numbers of livestock (see Table 38).

*Table 38**
The number of livestock (millions)

	1916	1941	1946	1961	1967	1970	1971
Cattle (including cows)	58.4	54.8	47.6	75.8	97.1	95.2	99.1
Cows	28.8	28.0	22.9	34.8	41.2	40.5	41.0
Pigs	23.0	27.6	10.6	58.7	58.0	56.1	67.2
Sheep and goats	96.3	91.7	70.0	140.3	141.0	135.8	143.2

*Source: *Narodnoe Khozaistvo, SSSR v 1969 g*, P 367 (Moscow, 1970)

Table 39 shows the productivity level of the animals has increased.

*Table 39**

	1940	1945	1950	1965	1970
Average annual milk output from one cow (kg)	1,124	987	1,137	1,987	2,302
Average annual wool clip from one sheep (kg)	2.5	1.9	2.2	2.9	3.3
Average number of eggs laid, annually by one hen	-	-	-	132	166

*Source: Vredenskii, ed: *SSSR 1917-1967*, P 235 (Moscow, 1967)

The production figures for the main products of animal farming are given in Table 40.

*Table 40**

	1913	1940	1945	1960	1965	1969	1970
Meat (the slaughtered weight, mlns. of tons)	5.0	4.7	2.6	8.7	10.8	11.8	12.3
Milk (million tons)	29.4	33.6	26.4	61.7	75.8	81.5	82.9
Wool (1,000s of tons)	192	161	111	357	357	390	415
Eggs (1,000 millions)	11.9	12.2	4.9	27.4	29.1	37.2	40.4

*Source: *Narodnoe Khozaistvo SSSR v 1969 g*, P 294 (Moscow, 1970).

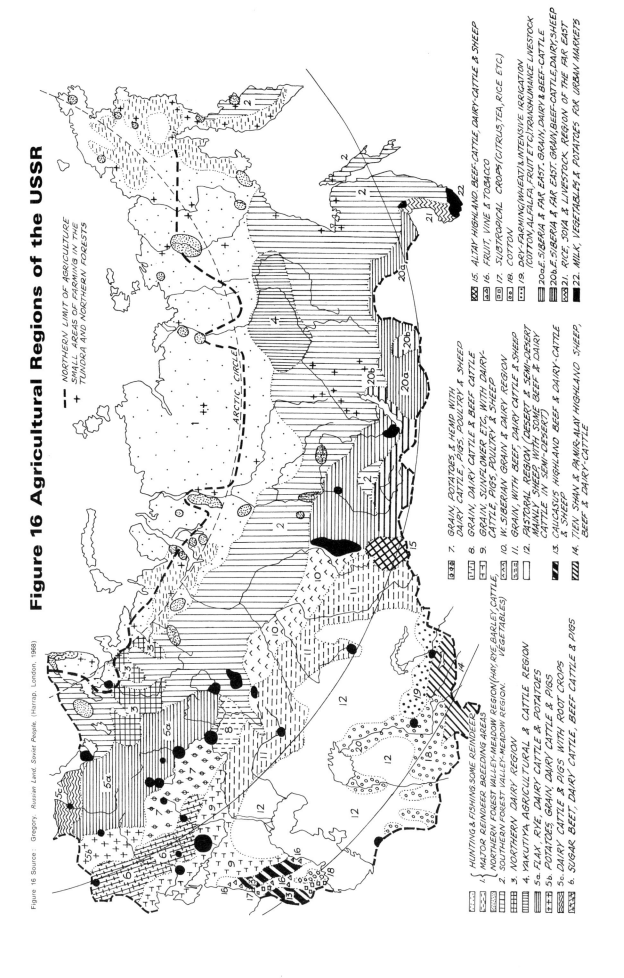

Figure 16 Agricultural Regions of the USSR

Figure 16 Source: Gregory. *Russian Land, Soviet People*, (Harrap, London, 1968)

The target production figures over the next few years are: for meat an annual output of 14 to 15,000,000 tons; for milk - 90 to 95,000,000 tons; for eggs - 45 to 50,000,000,000 units; and for wool - 480,000 to 500,000 tons.

According to Leonid Brezhnev in the speech he made to the Collective Farmers' Congress in 1969, there are a number of shortcomings in the sphere of animal husbandry. "Until now many farms have been sending cattle which are of very low weight and poorly fattened to the meat-packing plants. The simplest calculations show that if the weight of cattle as delivered averages between 350 and 400kg for the country then with the present head of cattle we shall be able to increase meat procurement in the country by about 2,000,000 tons .....To increase the productivity of livestock it is necessary in the first place to improve fodder production...the collective and state farms have been able to increase the expenditure of grain on feed by 35 per cent...but no serious headway has been made with respect to course and succulent fodder. Procurements of hay for livestock owned by collective farms have increased by only 9 per cent, and those of silage have even decreased. These figures point to serious shortcomings in the work done by collective farms and state farms to ensure a supply of fodder for livestock. We have to pay close attention to the numbers of livestock of all kinds. Certain farms are doing little to increase the numbers of livestock and are even allowing the numbers to decrease. This is an unhealthy and incorrect tendency. Today we need both an all-out increase of the production of livestock and a growth in the numbers of livestock (including poultry)."

(E) Labour

Some 27,000,000 people are employed in agriculture on collective farms, State farms and other State agricultural enterprises. This represents about 30 per cent of the working population. In 1928, the figure was 80 per cent.

(F) Finance

In March 1965 Brezhnev announced that investment in public agriculture was to be stepped up from 33,000,000,000 roubles in 1960-64 to 71,000,000,000 roubles in the following five-year period. (The overall investment in industry for the 1960-64 period was 71,136,000,000 roubles). In his speech to the Collective Farmers' Congress in 1969 he announced that capital investments in state and collective farms had increased by 19,000,000,000 roubles over the previous four-year period. Of the 71,000,000,000 roubles 41,000,000,000 were to come from centralised funds and 30,000,000,000 from the collective farms. The funds were to be allocated for capital investment in agriculture. The supply targets included 1,790,000 tractors, 1,100,000 lorries and trucks, 900,000 trailers and 550,000 combine harvesters. The funds were also to be used for land improvement, increased supplies of fertilisers and so on. The directives for the present five-year plan (1971-75) state that the total volume of capital investment in agriculture (including both state and collective farms) have been fixed at 129,000,000,000 roubles.

The economic measures approved by the 23rd Congress of the Communist Party of the Soviet Union in 1966 were aimed to help the growth of agriculture. Planning of agricultural production was improved. Collective and State farms receive stable plans of sales to the State of their agricultural products for six years in advance. This helps each one to develop production plans for several years in advance taking into account local conditions. Secondly the fixed prices paid by the State for state purchases were raised and a 50 per cent additional payment above the basic price was decided on for sales of grain over and above the plan.

In 1966 the value of kolkhoz production minus material costs, i.e. the gross value of their incomes, amounted to 20,200,000,000 roubles, 28.7 per cent higher than in 1964. A factor strengthening the economics of the collective farms and raising the peasant's standard of living is the introduction of a guaranteed wage, together with an increasing proportion of money paid into the general funds for distribution among collective farmers instead of payment in kind. In 1966 this proportion was 85 per cent compared with 59 per cent in 1959. In 1966 73 per cent of all the kolkhozes in the country changed to making only monetary payments for labour, leaving 27 per cent making payment in money and goods.

The State farms showed a profit at the end of 1966. Profit covers anything earned over and above the production target. (An excess of income over expenditure within the target laid down by the plan is not profit by planned growth). The profit earned over and above this is divided into four parts, one part going to the State and the other three parts being divided by the peasants. One of these can be divided between them as money payment; one part can be used for buying equipment not provided for in the plan; and the third part can be put towards such community services as new housing, schools, etc.

(G) Research and Development

(i) Research. The Ministry of Agriculture is the main body concerned with the organisation of agricultural research. It has some 98 higher educational establishments under its administration. It also has direct authority over the *All-Union Lenin Academy of Agricultural Sciences* and its sister Republican academies.

In 1963 "Measures Designed to Improve the Activity of the USSR Ministry of Agriculture and the All-Union Academy of Agriculture unified the network of agricultural research institutions. The USSR Ministry of Agriculture was entrusted with general guidance and co-ordination in this field as well as with the application of science to practical agriculture. The ministry directly controls the branch-specialised institutes and experimental stations, while the regional integrated research institutes are placed under the jurisdiction of the agricultural agencies of the respective union republics.

In the Soviet Union as a whole, scientific research work in forestry and agriculture is being conducted by more than 1,000 research and experimental establishments and 98 agricultural higher educational establishments where more than 700,000 scientists are working.

15(10) ALL-UNION LENIN ACADEMY OF AGRICULTURAL SCIENCES (VASHNIL)

The All-Union Lenin Academy of Agricultural Sciences is the supreme agricultural scientific institution of the USSR. It is a scientific methodological centre, responsible for the work in fundamental branches of agriculture of scientific research institutes, breeding and experimental stations, throughout the Soviet Union.

The academy was founded in 1929. In 1963 it was given the power to control institutes concerned with fundamental research in agriculture, basic agricultural theory and livestock raising, and mechanisation and use of chemicals in farming. It has six departments covering farming, animal husbandry, mechanisation and electrification of agriculture, silviculture and the improvement of agriculture and forestry, hydraulic engineering and soil improvement, and economics and organisation of agricultural production. These departments are responsible for the direction of scientific and technical research in their fields throughout the country. The academy takes a direct share in working out the national development plans in the areas covered by its department.

There are 73 academicians and 74 corresponding members. At the head of the academy is the presidium consisting of 12 members; the president, three vice-presidents, the scientific-secretary of the presidium, six departmental scientific-secretaries, and one other member.

The academy directly controls 35 research institutes, the central Scientific Agricultural Library, 16 central research institutes, numerous laboratories, experimental stations and so on, making some 95 scientific establishments in all. The number of scientists employed in academy scientific establishments was 4,017 in 1967.

There are six Republican academies of agriculture. These are in the Ukraine, Byelorussia, Uzbekistan, Kazakhstan, Georgia and Azerbaijan.

The bulk of the research effort of the agricultural academies falls into the applied research sector.

15(11) AGRICULTURAL SCIENCE TOWN

An agricultural science town close to and along the same lines as the famous Akademgorodok is being planned for the near future. It will house the Siberian branch of the Agricultural Academy, and its first five research institutes will cover stock breeding, mechanisation and electrification, agricultural economics, chemical applications and animal foodstuffs. The new town will be built on the opposite bank of the river Ob from Akademgorodok. There will be work for some 5,000 specialists, and the town itself will hold around 18,000 inhabitants.

The aim of the centre is to boost agricultural research and development in the eastern regions of the country, which at the moment are poorly served. It should also help to boost the level of agricultural research and production throughout the country.

(ii) Education and Training. The education and training of agricultural workers is being encouraged. More than one in three collective farmers now have secondary of higher education, and there are special facilities and encouragement for farm workers wanting to go to university.

With increasing mechanisation there is the need for more specialists. In 1969 there were 2,000,000 machine operators and some 330,000 specialists working on collective farms. There are 500,000 specialists working in agriculture as a whole. According to Brezhnev in his speech to the Collective Farmers' Congress in 1969, this is by no means enough.

"Serious attention needs to be paid to the training of personnel capable of coping with all the work connected with the use of new machines, the introduction of chemicals and the development of land improvement at a high level. In recent years large-scale work has been done to train personnel of this calibre for agriculture. However, their number is insufficient on a number of farms and a considerable number of machines are operated only on a one-shift basis....Matters should be organised so that machines are not idle and waiting for personnel and so that people are organising the use of chemicals in agriculture and working on land improvement."

(iii) Development. The main fields for technological development of agriculture are mechanisation, electrification, chemicalisation, research into plants, seeds, etc., irrigation and land reclamation, and specialisation.

(a) *Mechanisation.* Between 1966 and 1970 most agricultural machinery was planned to be replaced. The collective and State farms were to receive 1,790,000 tractors, 1,100,000 lorries, 550,000 combine harvesters. The general growth in the use of farm machinery is given in Table 41.

*Table 41**

	1928	1940	1945	1960	1966	1969
Tractors (1,000s)	27	531	397	1,122	1,700	1,821
Combine harvesters (1,000s)	2 only	182	148	497	540	581
Lorries (1,000s)	0.7	228	62	778	1,013	1,097

*Source: Vredenskii, ed: *SSSR 1917-1967*, P 233 (Moscow, 1967)

Much of the agricultural machinery gives a poor performance. There is a failure to prepare machinery for the spring sowing. The quality of repair work is low.

The supply organisation Selkhoztekhnika returned some 15 per cent of all purchased machinery to the factory for major repairs in 1965. Many of the machines have a large number of components which are not standardised from factory to factory.

According to Brezhnev in his speech to the Collective Farmers' Congress in 1969: "a start will be soon made on the construction of a new plant for the production of lorries and of other plants for agricultural machinery. Industry is going over to the production of more efficient tractors and combine harvesters. The output of tractor and motor-drawn trailers is increasing. In connection with the need for considerable expansion of equipment for land improvement and animal husbandry, we must see to it that a number of enterprises specialise in the production of this equipment and of facilities of this kind."

(b) Electrification. Use of electricity in agriculture has developed particularly fast in the post-war period - from 500,000,000 kilowatt-hours (kWh) in 1940 to 24,000,000,000 kWh in 1966. This quantity is planned to expand to 60,000 to 65,000,000,000 kWh by 1970.

In 1966 more than 60 per cent of all energy supplied was used for productive purposes (compared with the pre-war situation when practically all electric power supplied to the country was used for lighting purposes). During the five-year plan period (1966-70) the overall power supply to agriculture is planned to increase so that by 1970 there will be 128 hp available for every 100 hectares of sown land compared to 103 hp in 1965, and 18 hp per agricultural worker compared with 9.1 hp in 1965.

(c) Chemicalisation. The chemical industry is steadily increasing its output of mineral fertilisers. In 1968 it provided agriculture with 36,200,000 tons and in 1969 the figure reached 39,300,000. The planned target for 1970 was 46,000,000 tons. This compares with an output of 11,400,000 tons in 1960.

The Ministry of Agriculture has been experimenting for some years with comprehensive chemicalisation in 26 districts, situated in different climatic and natural regions of the country. These districts are supplied with more mineral fertilisers, (in accordance with quotas planned for the near future) and with more machinery than other areas.

As a result of the 1962 Party Plenum which requested the Ministry of Agriculture, the All-Union Lenin Academy of Agriculture and other research institutes to investigate, among other things, the need for improvement in the use of mineral fertilisers, the USSR Academy of Sciences set up a "Special Council for Chemical Processes in Agriculture."

(d) Irrigation and land reclamation. The May (1966) Plenary meeting of the Communist Party of the Soviet Union worked out a long-term programme of land reclamation and land fertility. In 1966 and the first half of 1967 the area of irrigated and drained land in the USSR was increased by more than 1,200,000 hectares. It is planned to enlarge the irrigated areas in Southern Russia, Southern Ukraine, the Kuban Areas and Asia over the next 10 years by some 7 to 8,000,000 hectares, and drained land by some 15 to 18,000,000 hectares. The collective and State farms invested more than 8,000,000,000 roubles in land improvement over the three years prior to 1969.

More progressive methods have been employed in recent years in construction projects associated with land improvement; open canals with earth banks are making way for canals lined with concrete and pipeline irrigation systems; extensive use is being made of closed drainage; projects are under way to automate the control of land improvement systems and mechanised watering is being introduced.

(Source: Vladimir Matskevich, Minister of Agriculture, *Soviet News*, March 31, 1970).

(e) Crop and stock improvement. The introduction of new and highly productive varieties and hybrids into Soviet farms is an important factor in intensifying agricultural production. Among the more important research centres for the development of grain production at the *Krasnodar Agricultural Research Institute,* the *Mironovsky Institute for Wheat Selection and Seed Production,* the *Agricultural Research Institute of the South-East* and others. One major success has been in the selection of the sunflower, the staple oil bearing crop in the Soviet Union. Research is being conducted into the effect of electrical and magnetic fields on plant growth. A number of electrical processes are used in agriculture to intensify existing plant growing techniques. Electrical processes include cleaning, sorting, de-infestation, treatment of seeds before sowing, and more recently electric stratification of vine grafts and electric protection of plants from pests.

(f) Specialisation. According to Vladimir Matskevich, the Minister of Agriculture: "Big farms for the fattening of cattle and pigs, groups of large greenhouses, commercial orchards and other specialised farms are now being created all over the country. This is a progressive process. At the same time we have many farms which will continue to develop a number of branches for some time to come. As far as these farms are concerned, they must, on one hand, decide which branches they can best combine, and on the other hand, set up big production departments and work teams so as to concentrate production, and achieve specialisation within their farm."

(Source: *Soviet News,* March 31, 1970).

Journals: The main agricultural journals include: *Agrokhimiia* (Agrochemistry); *Gidrotekhnika i Melioratsiia* (Hydraulic Engineering and Reclamation); *Zhivotnovodstvo* (Livestock Breeding); *Zemledelie* (Agriculture); *Selektsiia i Semenovodstvo* (Selection and Seed-Growing); *Selskokhozaistvennaya Biologiia* (Agricultural Biology; *Khlopkovodstvo* (Cotton Growing); *Doklady Vsesoyuznoi ordena Lenina Akademii Selskokhozaistvennykh Nauk im. V.I. Lenina* (Transactions of the All-Union Lenin Academy of Agricultural Sciences); and *Izvestiia Timiriazevskoi Selskokhozaistvennoi Akademii* (Proceedings of the Timiriazev Agricultural Academy).

(H) Policy and Plans

In 1953 the September meeting of the Communist Party of the Soviet Union (CPSU) passed a resolution on "Measures for the further development of agriculture". This planned for a strengthening of the material-technical base of agriculture and as a result output rose significantly. Between 1954 and 1958 grossoutput rose by 51 per cent, giving an average annual growth rate of 8.5 per cent.

This agricultural progress was halted by reductions in capital investment in agriculture and by increases in the cost of industrial goods needed by agriculture. The 1959-65 seven-year plan for agriculture was not fulfilled. Gross output was meant to increase by 70 per cent and actually increased by only 15 per cent, giving an annual growth rate of 2.1 per cent.

At the 1965 March meeting of the CPSU a resolution was passed "on the necessary measures for the further development of agriculture in the Soviet Union". This outlined the need for a sharp increase in capital expenditure, for the introduction of technological methods, for raising the price of agricultural goods, and for lowering the cost of industrial goods needed by agriculture. The 1966-70 five-year plan for agriculture was drawn up along these lines.

In 1969 Brezhnev outlined the main trends of agricultural policy in the following way: ".....first of all, the further consolidation and improvement of the machine and technical basis, extensive land improvement and chemicalisation of agriculture. An important feature of this policy is the new system of planning the procurement of agricultural products and the encouragement by economic means of agricultural production."

The directives of the present five-year plan (1971-75) outline the need for the development of all branches of agriculture. Grain production is a key problem with a planned average annual increase of 27.5 per cent. Agricultural power capacities are to be increased by 50 per cent (161,000,000 horse power) and the tractor fleet by 27 per cent (540,000 machines). The consumption of electricity will be doubled to reach 75,000,000,000 kWh. The supply of mineral fertilisers is to increase from 46,000,000 tons to 75,000,000 tons. In stock farming the basic problem will be to increase the supply of fodder. Economic levers and the transfer of all state farms to operation on a profit and loss basis are seen as important organisational means for achieving the aims set out in the plan.

16 Medical Services

(A) Introduction

The medical services in Russia prior to the revolution were very limited, in particular in the outlying regions of the empire. The total number of doctors operating in 1914 was 23,000 and most of these were in the towns. The mortality rate was high and life expectation low.

In 1918 the Soviet government organised the People's Health Commissariat, the forerunner of the present Ministry of Health. Since then the State has assumed responsibility for the health of the population, laying particular stress on the need for preventive medicine.

(B) Organisation

All the health services come under the control of the USSR Ministry of Health. The Ministry's tasks include the planning, co-ordination, control and guidance of all the work carried out by the ramified network of public health bodies and medical institutions, as well as of the medical supplies and pharmaceutical industries.

Each of the 15 union republics has its own ministry of health which is in charge of health services within the republic. The central Ministry of Health comes under the Council of Ministers of the USSR, and the ministries of union republics under the councils of ministers of their republics. There are regional, territorial and city public health departments of local soviets. They are subordinated both to the local soviet and to the higher public health body. All these public health bodies are headed by doctors. Urban and rural medical problems are dealt with by heads of the urban district health departments, or by chief physicians of rural district hospitals. They are responsible for all aspects of medical and public health work within a district or town.

Medical services are administered through hospitals, polyclinics, specialised clinics, consultation centres and first aid posts. These are organised on a territorial basis. Each region, catered for by a hospital, polyclinic and consultation posts, is further divided into smaller areas, which come under the administration of that area's doctor. In the towns each such area covers 4,000 patients, and has its own local doctor, and nursing sister.

Medical services are also organised within the factories. Industrial enterprises where the number of workers exceeds 800 (or at enterprises of the coal, oil, mining and chemical industries, exceeding 500) have medical posts. Nearly all large industrial enterprises have their own medical and public health units consisting of a hospital polyclinic, a number of first aid posts and an overnight sanatorium. Factory polyclinics function on the same territorial principle as ordinary polyclinics. One general practitioner usually attends 2,000 workers while at enterprises of the chemical, coal mining and oil processing industries he serves 1,000 workers.

The health services also provide creches and kindergartens for pre-school age children, and sanitoria and health rest homes.

(C) Medical Services

Table 42 illustrates the growth of the hospital network:

*Table 42**

	1913	1940	1958	1965	1969
The number of hospitals, maternity homes, specialised clinics, etc. (not including military hospitals) - 1,000s	5.3	13.8	26.0	26.3	26.4
The number of hospital beds - 1,000s	208	791	1,533	2,226	2,567
The number of hospital beds per 10,000 of the population	13	40	73	96	106.5

*Source: Vredenskii, ed: *SSSR 1917-1967*, P 271 (Moscow, 1967) and *Narodnoe Khozaistvo SSSR v 1969 g*, P 727 (Moscow, 1970)

There were 642,500 qualified doctors in 1969, which represents some 24 per 10,000 population. In addition the number of paramedical personnel (nurses, midwives, medical assistants) was 2,029,700 in 1969. Table 43 shows a breakdown of the special interests of qualified doctors:

*Table 43**

	1940	1965	1969
General practitioners	42,600	114,900	128,500
Surgeons	12,600	52,500	63,000
Pediatricians	19,400	71,700	76,200
Obstetricians and gynaecologists	10,600	35,400	39,600
Opthalmologists	3,600	13,100	15,400
Psychiatrists	2,400	10,000	13,300

*Source: Minister of Health, Boris Petrovsky, in his pamphlet, *Public Health in the USSR*, (Novosti, 1968).

These figures do not account for some 300,000 qualified doctors. These may be medical research workers; for example, there are 36,000 epidemiologists at 4,900 bacteriological and virological laboratories working on the control of dangerous infections.

(D) Disease Prevention and Eradication

Since 1917 the incidence of disease and the mortality rate have decreased considerably. Certain illnesses, such as the plague, cholera, small pox and relapsing fever have been completely eradicated. Malaria which was a serious problem both before and immediately after 1917, is no longer prevalent. Diphtheria and poliomyelitis are both nearly eradicated. The mortality rate in 1965 was 7.3 per 1,000 population; infant mortality in that year was 27 per 1,000. The average life expectancy is 70 years compared with 32 years in 1896/7.

Cardiovascular and oncological diseases stand high among the causes of death, thus placing the USSR in the developed nations disease league. In 1958 cardiovascular diseases were responsible for 31 per cent of all fatal cases (25 per cent of the men's deaths and 37 per cent of the women's). Malignant tumours account for 15 to 20 per cent of all deaths.

(E) Finance

All expenditure on public health and physical culture comes out of the State budget and the State bodies, or co-operative, trade union and other mass organisations, as well as the funds of co-operative farms. In 1965, the State allocations for this field amounted to 6,700,000,000 roubles, or 28.9 roubles per head of the population.

The medical services - hospitalisation, medical advice, tests, antenatal and postnatal care, and so on - are all free. A small fee has to be paid for medicines, Places in holiday homes and sanatoria are either free, paid for by the trade union or other mass organisation, or paid for by the visitor at 30 per cent of the cost.

(F) Medical Supplies Industry

The medical supplies industry comes partially under the Ministry of Health and partially under the Ministry of Medical Equipment. The Ministry of Health administers some 130 establishments, which put out close to 5,000 items of medical supplies.

The production of medical supplies is increasing steadily in volume. During the 1959-65 seven-year plan progress was greatest in the manufacture of antibiotics and synthetic vitamins, both of these nearly quadrupling their production. During the last five-year plan 1966-70 500,000,000 roubles was allocated (50 per cent more than for the previous seven-year plan period) for the construction of new factories and the extension of existing ones, for scientific research institutes and design offices with experimental departments.

Medicinal preparations are put into production after being approved by the Pharmological Committee of the USSR Ministry of Health.

(G) Research and Development

16(1) ACADEMY OF MEDICAL SCIENCES OF THE USSR

The highest scientific research body in the field of medicine and public health is the *Academy of Medical Sciences of the USSR*. The main tasks of the academy are:

(a) to elaborate and solve the most important problems facing medicine;

(b) to co-ordinate and plan research work in the field of medicine, and to keep a check on the way in which it is carried out;

(c) to prepare research workers in the field of medicine and related sciences.

The academy has 106 full members of academicians and 150 corresponding members. It is directed by a presidium, elected for four years at a general meeting of the academy. The learned secretaries of different sections, the vice-presidents and the president are also elected for terms of four years.

The general meeting of the Academy of Medical Sciences, held annually, is the policy making body. It discusses the development of scientific research in the medical field, the work of the various sections of the academy, the application of the results of research to practice, and the preparation of medical workers.

The presidium is made up of the president, two vice-presidents, the chief scientific secretary, three academicians who are secretaries of the three sections, and two others. The presidium is the chief executive body. It plans and co-ordinates the research work of the different sections of the academy and the different institutes, laboratories, clinics and the like; it organises scientific conferences, synposia, expeditions and so on; it keeps up scientific links with the USSR Academy of Sciences and the Republican academies, and also with scientific organisations abroad; oversees the education of medical research workers in its institutes; and it is concerned with questions of introducing new technology and the result of scientific research.

The Academy has three sections; medical and biological sciences; clinical medicine; hygiene, microbiology and epidemiology. Each section administers those research institutes, laboratories and clinics concerned with its particular field. In 1970 there were 28 institutes attached to the academy; of these 23 were in Moscow, 2 in Leningrad, 1 in Kiev, 1 in Sukhumi and i in Obninsk. The total number of people working in these institutes was 15,355, of whom 3,424 were qualified scientists or doctors.

A certain amount of medical research is also carried out in the USSR Academy of Sciences in its Departments of Physiology and of Biochemistry, Biophysics and Bio-organic Chemistry.

In addition to the USSR Medical Academy of Sciences, the Ministry of Health is also a key body in medical research and development. It runs a large number of research and teaching institutes. These include the *All-Union Research Institute of Medical Instruments and Equipment*, the *All-Union Chemico-Pharmaceutical Research Institute*, and the *All-Union Research Institute of Medicinal Plants*.

(i) Medical science in the USSR. A number of scientific trends and schools have evolved in the USSR and are recognised throughout the world. Ivan Pavlov's work on the physiology of the higher nervous system is one example. Centres of research include the *Pavlov Institute of Physiology* and the *Sechenov Institute of Evolutionary Physiology and Biochemistry*, both in Leningrad. The biochemists, A. Bakh, A. Palladin and V. Engelhardt, have studied the complex biochemical processes of living organisms, at the *Bakh Institute of Biochemistry and the Institute of Molecular Biology*. Developments in heart surgery are connected with the names of the surgeons - P. Kupriyanov, A. Vishnevsky and others. Academician Vishnevsky is a director of the Medical Academy of Sciences *Institute of Surgery*, in Moscow.

Other areas of medical research include the study of vascular hypertension, atherosclerosis and cardiology, at the *Myashikov Institute of Cardiology* and the *Bakulev Institute of Cardiac and Vascular Surgery*, both in Moscow; plastic surgery and corneal transplantation (under Academician Filatov at the Institute of Surgery); helminthiases (under Academician Skryabin at the *Skryabin Institute of Helminthology*). Medical scientists have recently become interested in viral infections that have proved difficult to control. (Research centres include the *Institute of Virology* and the *Institute of Poliomyelitis and Virus Diseases*, both in Moscow).

The problem of cancer is also very important. In this field Professor L. Shabad of the *Institute of Experimental and Clinical Oncology* has received an international prize from the United Nations for his work on cancerogenic agents in the environment. Space medicine is also a new and developing branch of the medical sciences. In the medical instruments field recent developments have included an artificial kidney apparatus, lasers and laser coagulators for possible use in the treatment of diseases of the eye, brain, mouth etc.

Journals: There are some 90 medical journals in all published in the Soviet Union. These include: *Byulleten Eksperimentalnoi Biologii i Meditsiny* (Bulletin of Experimental Biology and Medicine); *Vestnik Akademii Meditsinskikh Nauk SSSR* (Journal of the USSR Academy of Medical Sciences); *Antibiotiki* (Antibiotics); *Vestnik Rentgenologii i Radiologii* (Journal of Roentgenology and Radiology); *Voprosy Virusologii* (Problems of Virusology); *Voprosy Neurokhirurgii* (Problems of Neuro-surgery); *Voprosy Onkologii* (Problems of Oncology); *Zhurnal Mikrobiologii, Epidemiologii i Immunobiologii* (Journal of Microbiology, Epidemiology and Immunobiology); *Kardiologiya* (Cardiology); *Neurofiziologiia* (Neurophysiology) and *Patalogicheskaya Fiziologiia i Eksperimentalnaya Terapiia* (Pathological Physiology and Experimental Therapy).

(H) Medical Education

In 1969 there were 76 medical schools and nine universities with medical faculties. Some 25,000 doctors graduated from them each year. Each republic has at least one medical school. The courses last for six to seven years. The seventh year, introduced in 1968, is a postgraduate residential year when the young doctor works at the institute clinic, studying a speciality under an experienced professor.

All doctors are expected to attend regular postgraduate courses. For those working in the country it should be once every three years, and for those working in the city once every five years. There are 13 state institutes for postgraduate medical training and 14 special departments attached to medical institutes. In 1965 these were attended by 34,600 doctors. And another 13,200 attended postgraduate courses at local clinics and large hospitals.

In the field of pharmacy, Russia possesses six teaching institutes, 19 faculties of pharmacy, and 22 pharmaceutical medium-grade schools. In 1968 35,700 pharmacists had higher education qualifications and about 90,000 specialists secondary pharmaceutical training.

(I) Plans and Policy

According to the Minister of Health, Boris Petrovsky, "The health service is based on centralised guidance, planning and financing; its main features are as follows: it is State run, there are close connections between medical science and practice, emphasis is placed on the prevention of disease and the active participation of the population in measures aimed at protecting their own health and that of all members of society." (Petrovsky: *Public Health in the USSR*, P 3, Novosti, Moscow, 1968).

The Ministry of Health is also responsible for physical culture in the country.

17 Transport

(I) RAILWAYS

(A) Introduction

The first railway was completed in Russia in 1851. It ran between Moscow and Saint Petersburg. Widespread building of railways began in the middle of the 1860s, and by 1913 Russia occupied second place in the world in length of track, with a total of 71,700km, after the United States. However, the network was quite insufficient for a large country, its density being only 0.32km per 100 sq.km of territory (lower than in any other industrialised country). In addition, it was spread out very unevenly 83 per cent of the network was in the European part of the country and only 17 per cent in the Siberian and Asiatic part. There was a single line across Siberia to Vladivostok and a single line also to Tashkent in Central Asia.

The width of track was equal to 1,524mm, which was wider than the tracks laid in Western Europe (1,435mm). This led, and still does, to certain difficulties at border crossings.

During the civil war and wars of intervention which followed the 1917 Revolution, the railway lines were often scenes of battle and they consequently suffered. Some 4,000 railway bridges, numerous stations, depots and railway lines were destroyed. However, the railways had recovered to their pre-war (1913) level by 1926 as far as freight traffic was concerned. It is interesting to note that it was on the railways that the first *Subbotnik* took place on May 10, 1919. *Subbotniki* were Saturdays that the workers donated to the state; they worked overtime for nothing to help the country get back onto its feet again.

In 1931 the Central Committee of the Communist Party confirmed a development programme for the reconstruction of transport. Plans for the railways included electrification, the introduction of large locomotives and big freight wagons equipped with automatic coupling and breaking, and mechanisations of the loading and unloading operations.

Reconstruction of the old lines took place along the busiest and most important routes. At the same time many new lines were built. By 1941 the railway network had grown by 30,000km compared with its size before the revolution. The new lines included; the completion of one to the ice-free port of Murmansk; the Turkestan-Siberian line; and the addition of a track to the Trans-Siberian line making it double track along its whole length.

During the Second World War, the railways again suffered enormous destruction. 65,000km of track, 13,000 railway bridges, nearly 4,100 stations, 317 engine depots, 129 factories, some 16,000 steam engines and 428,000 wagons were destroyed. Normal conditions were restored by 1948 and since then the railways have been expanding all the time. Table 44 gives a good idea of the development of the railways since 1917.

*Table 44**

	1913	1940	1950	1960	1965
Freight density (in millions of tons-km per 1km exploited length)	1.5	5.3	5.9	13.4	16.6
Average speed of goods train (km/hour)					
district	13.6	20.3	20.1	28.3	33.6
technical	22.0	33.1	33.8	40.4	45.3
Average daily run of a goods locomotive (km)	119.1	256.8	245.5	367.2	476.5
Average daily run of goods wagon (km)	72.0	139.9	146.4	227.0	259.7
Average weight of goods train (brutto-ton)	57.3	1,301	1,430	2,099	2,368
Mechanisation of loading-unloading work					
on approach tracks of industrial enterprises (in percentage)	–	45	70	87	89
on loading bays of main railway (in percentage)	–	12	31	67	79

*Source: Vredenskii, ed: *SSSR 1917-1967*, P 243 (Moscow 1967)

(B) Administrative and Operational Organisation

There are two main ministries concerned with the administration and policy of the railways: the Ministry of Railways and the Ministry of Transport Construction.

17(1) MINISTRY OF RAILWAYS

The Ministry of Railways runs the railways within the broad planning and investment targets determined by the central planning agency. It has a number of different departments responsible for different functions: commercial, operating, financial, technical, planning and international relations. The overall direction of the railways is conducted by an executive council. This has 15 members, headed by the minister and made up of deputy ministers and certain chief officers. The council is supported by a research advisory committee composed of chief officers and outside university professors. The council and minister normally act in concert. The minister can, however, decide against the majority of the council but in this event he has to report the fact to the central government.

The ministry is responsible for the training of all their own engineering and technical staff. They have 12 railway technical institutes which are substantially of university status and have an output of 6,000 qualified engineers each year. The railway also has more than 80 technical colleges throughout the country with an output of 240,000 technicians each year. Most of these are then employed on the railways, though some work in the railway construction industry which is the responsibility of the Ministry of Transport Construction.

The railways are divided into 26 self-accounting regions, each controlled by a general manager. Each region is organised in a similar manner to the Ministry, with separate departments for operations, locomotives, rolling stock, track and structure, signalling, electrification and power, planning etc. Each regional department is administratively responsible to the chief of railways in that region. They also have a responsibility to the head of the corresponding department at Ministry level. The size of regions varies. The Moscow one, for example, is 10,000 route km, and is split up into separate divisions of about 1,000 route km each.

17(2) MINISTRY OF TRANSPORT CONSTRUCTION

The structure of the Ministry of Transport Construction is believed to be similar to that of the Ministry of Railways. Within the railway industry it is responsible for such construction work as the laying of new tracks. There has been discussion about the need for more liaison between it and the Ministry of Railways as the division of responsibility between construction and operation of the railways does not necessarily help efficiency. *(Pravda,* January 13, 1970).

(C) The Railway Network (see Figure 17)

The Soviet railways had 134,600 route km in 1969, excluding branch lines. They constitute 10 per cent of all railways in the world. Over 60,000km of track has been laid since the revolution. Much of this has been in the Eastern and Central Asian parts of the country. In addition there are 114,000km (1965) of branch and approach tracks from industrial enterprises. These lines account for some 75 per cent of the total loading and unloading operations carried out on the railways.

Of the network 70 per cent is single track and of this 12,000km is converted to Centralised Traffic Control (CTC). Any line needing to convey 30 or more pairs of trains per day is regarded as requiring CTC operation. An average of 87 per cent of the traffic ie either electric or diesel hauled. Over 30,000km of track is electrified. Of this 10,500km are equipped for 25kV AC; this is mainly in the newer networks of the Ukraine, Siberia, Central Asia, Caucasus and Far East. The earlier lines, lying between Moscow, Leningrad and Rostov, are 3kV DC. The 25kV AC electrification capital costs are some 22 per cent less than for 3kV DC. Approximately 75 per cent of all overhead line maintenance is carried out under live conditions because of the very heavy traffic. Electrification is proceeding at the rate of 1,800 to 2,000km annually.

The last five-year plan (1966-70) provided for the construction of a large number of new lines. Many of these were in areas where railways are necessary to help exploit the natural resources. For example: the Ivdel-Ob will help to start tapping the extensive forest reserves in the Ob area; the Divnoye-Elista line has linked the capital of the Kalmyk Autonomous Republic with the railway network; the Tavda-Sotnik line is important for the oil fields and timber industry of the Tyumen (Western Siberia) region. All these lines came into service at the beginning of 1970. More than 1,200km of new lines and double tracks were planned for the final year of the five-year plan. One of the major lines under construction is the Baikal-Amur line, running from Ust-Kut on the river Lena to Komsomolsk-on-Amur. It is 3,000km long. During the 1971-75 plan, between 7,000 and 8,000km of the railway network will be double tracked, and between 5,000 and 6,000km will be built.

The speeds achieved by goods trains compare favourably with those of other countries.

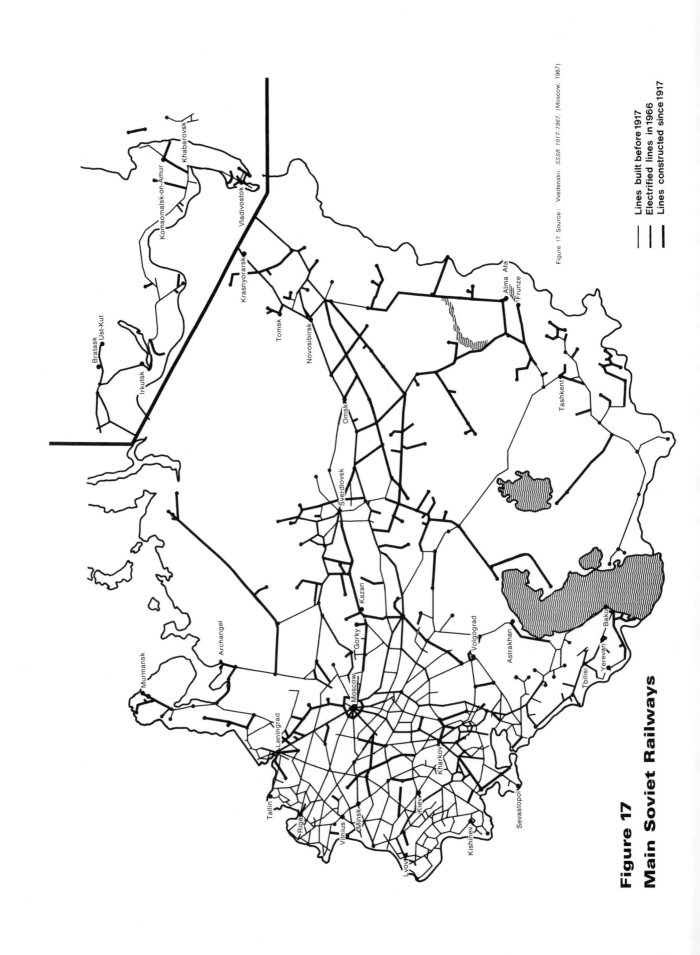

Figure 17
Main Soviet Railways

However, the average speeds of passenger trains are considerably lower. The main reasons for this are the large number of single track lines and the density of the traffic. However, passenger train speeds are being increased. On a number of lines, for example, Moscow-Leningrad, trains are used with speeds of up to 160km/hour and the average speed is 107km/hour. On main routes, such as Moscow-Brest, Moscow-Lvov, Moscow-Sevastopol, the average speeds are 65-80km/hour.

(D) Freight and Passenger Traffic

The volume of freight carried by the railways has grown from 70,400 million ton-km in 1913 to 2,367,100 million ton-km in 1969. The railways carry more than two-thirds freight traffic in the USSR. Similarly passenger traffic has increased from 30,300 million passenger-km in 1913 to 261,300 million passenger-km in 1969. This represents well over half the total passenger movement.

In 1967, 2,500,000,000 tons of goods were carried on the railways. The average load per track is 26,000,000 tons per annum, but on some routes it reaches 140,000,000 tons. However, more than a quarter of total rail freight was transported over distances less than 100km.

A growing volume of freight is being transported in containers. At present 1,000 stations serving 4,000 towns and villages are classed as container depots. Smaller goods stations with three or less wagons a day are being progressively closed down.

The Railway Plan provides for the introduction of 20-ton containers of International Standard Organisation (ISO) standard where practicable. At present the majority in service are of small capacity (up to five tons). There are 725,000 containers in service, and about 20 different types. Daily loadings on containers have grown from 3,400 in 1950 to 42,200 in 1968. Loaded journeys comprise 80 per cent of container movement. In 1968 28,400 tons of freight was handled in containers; the unit cost is 58 per cent lower than for equivalent consignments carried in covered wagons.

Collection and delivery of freight by road is undertaken either by railway-owned road vehicles or by a road delivery association acting as agents for and paid for by the railway.

(E) Production and Labour

The railway employ more than 3,500,000 people.

Since the early 1930s the Russians have developed mechanisation on the railways. They have reduced the number of man-days required to renew one km of track from 1,700 to 400. They are introducing automation into the marshalling yards - for example, the Losinoostrovskaya Marshalling Yard in Moscow. Computers are also being introduced to deal with the timetables. In Leningrad they have installed a system where all the data necessary for the timetable - times of arrival and departure, routes, platforms, removal to sidings, etc. - is reproduced on punch cards. The cards are fed in order into a punched card reader which then automatically sets routes and signals at the times recorded, without any manual operation by the staff.

The past few years have also seen changes in the operation of the railways as a result of the economic reforms introduced in 1965. Two railways and 14 railway transport enterprises changed over to the new system of planning and incentive in 1966. In 1967, 16 more railways changed over to the new system, and the rest followed in 1968. The Minister of Railways has claimed that the new system is a success and operations on the railways have improved, (*Pravda*, August 2, 1968). It has led to more efficient use being made of fixed assets. Railwaymen are interested in both their own work and that of the enterprise as a whole. The railways have to work in with the industrial enterprises also operating under the new economic system. The railways consignors and consignees are more aware of their mutual responsibilities - their transport obligations, the need to deliver freights on time, the question of damages suffered during transportation, and the need for complete utilisation of rolling stock.

The railway car depot Moskova was the first to start employing continuous flow and conveyor lines for the repair of railway wagons. Now 170 other depots have introduced this method and their productive capacity has increased by 15 to 50 per cent depending on local conditions. Labour productivity has gone up 25 per cent and the time taken to repair railway cars has gone down 25 per cent on average.

(F) Technology and Research

The main research centre for the railways is the *Railway Research Institute*. There are other research institutes associated with the manufacture of items such as locomotives and wagons. The Ministry of Railways had a total of 12 higher educational establishments subordinated to it in 1962.

17(3) RAILWAY RESEARCH INSTITUTE

The Railway Research Institute comprises three establishments: the largest is in Moscow and is called the Institute; then there is a test track and associated laboratories, called the Belt Railway, which is at Shervinka some 40km outside Moscow; and finally there is a laboratory in the Urals devoted to the problems of railway operation in extreme climatic conditions. There are 21 scientific departments employing 5,067 people of whom 435 have university degrees and 30 hold doctorates. The functions of the Railway Research Institute are research and development and to some extent design. It has three pilot plants engaged in prototype manufacture, each plant employing some 300 people, mainly skilled fitters and machine operators.

The Shervinka Belt Railway has three circular tracks, each 6km in circumference. One is electrified and two are for diesel traction. They are used to test new designs for rails, sleepers, fastenings and other track components. An 8,000-ton train runs each day continuously for 16 hours at a speed of 60 to 70 km/hr. This is equivalent to 400 20-ton axle loads every five minutes, or ten years' fatigue life in six months. There are 35 laboratory coaches for recording test results.

The scientific departments of the Research Institute include:

> the gas turbine and power equipment laboratories;
>
> the track maintenance and machinery department;
>
> the vehicle braking laboratory;
>
> the signalling laboratory;
>
> electrification development laboratory (at Shervinka);
>
> brake test facility;
>
> electric traction laboratories (at Shervinka);
>
> diesel test laboratories (at Shervinka);
>
> metallurgical laboratories;
>
> fatigue laboratories;
>
> high voltage laboratory;
>
> rolling stock department;
>
> computer department.

The budget of the Ministry of Railways' research department is between 11 to 12,000,000 roubles per year and this has been increasing progressively each year. In addition, the ministry provides locomotives, rolling stock and other services free of charge.

Journals: Journals concerned with railway transport include: *Elektricheskaya i Teplovoznaya Tiaga* (Electric and Diesel Traction); *Vestnik Vsesoyuznogo Nauchno-Issledovatelskogo Instituta Zheleznodorozhnogo Transporta* (Journal of the USSR Railway

Transport Research Institute); *Put' i Putevoe Khozaistvo* (Railway Track and Track Equipment); *Zheleznodorozhny Transport* (Railway Transport); and *Gudok* (Whistle).

(G) Equipment

Every three years the Ministry of Railways regulates the mechanical equipment on the railways by issuing a booklet with specifications of the equipment to be used including new equipment that can be designed either centrally or in the regions. The technical specifications of the machine is first put out; then a prototype is manufactured at one of the ministry's pilot plants; mass production is then taken over by one of the ministry's works.

General development of machinery is controlled by a state committee covering transport, commerce, building, the ministries of Power, Railways and Construction, and the Railway Research Institute. Forty seven per cent of machines are off-track; 32 per cent are portable; and 21 per cent run on track. Present work on equipment includes:

(a) *Rails*: Tests are being made of rails with varying proportions of chrome, molybdenum, vanadium, nickel, etc. Fifty different types have been tried in the past few years. They are also working on the hardening of rails with a carbon content of 0.8 per cent by heat treatment followed by oil quenching.

(b) *Sleepers*: Seven million concrete sleepers are laid per annum, and this number is to go up to 12,000,000. The Russian minimum is 1,800 sleepers/km. There are seven new factories being built jointly with the Hungarians for the manufacture of pre-stressed concrete sleepers.

(c) *Track laying*: Before the Second World War most tracks were ballasted on sand. There are 80 different track machines for permanent way work and 59 general types in the equipment catalogues. 37 other machines are being designed. The VPO 3,000 machine was developed a few years ago for track modernisation work. It is a levelling, tamping and lining machine, hauled by a locomotive and working at two to three km/hour.

(H) Future Plans

Between 1966 and 1970 (the period of the last five-year plan) the main problem for the development of railway transport was to increase their traffic capacity. This was and is being done by changing over from steam to electricity and diesel power, by increasing the speed of trains, and by laying more tracks. The productivity of the railways was to increase by 23 to 25 per cent.

As has been mentioned previously the main trend on the Soviet railways is to increase mechanisation and automation. Improvement of freight and passenger transport also calls for increased train speeds; more double tracks and larger weights of trains. Some of the developments include the following:

(a) *The Russian Troika*: This is a fast express train with a maximum speed of 200km which is being built at the Kalinin carriage building works. A pilot model of the train was due in 1970. Its carriages will be three-metres longer than the conventional model; its bogie is fitted with pneumatic axle suspension springs which will automatically keep the car body level. The Russian Troika can be used on existing tracks and is planned for the Moscow-Leningrad, Moscow-Brest, and Moscow-Gorky lines.

The Kalinin works are also testing a model of an express car designed for speeds of 300 to 400 km/hour. But in this case new lines would have to be laid or the old ones improved for these speeds.

(b) *The Monorail*: One of the first monorails to be built in the USSR was due to go into service in 1970 between the centre of Kiev and Kiev airport (a distance of 26 miles). The Kiev Polytechnical Institute is responsible for its development.

(II) AUTOMOBILES

(A) Introduction

Prior to the revolution there was no automobile industry in Russia, only 24,000km of roads had hard surfacing, and the 8,000 automobiles in the country were all imported. After the revolution most of these were in unserviceable condition and the first task was to organise reconditioning. The 1920s saw the beginning of small-scale production of three and five-ton lorries and of the first home-produced small-capacity four-seater light cars. During the first five-year plan (1929-33) mass-production factories were constructed. For example, the AMO works (formerly the Likhachev Automobile Works in Moscow) was reconstructed to produce an annual output of 25,000 2 to 5-ton lorries. The Gorky Vehicle Plant was established at the same time to produce 1 to 5-ton lorries and four-seater light cars. The assembly of vehicles, engines and units was organised on the flow-line conveyor principle. The production technology was the most advanced in the country.

The pre-war period in the development of automobile engineering was characterised by the organisation of mass-production technology and by an increase in vehicle production, mainly in the two large plants - the Likhachev Works in Moscow and the Gorky Plant. The main aim of the industry at this time was to increase the output of commercial vehicles, and by 1937 the USSR occupied second place in the world (after the United States) in the production of commercial vehicles. In 1940 production exceeded 145,000 vehicles a year. At the same time in the road network the length of hard surfaced roads was increased several fold.

The Second World War slowed up the development of vehicle technology. At the same time, production was started in a number of new places, due to the evacuation of the Likhachev Works from Moscow. After the war work was immediately begun to restore and develop further the automobile industry. By the end of the fourth five-year plan (1946-50) production was treble the 1940 level, and it has been growing ever since.

(B) Administration

Overall administration of policy and finance is carried out within the general directives of the central planning agency by the *Ministry of the Automobile Industry*. It seems likely that this ministry is organised along the same lines as the Ministry of Railways with different departments responsible for different functions, and overall direction being in the hands of an executive council headed by the minister.

(C) Road Network

The length of hard surface roads increased from 24,300km in 1913 to 483,200km in 1969. However, of this only 191,000 were metalled roads and over half of that had been built in the period 1960-69. At the same time the principal road network is rudimentary consisting of dirt tracks that are impassable in mud or snow.

The road network is expanding at a rate of 4 per cent a year (compared with 7 to 14 per cent for road traffic, 400 per cent for car production and 150 per cent for lorry production). Up to 1970 the level of road construction remained the same. However, the next five-year plan (1971-75) provides for an investment of 1,800,000,000 roubles a year in road construction compared with 440,000,000 roubles before. Up to 1958, roads were financed from the central budget. Since then a tax has been levied on the profits of enterprises amounting to some 0.05 per cent of the annual value of their production. It is now proposed to step this tax up to 0.5 per cent and also to impose an annual levy on every vehicle.

(D) Freight and Passenger Transport

Freight and passenger transport by car and lorry still represents a very small percentage of the whole (for example, some 5.3 per cent for freight in 1966). However its share is now increasing fast (see Table 45).

*Table 45**

	Freight transport (1,000 million ton-km)	Passenger transport (1,000 million pass-km)
1913	0.1	-
1940	8.9	3.4
1950	20.1	5.2
1960	98.5	61.0
1966	154.0	136.1
1969	200.1	183.0

*Source: Vredenskii ed: *SSSR 1917-1967*, P 239 (Moscow, 1967) and *Narodnoe Khozaistvo SSSR v 1969 g*, P 469 and 477 (Moscow, 1970).

(i) Freight. The recent growth of the number of vehicles means that automobile transport carries about four to five times more freight by weight than all the other types of transport taken together. However, the average distance transported is only 13km and so the overall share of automobile transport remains small. Transport prices have generally been fixed to encourage the use of railways at the expense of roads. In 1965 almost a quarter of total rail freight was transported over distances less than 100km, although road transport can be more efficient on short journeys.

Freight transport by road is now being developed. Heavier ehicles are being manufactured which will greatly increase the overall freight-carrying capacity of road transport. At present there are a lot of vehicles being used with a small load capacity. The new vehicles should release up to 800,000 drivers (according to Dmitry Velikanov, corresponding member of the Academy of Sciences, *Soviet News*, September 23, 1969), which should help in those parts of the country, particularly Siberia and the Far North, where there is a great shortage of drivers.

The number of vehicles in each transport depot is being increased. The productivity of vehicles in a depot with 100 lorries is up to 80 per cent higher than in depots with only 10 to 24 vehicles. There is more mechanisation and, as on the railways, the use of containers is increasing.

(ii) Passenger. In 1968 there was one car for every 400 people. The policy is to expand car ownership, and therefore passenger transport by car, over the next few years.

(E) Production

(i) Output. In 1970 there was an output of 916,000 automobiles, of which 572,000 were lorries and buses. This compares with a 1966 output of 675,000 automobiles of which 445,000 were lorries. This is an increase in passenger cars of some 150 per cent during the past five years. Future plans envisage further increases in the annual production of passenger cars.

This scale of development is unprecedented in the Soviet motor industry. It is being brought about by both expanding and increasing the number of plants, and by raising labour productivity. More than 80 per cent of recent production increases have been due to improved labour productivity. *(Pravda,* February 6, 1970*)*.

The main role in the production of trucks is now assigned to the Gorky motor works, though there are preparations for the construction of a number of works which will annually manufacture 150,000 three-axle trucks with a big load capacity. The output of heavy vehicles is concentrated at the Byelorussian motor works, which is turning out 40-ton and 75-ton tip-lorries. The Kama lorry factory, which is under construction, will be able to turn out 150,000 medium size vehicles a year. The main contribution to the manufacture of motor cars will be made by the works at Togliatti which will produce 660,000 cars a year at full capacity. The Zaporozhye works in the Ukraine will continue to produce small cars. The output of Moskvich cars will be doubled in 1971 at the Lenin Komsomol Works in Moscow. Medium class vehicles are represented by the Volga car produced by the Gorky works. The Ministry of Automobile Industry was in the process of building 21 new plants and reconstructing some 100 enterprises during 1970.

(ii) Labour and Administration. The rapid development and modernisation of the car industry means that the factories are being reorganised which in turn affects the car workers. The productivity of the car workers has been rising; for example, it rose by 43.5 per cent in the first four years of the last five-year plan at the Likhachev works in Moscow.

(F) Technology and Research

The research organisation within the automobile industry is complicated by the dispersal of authority between the central ministerial organisation and the separate automobile factories. Factory research enjoys an exceptional position within the industry.

17(4) CENTRAL SCIENTIFIC RESEARCH INSTITUTE FOR AUTOMOBILES AND ENGINES

The chief research organisation is the Central Scientific Research Institute for Automobile and Engines in Moscow. In 1963 this had a staff of 2,000 of which 60 per cent worked on research and development and 40 per cent on project work; it is also supposed to have good testing grounds. Its function is to prepare forward designs for engines, transmissions, and suspensions for the whole Soviet automobile industry.

The Institute's position becomes ambiguous, however, because the larger factories often carry out their own research. For example, the Moskvich plant has its own design office of 500 people and assumes responsibility for designing a whole vehicle, producing prototypes and carrying out extensive testing. The plant also designs and builds some of its own machine tools, and it only occasionally calls in the Central Scientific Research Institute for consultation.

In 1962 there was a total of 40 design bureaux within the automobile industry.

The period of the last five-year plan (1966-70) was one of vigorous development in the technology of automobile engineering. It represents a new stage in the qualitative and quantitative growth of production using the most advanced technology and progressive methods of complex mechanised and automated production processes. The quantity of high-capacity machinery in basic manufacture was planned to increase by 150 per cent over the five-year period. The number of automatic lines doubled, and the total number of automatic and semi-automatic machines increased by 15 to 20 per cent. The proportion of plant more than ten-years old was cut by about half. The development and introduction of economic-mathematical methods and the application of computer technology and other methods of production control are playing an important part. It is proposed to set up an information computing centre in the Ministry of Automobile Production, together with a whole series of factory systems in the Likhachev, Gorky and Yaroslavl Works. Increasing attention is being paid to equipping industrial plant with various conveyor systems, including automatic dispatch of components to assembly points. The extent of mechanisation and automation in assembly has been raised to 85 per cent.

Technical re-equipping for casting production is being achieved by introducing automatic lines, automatic and semi-automatic machines for producing moulds under high specific pressure, sand-blasting and sand-jet machines for core making, automatic rigs for knocking out and cleaning castings, and methods for mechanisation and automation throughout in the production of mould and core mixtures and of metal melt. Examples of high-efficiency casting processes are the automatic lines for casting brake drums and the automated foundry for precision casting using the lose wax process. These developments won Lenin Prizes in 1965 and 1966. In modern automobile design wide use is being made of light alloys particularly of aluminium. The Zavolzhsk Motor Works has reached the final stages for setting up production of the largest aluminium die and chill-castings in Europe. Improvements in forging production include mechanisation, automation and the installation of hot-stamping crank presses.

Increasing the quality and stabilising the strength properties of vehicle components represents the main trends in improving heat treatment and chemico-heat treatment. The five-year plan envisaged a trebling or quadrupling of the treatment of parts in controlled atmospheres; induction heating processes were developed; and about one-tenth of all components will be heat-treated directly in the automatic and flow lines used in mechanical working.

Mechanical working accounts for up to 30 per cent of total vehicle manufacture. Attention is being paid to renewing and improving machine-tool installations. A new process was developed for rolling the teeth of helicalconical gears in automobiles and tractors. Techniques involving plastic deformation are being widely adopted. Advanced broaching with hard-alloy broaches, continuous broaching, automatic honing processes, working with new hard-alloy tools, complex design processes for working the main vehicle parts, abrasive shaving, new processes for surface finishing, working with diamond tools are some of the high-efficiency processes widely used in the motor industry.

The five-year plan provided for the introduction of new processes for welding wheels and the wide application of multi-electrode machines, high-capacity automatic and semi-automatic plant, and also the development and establishment of processes for welding galvanised and painted components, and new processes of electron-arc and plasma welding. The use of pressing work and cold-extrusion processes is being increased. Plans scheduled the introduction of several plants for glaze or dip-welding with spreading of the paint in solvent vapours, painting by electro-deposition, etc. About 80 per cent of all painting will be done by high-productivity methods. (Source: "Fifty Years of the Motor Industry", *Avtomobilnaya promyshlennost,* 1967 (10); translated in *NLL Translations Bulletin,* PP 517-531, volume 10, no 5, May 1968).

(i) Electric cars. Some Soviet experts, including Academician Frumkin, believe that electrically powered motor vehicles will begin to be widely used by the end of the 1970s. *(Soviet News,* September 23, 1969). Research is being carried out on electric vehicles. A prototype of a ten-seater electric minibus was tested in Moscow in 1969. This combined a small petrol engine with storage batteries. The engine runs at low constant speed and drives the generator which supplies current to the electric motor and to the lead storage batteries, which, in turn, provide the extra power required for starting and climbing. The most optimistic forecasts see 1,500,000 electric cars and about the same number of electric lorries in the Soviet Union by 1980. By the year 2,000 their number will have increased to some 14,000,000. This will still only account for a fifth of the total number of vehicles in the country at that time (*Ekonomicheskaya Gazeta,* No 31).

(ii) Exhaust pollution. The Central Laboratory for Neutralisation and for Automobile Energetics Problems works on various aspects of exhaust pollution. It has set up special control posts, which measure the amount of harmful substances given off by a car exhaust over an eight to nine hour drive around town taking into consideration stops, starts, slowing down etc. Between 20 and 25 such posts are considered to be sufficient to check all the vehicles in Moscow twice a year. The laboratory has also devised a soot-meter for diesel motors, and developed a wide number of neutralisers for installing

on vehicles which directly reduced the amount of toxic substances in the exhaust.

Journals: Journals of the automobile industry include *Avtomobilnye Dorogi* (Motor Roads); *Avtomobilny Transport* (Motor Transport); *Avtomobilnaya Promyshlennost'* (Automobile Industry); and *Stroitelnye i Dorozhnye Mashiny* (Construction and Road Building Machines).

(G) Plans and Policy

At present the automobile industry in the Soviet Union is enjoying boom conditions. While previously priority was always given to the manufacture of trucks and buses, now there is an equal emphasis on passenger cars. The increase in manufacture is being effected by introducing mechanisation and automation, the construction of new plants, and the reconstruction of old, the introduction of new technology, the specialisation of factories and by increasing labour productivity. This policy is associated with an increase in road building.

The development of road transport will be particularly important for both the agricultural and consumer sectors of the economy. The farmers need hard surface roads for farm-to-market transport and also for station-to-farm carriage of fertilisers, machinery and fuel. The consumer industry needs rapid door-to-door deliveries of small consignments.

Long-term plans see the gradual transformation of the country's motor transport fleet from its present 70 per cent of lorries and 26 per cent passenger cars, to 70 per cent cars and 25 per cent lorries by 1980.

(III) SEA AND RIVER TRANSPORT

(A) Introduction

The development of water transport was slow before the revolution. Of the 500,000km of navigable rivers in Russia only some 65,000 were in regular use. The possibilities of river transport were limited by the shallow transit depths, the separation of rivers of different basins, and the small size of the marine fleet. On many rivers movement was possible only during the day because of the absence of necessary signal systems. There were only 2,000km of canals. The river fleet was privately owned. The rivers Volga and Kama were used quite intensively, while the Siberian rivers were hardly used at all. In 1914 there were 1,103 steam or paddle sea ships with an overall tonnage of 894,000 tons; sailing ships accounted for 202,000 tons. The total represented between one and two per cent of the world tonnage.

River and sea transport has been developed since the revolution. The northern and far-eastern seas, the Siberian rivers and other waterways have come into greater use. The northern sea route is now open for several months of the year. Important ports have been reconstructed and new ones built. The byelomorskii-Baltiskii canal was built in 1933, the Moscow-Volga in 1937, the Volga-Don in 1952, and the short deep-water Volga-Baltiskii (1964) since the war.

(B) Administration

Overall administration of policy and finance is carried out, within the general directives of the central planning agency, by the *Ministries of the Merchant Marine* and of the *River Fleet*. These would appear to be organised along the same lines as the Railways Ministry with different departments responsible for different functions, and overall direction being in the hands of an executive council headed by the minister. The *Ministry of the Shipbuilding Industry* is responsible for shipbuilding activities.

(C) River Network and Sea Routes

Between 1913 and 1965 the length of navigable inland waterways has more than doubled - from 64,600km in 1913 to 142,700km in 1965. Canals have been built to connect the waterways of the European part of the Soviet Union into a single network. Sea routes have likewise expanded since the revolution. New ports have been built in the Far East, on the Black Sea and along the northern sea route. This latter route is very important, despite its seasonal nature, and forms the basic transport system for many of the northern regions of the country. (See Figure 18)

The Soviet Union also has ships sailing to all parts of the world.

(D) Freight and Passenger Transport

Table 46 gives the growth in freight and passenger traffic.

*Table 46**

	1913	1940	1950	1960	1965	1969
Freight						
river (mln. ton-kms)	28,900	36,100	46,200	99,600	133,900	160,100
sea (mln. ton-miles)	20,300	12,800	21,400	71,000	209,900	324,700
Passenger						
river (mln. passenger-kms)	1,400	3,800	2,700	4,300	4,900	5,500
sea (mln. passenger-miles)	500	479	671	715	789	935

*Source: Vredenskii ed: *SSSR 1917-1967*, P 239 (Moscow 1967) and *Narodnoe Khozaistvo SSSR v 1969 g*, (Moscow 1970).

In sea transport the USSR occupied sixth place in the world for general tonnage of her sea fleet in 1965. The policy now is to increase river and sea freight and passenger transport, and the emphasis is on greater mechanisation, faster turn round, modernisation of ports, and containerisation. In general river transport is cheaper than railway, but it suffers from its seasonal nature. The bulk of river freight ships are diesel carriers with a speed of 12 to 18 km/hour. However, in recent years hydrofoils have come into use and could revolutionise transport on rivers. The bulk of the freight carried consists of timber and firewood, oil, oil products, minerals and building materials.

(E) Production and Output

During the past ten years or so, the merchant marine has been reinforced by 800 freight boats with an aggregate carrying capacity of 6,500,000 tons. At the end of the 1960s the fleet had some 1,300 boats in all with a dead weight of nearly 10,000,000 tons.

Nearly 90 per cent of the ships are powered with diesel or steam turbine engines.

(F) Technology and Research

Research institutes concerned with problems of marine and river transport include the *Central Scientific Research Institute of the Maritime Fleet* in Leningrad (works on the improvements of ships and ships' equipment, particularly interested in corrosion prevention,

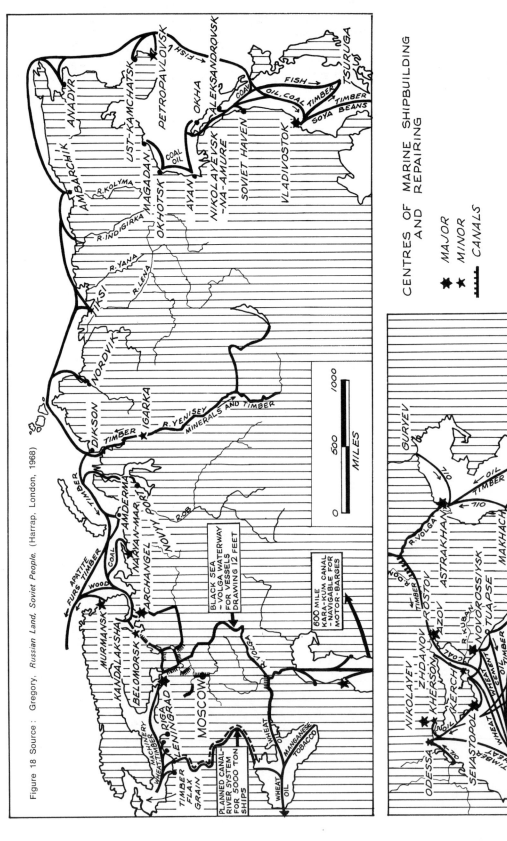

Figure 18 Source: Gregory, Russian Land, Soviet People, (Harrap, London, 1968)

Figure 18 Main Ports and Shipping Routes of the USSR

radio communications, radar systems and radio navigation systems); the *Central Scientific Research Institute of the River Fleet*; the *Ship Building Institute im. Admiral S.O. Makarov* in Nikolayev (specialises in theoretical research into ship structures); the *Odessa Institute of Engineers of Merchant Fleet*; the *Odessa Institute of Marine Engineers*; the *Vladivostock Higher Engineering Maritime School*; and the *Novosibirsk Institute of Marine Engineers*.

Interesting developments in the marine transport field include hovercraft and hydrofoils, nuclear powered ships (one ice breaker has been built, and two more are now commissioned) and catamaran fishing vessels (the prototype called 'Experiment' carried out successful trials in the winter of 1970/71).

Journals: Journals concerned with river and sea transport problems include *Morskoi Flot* (Merchant Marine); *Rechnoi Transport* (River Transport); *Subostroenie* (Ship Building); and *Vodnoi Transport* (Water Transport);

(G) Plans and Policy

The 1966 and 1970 five-year plan envisaged an increase in the general tonnage of the sea fleet of 150 per cent; and the river fleet was to be largely replenished so that it could deal with all freight moving along river routes; however, the greatest effort was to go into developing the Siberian rivers.

(IV) AVIATION

(A) Introduction

Prior to the revolution in 1917 Russia only had a few primitive aircraft-building plants, and these depended solely on foreign made parts. The famous aviation pioneers, such as N.E. Zhukovsky, S.A. chaplygin and K.E. Tsiolkovsky, received little government help. During the First World War most of this primitive industry was destroyed. However, only four days after the revolution, the Bolshevik government set up the Bureau of Commissars for Aviation and Aeronautics and by December of the same year the All-Russian Board was organised to manage the Russian Republic's air fleet. The aircraft-building plants were among the first to be nationalised, (between December 1917 and January 1918). In October 1918 the Central Aerodynamic Institute (TsAGI) was established with Professor N.E. Zhukovsky at its head. The main intention was that it should provide the scientific and technological research needed for the development of the aviation industry. Since then the aviation industry and air transportation have expanded fast, as Table 47 shows.

Table 47[*]

	1913	1940	1950	1960	1966
Air freight transport (million ton-km)	–	20	140	560	1,450
Air passenger transport (million passenger-km)	–	200	1,200	12,100	45,000

[*] Source: Vredenskii, ed: *SSSR 1917-1967*, P 239 (Moscow, 1967)

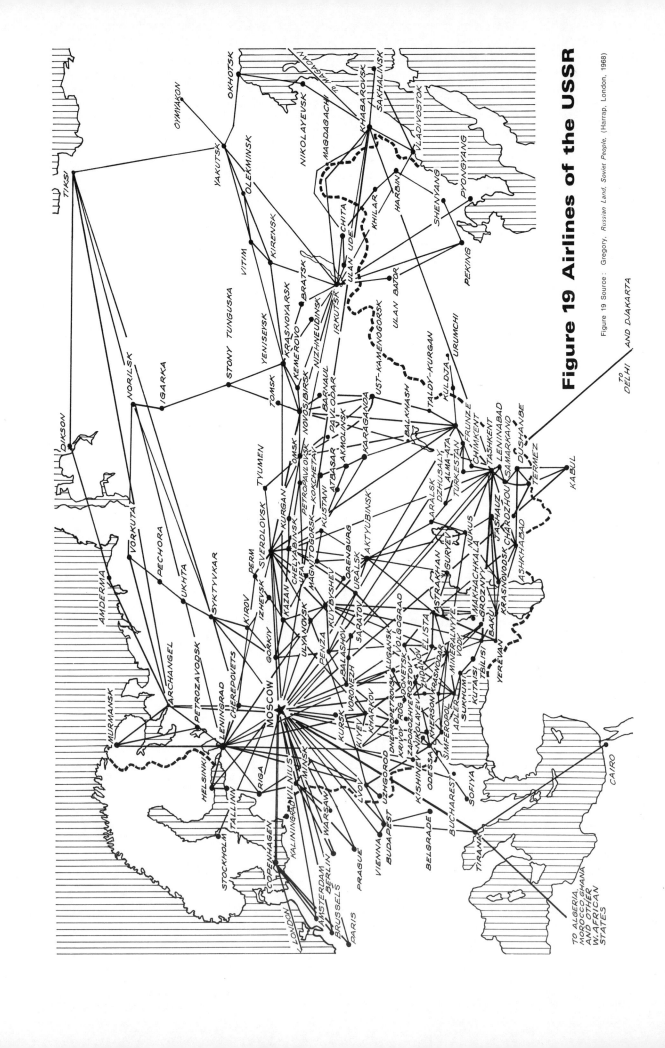

Figure 19 Airlines of the USSR

Figure 19 Source: Gregory, *Russian Land, Soviet People*, (Harrap, London, 1968)

(B) Administration

The Ministries of the Aviation Industry and Civil Aviation, and the Aeroflot are the three main bodies responsible for administration of the aviation industry and air transport. Aeroflot is the State civil airline which operates the air passenger and freight services both within the USSR and on the international routes.

(C) Network and Routes

The air network within the USSR extended to some 435,000 km. in 1965 not counting overlapping lines. Moscow is connected by air with the capitals of all the other republics and with a large number of foreign countries. (See Figure 19). Recently a number of new international routes have come into service. For example, there are now flights from Paris to Tokyo via Moscow, and from London to Japan via Moscow and Siberia; new lines are being started to Kuala Lumpur, Nairobi and Dresden and so on.

To help develop the oil industry of Western Siberia a new airfield was due to be opened at Surgut in 1970 and at Nizhevartovsky in 1971.

(D) Freight and Passenger Transport

Air transport only accounts for a small percentage of the total freight carried in the USSR. In 1966 the total carried by air was 1,450,000,000 ton-km. out of a total of 2,916,900,000,000 ton-km.

However, its role is growing; in 1969 the total air freight was 1,946,900,000 ton-km. More big freight-carrying planes are being introduced including large helicopters capable of carrying more than 10 tons. (The largest Soviet model broke the world record early in 1971 by lifting 40 tons, a mile high).

In passenger traffic, air transport accounts for 11 per cent of all movement between cities and is one of the fastest growing forms of transport for this purpose. In 1969 it accounted for 71,500,000,000 passenger-km, compared with 12,100,000,000 in 1960 and 1,200,000,000 in 1950.

(E) Production

In 1960 the Soviet Union had about 1,800 aircraft in commercial operation.

The flagship of the civil air fleet is the four engine 12-62 for 200 passengers. It cruises at 900km.p.h., and is used on domestic and international flights.

Three other important planes used by Aeroflot are the TU-104, and IL-18 and the AN-10. The TU-104 has a payload of 12 tons cruising speed of 850-900km.p.h. a ceiling of 10,000m and can fly non-stop for 4,000km. It seats 70 to 90 passengers. The IL-18 is a reliable and economic plane, designed under the direction of S. Ilyushin. It carries from 89 to 118 passengers and flies at 650-680 k.p.h and can cover 5,600km. non-stop. The AN-10 comes from O. Antonov's design bureau. It has four turboprop engines, develops a speed of 650-680 km.p.h. but has a shorter range than the IL-18. One of its chief advantages is that it can use short earth runways. These three planes are to be gradually replaced by the TU-134, Yak-40 and TU-154.

The number of piston engine planes is being cut down. Even local lines are serviced by small jets such as the TU-124 (56 passengers) and the AN-24 (44 to 52 passengers). During the 1971-75 period the 150-ton cargo jet plane 16-76 and the supersonic TU-144 will be introduced into the Aeroflot fleet.

(F) Technology and Research

According to Professor Grigori Tokaty (*Aeronautical Engineering Education and Research in the USSR*, Lawrence, Kansas, 1960). the aeronautical research establishments come under the administration of four principal bodies: the Ministry of Defence, the State

Committee of the Aviation Industry; the State Committee of Armaments; and the Chief Managing Department of the Civil Air Fleet.

The most important research establishments mentioned by Tokaty are: the *Central Aero and Hydrodynamic Institute (TSAGI)*; the old *TSAGI in Moscow* and the new *TSAGI in Zhukovskaya*; the *Institute of Hydrodynamics* at Novosibirsk; the *Flight Research Institute* at Zhukovskaya; the *Scientific Testing Institute of the Air Forces (NIIVVS)* the *Scientific Testing Institute of Aviation Armaments (NIIAV)*; the *Scientific Testing Institute of Aviation Instruments (NIIAP)*; the *Zhukovsky Aeronautical Engineering Academy* in Moscow; the *Central Institute of Aviation Engines (TsIAM)*; in Moscow; and the *All-Union Institute of Aviation Materials (VIAM)* Further details of the equipment and particular lines of research of these institutes can be found in Chapter 6 on Space Research P 85.

Experimental Design Bureaux are responsible for the design and production of new aircraft. Each bureau has two main parts: the design bureau which designs, and an experimental factory which produces the prototype. Both organisations come under the administration of one man called the Chief Designer and Director. Tokaty describes the process by which a new aeroplane is produced. The bureau may have a new idea which its chief designer then has to get accepted by the appropriate ministry and state committee officials. Once this has been done, the Minister of Aircraft Production issues a special order authorising the Chief Designer and his bureau to proceed to the design and development of the machine, and he also authorises the provision of the necessary money, materials, engines, instruments and so on. All expenses are met by the State. When the bureau has produced a prototype, this has to pass state tests carried out by the Scientific Testing Institute of the Air Forces. Once the new aircraft has been accepted for serial production, the design bureau loses any responsibility for it. The responsibility passes to the design bureau of the factory where it is being produced.

The aircraft design bureaux are often linked with the name of their chief designer, who in turn lends his initials to the planes developed within his bureau. There are manu examples: Tu - Tupolov (aeroplanes); Yak - Yakovlev (aeroplanes and helicopters); La - Lavochkin (aeroplanes); MiG - Mikoyan and Gurevich (aeroplanes); Il - Ilushin (aeroplanes); Su - Sukhoy (aeroplanes); Pe - Petlyakov (aeroplanes); An - Antonov (aeroplanes); Mi - Mil' (helicopters); Ka - Kamov (helicopters); Bra- Bratukhin (helicopters); Ar - Arkhangelsky (aeroplanes); LaGG - Lavochkin, Gorbunov and Gudkov (Aeroplane); Po - Polikarpov (aeroplanes); VK - Viktor Klimov (engines); NK - Nikolai Kuznezov (engines); ASh - Aleksandr Shvezov (engines); and AM - Aleksandr Mikulin (engines).

Every aeroplane and aeroengine has its code name consisting of the initials of the chief designer of the bureau where it was developed together with the number of the project. Fighters have odd numbers; and all other aircraft - bombers, airliners, helicopters, and so on - have even numbers.

Experimental designs that are being tested at present include the supersonic TU-144 which is due to go into passenger service by about 1973/74; the jumbo IL-62 which seats 200 and flies non-stop from Moscow to New York at 900km.p.h.; the AN-22 (or Antei) which will carry 700 passengers and covers 11,000km non-stop.

The question of airbuses is raised fairly frequently in the Soviet press, particularly in connection with developing air freight transport in the difficult conditions of the north and east of the country.

Tokaty lists the following higher educational establishments which provide graduate and post-graduate training in aeronautics; the *Zhukovsky Military Aeronautical Engineering Academy* in Moscow; the *Mozhaisky Military Aeronautical Engineering Academy* in Leningrad; the *Ordzhonikidze Moscow Aviation Institute*; the *Moscow Aviation Technological Institute*; the *Kharkov Aviation Institute*; the *Kuibyshev Aviation Institute*; the *Kazan Aviation Institute*; the *Ufa Aviation Institute*; the *Alma-Ata Aviation Institute*; the *Kiev Institute of the Civil Air Fleet*; and the *Leningrad Institute of the Civil Air*

Fleet. In addition to these specialised aviation colleges, there are 12 departments and chairs of aeronautics in non-aviation universities (the Moscow State University, the Moscow Higher Technical College, the Leningrad Polytechnical Institute and the Saratov State University are the best known), as well as four aeronautical engineering technikums, and four military establishments providing aeronautical training.

Journals: Journals of the aviation industry include *Grazhdanskaya Aviatsiia* (Civil Aviation); *Aviatsiia i Kosmonavtika* (Aviation and Cosmonautics); and *Iz. VUZ. Aviatsionnaya Tekhnika* (Proceedings of Higher Educational Establishments. Aviation Technology).

(G) Plans and Policy

Air transport is presently showing a faster growth rate than any other type of transport in the Soviet Union. According to V. Petrov and S. Ushakov, writing in their booklet on Soviet Transport (Novosti Press, Moscow), the main tasks facing civil aviation are to make flights more regular, and to reduce dependence on meteorological conditions. They see these problems as being solved by greater use of automatic landing control systems.

18 Communications Industry

(A) Introduction

Before the revolution Russia had a very poor communications network, particularly in the rural areas. The first inter-town telephone link was organised in 1882 between St. Petersburg and Gatchina. The Moscow-St.Petersburg line went into operation 16 years later in 1898. By 1917 there were 232,000 subscribers on the city telephone exchanges, and of these about a half came from Moscow and St. Petersburg. There was no proper telephone service in the rural areas. Radio communications were equally poorly developed before the revolution, despite the fact that a Russian scientist, A.S. Popov, was one of the early pioneers of radio.

(B) Organisation

Because of its nature, the communications industry is one of the most highly centralised industries in the whole country. The means of communication come under the administration of the all-union *Ministry of Postal Services and Telecommunications* and its republican counterpart ministries. They organise and control and local postal, telegraph and telephone services and the technical equipment needed for radio and television transmission.

There is also a *Ministry of the Radio Industry* which is concerned with the development of the radio and television transmission and with the technological problems this involves.

(C) Network and Services

(i) Telephone and Telegraph Services. The inter-city telephone and telegraph communications network has developed on a radial basis, with each local station having a direct link with the capital.

The first 18-channel tonal telegraph system was organised in 1939, between Moscow and Khabarovsk. By 1966 the length of such telegraph channels had reached 7,900,000 kilometres. In large centres the methods of sending telegrams are now completely automated. The number of telegrams sent has grown from 42,000,000 in 1913 to 300,000,000 in 1966.

Photo-telegraphic equipment has been installed in 200 cities. One important use of this is to transmit the central Moscow newspapers to cities in the Urals, Siberia, the Far east and Central Asia.

Telephone communications developed steadily after the revolution, gradually linking Moscow with the more distant regions of the country - Novosibirsk and Kuznatsk, Vologda and Archangelsk, and finally in 1939 the 8,600 km link between Moscow and Khabarovsk was built. Since the war, the telephone network has grown rapidly, chiefly on the basis of cable and radio-relay lines. During the 7 year period from 1959 to 1966 the length of cable lines was tripled, while the length of radio-relay lines increased 6.7 times. The longest overland cable trunk line in the world was built between Moscow Khabarovsk and Vladivostock. Other important cable trunk lines are Moscow-Kiev-Lvov and Moscow-Saravot-Tashkent; an important radio relay line runs from Moscow to Simferopol, Sochi and Tbilisi. 12, 24, and 60-channel systems have been developed for high-frequency symmetrical cables. For coaxial or concentric cables, equipment has been developed where up to 1,920 telephone channels can be carried along two tubes, while the two other tubes carry two television transmissions. The number of telephone calls made between towns rose from 163,000,000 in 1958 to 283,000,000 in 1966. Over the same period the overall length of the inter-city telephone network tripled.

By the end of 1965 there were 4,192,000 telephone subscribers connected to the urban telephone exchanges. Of these numbers 79.9 per cent (3,350,207) were operating on an automatic telephone system. By 1970 it was planned that this number should have risen to 3,700,000.

In the rural areas there were 1,113,600 telephone subscribers at the end of 1965. These took in 68.1 per cent of the state farms and 31.8 per cent of the collective farms. It was planned to increase the number of telephone subscribers by a factor of 1.8 by the end of 1970. Of these 1,000,000 were to be operating on the automatic system.

(ii) Radio and Television. In July 1918 a decree was passed centralising the research, development and organisation of the radio industry. The central research institute was the Nizhegorodskaya radio laboratory directed by M.A. Bonch-Bruyevich. The first Soviet radio station began transmission in the autumn of 1922 in Moscow. By 1929 there were 23 broadcasting stations in the USSR, including one with a power of 75 kilowatts, at that time one of the most powerful in the world. The first five-year plan (1929-32) saw the construction of a 500 kilowatt broadcasting station, and in 1935 work began on the first short-wave transmission centre. Since then and after the war the radio transmission network has expanded enormously. In particular, ultrashort wave broadcasting with frequency modulation has developed widely. The broadcasting facilities in the country allow for the transmission of five central programmes a day, including special programmes for Siberia and the Far East (taking into account their time differences), two programmes in each union republic, local programmes in the autonomous republics and other regional areas, and overseas broadcasts for foreign countries.

The Moscow and Leningrad television centres began regular transmissions in 1939, but the war interrupted the development of television. After the war television centres were built in Kiev, Riga, Sverdlovsk, Tallin, Minsk, Tashkent, Tbilisi and other cities. By 1959 there were 60 altogether. In 1971 the television network consisted of 127 programme centres, and over 1,000 re-transmitting stations. In 1965 the television network reached 115,000,000 people in the country. By 1970 this number had increased to 168,700,000 or 70 per cent of the population.

In Moscow and Leningrad three television programmes are transmitted; the capitals of eight union republics and five other cities have two programmes. The central television programme is transmitted to the capitals of 12 union republics, eight autonomous republics and 98 other towns. In 1971 there was colour television in Moscow, Kiev and Tbilisi. The number of television sets in the country in 1965 was 16,000,000 and this had risen to 35,000,000 by 1970.

(D) The Orbita Space Communications Network (see also Chapter 6)

The Orbita space communications network has 30 ground stations which are served by the Molniya communications satellites. It is used for relaying black and white and colour TV, radio, telegraph, telephone, meteorological charts and newspaper facsimilies. It was developed to help provide a full communications network over the whole of the USSR. A purely ground network on this scale would involve substantial engineering and logistics problems, because of the vast distances and difficult weather conditions.

The Orbita network operates for some 16 to 20 hours a day.

(E) Postal Services

The first effort after the revolution was to improve the postal services in the country areas. In 1969 there were 80,000 post offices operating in the USSR, including 59,000 in the villages. For long distances, the bulk of letters and newspapers is carried by plane.

Mechanised sorting methods are being introduced in Moscow, Chelyabinsk, Kiev, Tbilisi, Murmansk, Vladivostok and 15 other cities.

(F) Research and Development

Most of the communications research institutes, design bureaux and so on are administered by the Ministry of Postal Services and Communications.

18(1) LENINGRAD ELECTRICAL ENGINEERING INSTITUTE OF COMMUNICATIONS (LEIS)

The Leningrad Electrical Engineering Institute of Communications im. M.A.Bonch-Bruyevicha is one of Russia's leading communications research institutes. Areas of research include stereoscopic colour television and computer design and construction. The institute has developed a TV microscope operating in both the visible and invisible spectra, a high-speed storage device using condensers and semiconductors, a photo-television unit for submarine and subterranean observations, and an apparatus for direct transmission of photo telegrams.

18(2) MOSCOW ELECTRICAL ENGINEERING INSTITUTE OF COMMUNICATIONS (MEIS)

The Moscow Electrical Engineering Institute of Communications specialises in radio relay communication lines, radio communications and broadcasting, TV, automation and mechanisation of production processes, telephony and telegraphy, and long-range communications. The institute maintains a number of shops for carrying out contract work and for developing simpler types of communications equipment. Such equipment has included a 'ballistic' aerial, an electronic telegrapher operating on small batteries, an automatic magnetic tape symbol printing apparatus, and an ultra-small semiconductor receiver powered by flashlight batteries.

18(3) NOVOSIBIRSK ELECTRICAL ENGINEERING INSTITUTE OF COMMUNICATIONS

The Novosibirsk Electrical Engineering Institute of Communications has carried out research into the development of an electronic telegraph apparatus and a TV-synchronising generator using semiconductors and ferrites.

18(4) SCIENTIFIC RESEARCH INSTITUTE OF URBAN AND RURAL TELEPHONE COMMUNICATIONS

The Scientific Research Institute of Urban and Rural Telephone Communications is situated in Leningrad. It carries out research into switching devices, automatic control, electrophotography, facsimile systems, and high frequency studies related to ferromagnetic materials.

These four institutes come under the administration of the Ministry of Postal Services and Communications. However, some research related to communications problems falls outside that Ministry's research network. The *Moscow Physical-Technical Institute* and the *Tomsk Polytechnical Institute*, both of which come under the Ministry of Higher and Secondary Specialised Education, carry out research into radio communications and TV broadcasting respectively.

According to the Minister of the Radio Industry, one of the most important lines of research is into problems of broadcasting on millimetre and sub-millimetre wavelengths. Laser communications systems are also being studied.

Journals: Journals of research include *Vestnik Sviazy* (Journal of Communications); *Iz. VUZ. Radioelektronika* (Proceedings of the Higher Educational Establishments. Radioelectronics); *Radiotekhnika* (Radiotechnology) and *Radiotekhnika i Elektronika* (Radio Technology and Electronics).

(G) Plans and Policy

During the present five-year plan (1971-75) particular attention is being paid to the development of the space communications network, the construction of radio-relay and cable lines, and the reconstruction and technical modernisation of existing television centres. Most of these will be equipped with video-tape equipment which should make it easier to accommodate the regional time differences when relaying important programmes. It is planned to extend the transmission of colour TV programmes to another 17 cities.

19 Patents, Information, Libraries and Museums

(1) PATENT SERVICES *

(A) Introduction

The first patent law in Russia was passed in 1812. In 1919 after the Bolshevik Party came to power, the state decreed that inventions belong to society. This principle provides for the unimpeded circulation of patent information. Another specific feature introduced by the government into the patent system is a concern for the practical utilisation of inventions.

However, in practical terms, there was no system of patent information organised on a national scale until the beginning of the 1960s. Since then the development of a nation-wide organisation has proceeded rapidly.

(B) Organisation

19(1) COMMITTEE FOR INVENTIONS AND DISCOVERIES

The Committee for Inventions and Discoveries is the leading organisational body in the patent field. It is attached to the Council of Ministers.

The range of the committee's responsibilities is somewhat wider than that of many patent offices in the capitalist world. Its duties include: the provision of would-be inventors with information; assisting them in making their application; searching the applications for novelty; examining the applications for patentability; publishing the specifications and other information normally found in official patent gazettes; passing on information of authors' certificates with a technical recommendation to the appropriate State industrial enterprises; following through the use or otherwise of the inventions by the enterprises, and where use is made, ensuring that the inventor is rewarded; in certain areas where the inventions do not fall easily into the fields of established enterprises, the committee organises technical development work before passing them onto the appropriate enterprise; guiding the filing of patent applications abroad; co-operating with the USSR Chamber of Commerce (which acts as the professional patent agent for applications for

---

* Much of the information in this section, particularly parts (B) and (C), is based on Mathys: *A Commentary on a Visit to The Committee of Inventions and Discoveries under the USSR Council of Ministers in Moscow*, February 1968; and Artemies: "The System of Patent Information and Patent Services in the Soviet Union", *Industrial Property*, PP 222-226, 16 September 1969.

Soviet patents by foreigners and for foreign applications by Russians); co-operation with the All-Union Soviety of Inventors and Rationalisers; issuing orders, instructions, regulations and clarifications; and settling disputes.

The Committee for Inventions and Discoveries is developing a number of Industrial Patent Departments in the main industrial centres. There were some 25 in 1969. Each is provided with a comprehensive library of Soviet and foreign patent literature and with a staff of searchers, information officers and patent lawyers. They have close links with industry and are there to provide the industrial enterprises with information and to assist individual inventors by making searches, helping inventors prepare their applications, and following these applications through.

19(2) PATENT

Patent is a polygraphical organisation. It was started in 1964 with a staff of 40 and has expanded rapidly so that by 1969 its staff numbered 1,500, of whom 700 were workers and 800 specialist engineers. There are 11 branches in Moscow and 24 in other cities attached to the Industrial Patent Departments. All these are connected by teletype.

Patent microfilms patent specifications and other patent literature of the world, and keeps its branches and other customers (numbering some 20,000) informed. Every year it passes on some 80,000,000 copies of patent specifications, and in 1968 passed on 250,000,000 documents altogether. These numbers will fall, however, as the various libraries are built up.

Anyone may request Patent to prepare an application for an industrial design, trademark or invention, to undertake translations, to make drawings and so on. It provides considerable assistance to authorities in preparing the necessary documents with a view to patenting Soviet inventions abroad.

19(3) ALL-UNION SOCIETY OF INVENTORS AND RATIONALISERS

The All-Union Society of Inventors and Rationalisers is a trade union organisation, separate from the main committee for Inventions and Discoveries. It is financed by contributions from all industrial enterprises which use inventions and innovations. It has 50,000 branches spread across each state, region, city and industrial enterprise, with some 4,500,000 members. The society helps its members to draft specifications, and the like, it will argue with the Patent Office free of charge, and will sometimes give financial help to assist the early development of an invention. It also supervises the introduction of inventions and ensures that authors receive the full remuneration due to them.

19(4) USSR CHAMBER OF COMMERCE

The Chamber was founded in 1922, and is independent of the Committee for Inventions and Discoveries. It is not a government organisation but self-selecting. The function of the chamber is to facilitate overseas trade, for example, by exhibitions and trading centres. It also acts as an agent for all Russian patents applied for in foreign countries and for all foreign applications for Russian patents.

The chamber has 70 full-time and more than 30 part-time experts employed in two departments, one for incoming and one for out-going applications. In 1968 there were 4,500 incoming applications and 4,400 out-going applications. The staff comprise technical, legal and translating personnel, and they are also training up patent lawyers.

19(5) ALL-UNION TECHNICAL PATENT LIBRARY (VTPB)

The All-Union Technical Patent Library, situated in Moscow, holds the basic patent resources of the country. Before the present development drive, it was virtually the only patent library in the whole country, but the newly-formed libraries of the Industrial Patent Departments now hold large duplicate collections of patent information.

The all-Union Technical Patent Library has a staff of 500, and is at present increasing its collection by some 300,000 descriptions of Soviet and foreign inventions each year, as well as with other items of patent information and scientific and technological literature. The library works in close co-operation with Patent, which can supply copies of the library's documentation.

19(6) PATENT EXAMINATION RESEARCH INSTITUTE OF THE USSR (Patent Office)

The Patent Examination Research Institute of the USSR was founded in 1960 to take over the work previously done by a small staff of the Committee. It has now 1,500 employees, of whom 1,000 are technical experts and 500 are on the administrative side. The job of the institute is to examine patents, trade marks and industrial designs.

The institute has two deputies, one is an expert on examination, and the other an expert on general questions. Applications are subject to a pre-examination for formalities and classification, before full examination in one of the institute's 23 examining branches. Apart from the branches, the institute also has functional departments covering such subjects as methodology, organisation and service. It deals with more than 100,000 applications a year.

19(7) CENTRAL SCIENTIFIC RESEARCH INSTITUTE OF PATENT INFORMATION AND TECHNICAL-ECONOMIC STUDIES

The Central Scientific Research Institute of Patent Information and Technical-Economic Studies was established in 1963. It is the central body in charge of patent information, and as such is one of the key departments of the Committee for Inventions and Discoveries. It prepares descriptions of inventions and other Soviet and foreign information for publication. The actual printing is done by Patent. It also issues and circulates to its subscribers information for practical use, or ones that have already been successfully introduced into the production process.

The institute also conducts research with the aid of computers and by other means into ways of simplifying the transmission of information and its co-ordination and use. A particular aim is to highlight the economic effects of the Soviet patent system on USSR industry. A third function of the institute appears to be to act as an "organisation and methods" department for all the relevant activities under the Committee for Inventions and Discoveries.

19(8) LICENSINTORG

Licensintorg is the All-Union Export-Import Association which specialises in commercial operations in the field of licences for various inventions and scientific and technical achievements. It is an independent organisation carrying out its work on a self-supporting basis. It has the right to conclude trade agreements with foreign organisations and firms both in the Soviet Union and abroad. It also has the right to carry out various juridical actions connected with fulfilment of its operations. Licensintorg was set up in 1964.

19(9) COUNCIL OF EXPERTS

The Council of Experts is set up by the Committee for Inventions and Discoveries, and its main function is to settle disputes between applicants and the committee over the granting of author's certificates and patents, and protests by other interested parties against the grant of protection. Its decisions become valid after endorsement by the deputy-chairman or chairman of the committee.

The council has other functions such as making recommendations on important inventions, but 70 per cent of its work is on disputes and protests.

(C) Publications

Brief, preliminary information on Soviet inventions is published in the official bulletin of the Committee for Inventions and Discoveries, which comes out three times a month. The bulletin is called *Discoveries Inventions Industrial Designs Trademarks* and was first issued in 1924. It publishes the formula or claims of every discovery and invention as entered in the USSR State Register, as well as information in industrial designs and trademarks. It was intended that the bulletin should carry an appendix in English from around 1970.

A guide to the previous year's issues of the bulletin is published in five volumes. A brief handbook is also published called *Bibliographical Guide to Patents Valid in the USSR*.

Since 1968, a quarterly publication entitled *Introduction of Inventions* has been issued. This contains brief information on new solutions of particular technological problems which are already in use in industry, and the economic results obtained.

(D) Patent Procedure

The most widespread practice is to issue Soviet inventors with a Certificate of Authorship, not a patent. In accepting this certificate the inventor makes over the invention to the State and in return receives a payment after his invention has been introduced into the economy, based on the economic return which it gives. Payments are levied on the enterprise, or enterprises, benefiting from the invention; they also have to pay a sum equal to 35 per cent of the basic payment they make into a "bonus fund for assistance in the introduction of inventions and rationalising proposals", which is transferred to the appropriate organisation or managed by the enterprise as appropriate. The minimum payment an inventor may receive is 20 roubles and the maximum is 30,000 roubles. There is a basic scale for payments expressed as a percentage of the economic return. However, this basic scale can be increased in certain fields so as to stimulate inventions of a particular type. In addition to cash received, an inventor can be awarded a higher degree for his invention and may also be entitled to additional accommodation on a par with scientists.

The formal procedure for applications is the same whether for a patent or for an author's certificate, except fees are charged for the first and none for the second. Most Russians apply for the author's certificate, and most foreigners for patents. The individual inventor can be helped in his application by three organisations: the USSR Chamber of Commerce, the All-Union Society of Inventors and Rationalisers, and Patent. When an invention is made as part of the normal duties of the inventor at his place of work, the organisation concerned has to apply for the certificate of authorship, this is then granted in the name of the organisation with the inventors' names inserted.

Only some 32 per cent of Soviet applications are granted. The main reason for this is that the granting of author's certificates put an obligation on the Committee for Inventions and Discoveries to urge the use of the invention in Soviet industry. In 1968 110,000 new applications were made for author's certificates and patents; 70,000 applications were processed; 32 per cent of the applications were granted; and 20,000 inventions were put into industrial use (some of these came from author's certificates granted prior to 1968, owing to the time taken in processing applications).

Disputes between applicants and the Committee for Inventions and Discoveries over granting of author's certificates and patents, and protests by other interested parties against the grant of protection are dealt with by the Council of Experts. In certain circumstances disputes can be taken to the Courts.

(E) Research and development

The Soviet Patent Office (the Patent Examination Research Institute of the USSR) devotes particular attention to improving the professional standards of patent specialists. The Committee for Inventions and Discoveries operates permanent courses for the training of leading specialists, engineers and technologists engaged in patent and invention matters. More than 30,000 people from different organisations had attended such courses by 1969. In 1968 the Committee set up a central institute to provide further (and more intensive) training of responsible staff and specialists in economics as it concerns patent work.

With regard to the actual quality of inventions and their contribution to the development of industry: "the system appears to work reasonably well in the case of "service" inventions, which are normally part of the plans of the Ministry concerned or satisfy the needs of a particular factory. But the encouragement offered to the independent inventor by these arrangements is somewhat limited. He depends for his success on getting his invention included by a Ministry of factory in its plan, and has no simple way of trying out his invention for himself in factory conditions. The Committee for Inventions and Discoveries is required to assist the application of inventions by making recommendations about the use of inventions to appropriate Ministries, which then consider them for inclusion in their plans. But this procedure is necessarily clumsy." Source: Zalenski et al: *Science Policy in the USSR*, P 476, (Paris, 1969).

(II) INFORMATION SERVICES *

The scientific and technological information service is organised on both a centralised and regional basis. The most important centralised information body is the All-Union Institute for Scientific and Technical Information.

19(10) ALL UNION INSTITUTE FOR SCIENTIFIC AND TECHNICAL INFORMATION (VNITI)

The All-Union Institute for Scientific and Technical Information is attached to the Academy of Sciences of the USSR and its purpose is to process information on the natural and applied sciences, and on the humanities. Departments within the institute cover separate subjects such as mathematics, mechanics, physics, metallurgy and so on. Each department processes the published information concerning its subject. It also selects material for publication in the Institute's *Abstract Journal* (RZ). The Institute receives more than 23,000 publications from 102 countries each year. Its archives number more than a million documents and constitute the most important collection of foreign literature in the USSR.

In addition to its *Abstract Journal*, the Institute publishes express information series, which carry shortened versions of the most important articles of descriptions of the particular subject concerned put out during a certain period. It also produces bibliographical filing cards. The aim is that eventually all the information references will be transferred to a computer information retrieval system, which will have terminals throughout the country.

---

* This section is largely based on *Science Policy and Organisation of Research in the USSR*, P 48, (UNESCO, Paris, 1967).

VINITI carries out scientific research into methods of storing and retrieval of scientific and technological information, including ways of mechanisation, processing, selection and long-distant transmission of information. In 1965 it began publishing an Abstract Journal on scientific and technical information.

The institute carries out orders for photocopies, microfilms and translations of foreign sources into Russian. It has a special translation bureau to deal with this side of its activity.

Institutes for Scientific and Technical Information operate within the union republics. There they are the main information centre co-ordinating the work being done within individual enterprises and other organisations within the republic. They control the use of information materials, and publish collections of information on important industries. There are a number of other centralised information bodies.

19(11) CENTRAL INSTITUTE FOR SCIENTIFIC INFORMATION ON CONSTRUCTION AND ARCHITECTURE (CINIS)

The Central Institute for Scientific Information on Construction and Architecture was formed in 1943. As its name implies it processes information on scientific research in the fields of construction and architecture.

19(12) ALL-UNION SCIENTIFIC RESEARCH INSTITUTE OF MEDICAL AND MEDICO-TECHNICAL INFORMATION

The All-Union Scientific Research Institute of Medical and Medico-Technical Information was organised in 1965. It developed out of the department of Scientific Medical Information of the USSR Academy of Medical Sciences. The institute publishes a *Medical Abstract Journal* which covers information from over 2,500 periodicals, and is also collecting material on the development of the medical sciences. It controls the activities of all information bodies within the public health service.

19(13) ALL-UNION SCIENTIFIC RESEARCH INSTITUTE FOR TECHNICAL INFORMATION, CLASSIFICATION AND CODIFICATION (VNIKI)

The All-Union Scientific Research Institute for Technical Information, Classification and Codification was created in 1965. It is administered by the Committee of Standards, Measures and Measuring Instruments (attached to the Council of Ministers), and it published information concerning standards and norms.

In addition to these institutes of information, most specialised branches of industry are served by some kind of central specialised information body. For example, there are specialised information organisations for rail transport, the electrotechnical industry, tool construction, the food industry and so on. All these organisations have their central specialised scientific, and technical libraries, they process information on development in the USSR and abroad within their field, and they carry out research into ways of improving the information services.

There are three kinds of regional information bodies:

(i) The *department or bureau of scientific and technological information* is attached to a scientific research institute, design office or project organisation. Its job is to search for information concerning the research work of the organisation to help in the exchange of ideas and methods between different departments of the same institute or between different institutes, and to prepare information about the results of the institute's research work for the central information bodies.

(ii) The *bureau or department of technological information* is attached to an industrial enterprise. Its main job is to provide information and patent material on the latest developments in the particular branch of industry concerned, to encourage the introduction of new techniques, and to prepare information for the central information bodies. The bureau appears to have a quite active role in encouraging the introduction of new ideas into industry. It works closely with the All-Union Society of Inventors and Rationalisers (see Section 19(3), organising meetings and seminars on new production techniques. It uses film, radio and the local press to publicise advanced production methods, and publishes its own information bulletins and posters.

(iii) The *Central Bureaux of Technical Information* (CBTI) are local information organisations. They are concerned with the efficient use of the available information and reference materials, and with the dissemination of information on advanced production and technological methods.

(III) LIBRARIES *

In 1970 there were 370,000 libraries in the Soviet Union. Of these 51,000 were scientific, technical and other specialised libraries, 182,000 were school libraries, 127,000 were public libraries and 50,000 were children's libraries. The largest library in the whole country is the Lenin Library of the USSR in Moscow. Each republic and autonomous republic also has at least one large library in its principal city.

The scientific, technical and specialised libraries include those of the Academy of Sciences of the USSR and Republican academies of sciences, the universities, higher educational institutes and research institutes. The State and trade unions provide the necessary finance to maintain the libraries and pay staff salaries.

The central libraries, apart from the Lenin Library already mentioned, include:

(i) Ushinsky State Scientific Library of the Academy of Pedagogical Sciences in Moscow, which covers all fields of education and pedagogics, and has 1,450,000 volumes.

(ii) State Central Scientific Medical Library in Moscow has over 1,450,000 volumes.

(iii) All-Union Geological Library in Leningrad, has about 939,019 books, monographs, periodicals and maps.

(iv) All-Union Patent and Technical Library in Moscow (see section 19(5)).

(v) State Public Scientific and Technical Library in Moscow is attached to the All-Union Institute for Scientific and Technical Information. It houses 5,500,000 books, periodicals and documents and has special collections of industrial firms' catalogues, and scientific translations. It co-ordinates bibliographical activities of 20,000 technical libraries throughout the country.

(vi) State Central Polytechnic Library in Moscow has 2,500,000 volumes.

(vii) State Historic Library, in Moscow.

(viii) State Foreign Languages Library, in Moscow.

---

* Information in this section is taken mainly from *The World of Learning 1969 70*, 20th Edition, (Europa Publications, London, 1970).

Libraries of the Academies of Sciences include:

(a) *Library of the Academy of Sciences of the USSR* in Leningrad. The Library has 14,800,000 volumes, collections of manuscripts, incunabula and early works on the natural sciences and mathematics. It acts as inquiry, loan and reference centre for the network of 100 libraries attached to the academy's departments and institutes.

(b) *State Public Scientific and Technical Library* of the *Siberian Department of the USSR Academy of Sciences* in Novosibirsk. The library has 5,000,000 volumes and acts as co-ordinating centre for 64 academy institutes situated in Siberia and the Far East.

(c) *Central Scientific Agricultural Library of the All-Union Lenin Academy of Agricultural Sciences* in Moscow. The library houses 2,500,000 volumes, and acts as a co-ordinating centre for 1,300 libraries in agricultural research institutes, laboratories and experimental stations.

(d) *Fundamental Library of the Academy of Medical Sciences*, in Moscow. It has about 500,000 volumes and acts as a co-ordinating centre for 42 libraries in the institutes and laboratories of the Medical Academy of Sciences.

Scientific libraries are attached to each of the 14 Republican academies of science.

Libraries attached to polytechnic and industrial higher educational institutes include:

(1) *Central Scientific Library of the Moscow Timiriazev Agricultural Academy* in Moscow. The library's collection includes rare books on agriculture and the private collections of many prominent agriculturalists.

(2) *Library of the Kharkov Lenin Polytechnic Institute* in Kharkov. The library's collection includes autographs by prominent nineteenth century scientists and charts and drawings of the Donetsk coal basin by the brothers Nosov.

(3) *Scientific Library of the Kiev Polytechnic Institute* in Kiev. The library has complete files of Russian and foreign technical periodicals since about 1900; its special interests are chemistry, chemical technology, mining and metallurgy.

(4) *Scientific Library of the Lvov Polytechnic Institute* in Lvov. The library has more than 625,000 volumes including rare editions on mathematics.

(5) *Fundamental Scientific and Technical Library of the Moscow Bauman Higher Technical School* in Moscow. The library has more than 1,287,000 volumes including complete files of Russian and foreign technical magazines.

(6) *Library of the Riga Polytechnic Institute* in Riga. The library has a good collection on engineering, mechanics and architecture.

(7) *Scientific Library of the Tomsk Polytechnic Kirov Institute* in Tomsk. The library has 628,000 volumes including complete files of Russian and foreign technical magazines.

(8) *Library of the Ural Kirov Polytechnic Institute* in Sverdlovsk. The library has more than 1,548,000 volumes including rare eighteenth century volumes as well as contemporary Soviet and foreign scientific magazines and books.

(IV) MUSEUMS *

There are some 30 museums devoted to the natural sciences and technology. The majority of these are in Moscow and Leningrad.

---

*Information in this section is taken mainly from *The World of Learning 1969-70*, 20th Edition, (Europa Publications, London 1970).

(A) Natural Science

   (i) *Museum of the Arctic and Antarctic* in Leningrad. It has documents of and the original equipment of all Soviet expeditions to these areas.

   (ii) *Chernyshev Central Scientific Geological and Prospecting Museum* in Leningrad. It has 1,000,000 geological specimens from all over the Soviet Union.

   (iii) *Mining Museum of the Plekhanov Mining Institute* in Leningrad. It traces history of the mining industry in the nineteenth and early twentieth centuries.

   (iv) *Dokuchayev Central Soil Museum* in Leningrad. It contains about 4,000 specimens of soil from nearly every soil zone in the world.

   (v) *Anuchin Anthropological Institute and Museum* in Moscow. It contains collections of Russian explorers in Africa as well as items found on Soviet territory.

   (vi) *Timiriazev State Museum of Biology* in Moscow. It has some 30,000 exhibits tracing the origin and evolution of life on Earth.

   (vii) *Museum of the Earth Sciences of Moscow State University*. It covers geography, geology and biology of the Earth and is used for teaching in the university.

   (viii) *Fersman Mineralogical Museum of the USSR Academy of Sciences* in Moscow. It contains 110,000 mineral samples from all over the world.

   (ix) *Pavlov Museum of Geology and Palaeontology* in Moscow. It was founded in 1861 and has 300,000 exhibits.

   (x) *'Belovezhskaya Pushcha'Museum* in Byelorussia. The museum is attached to a game reserve and shows the work being done to preserve the almost extinct European bison.

   (xi) *Ilmen Mineral Preserve Museum* in the Cheliabinsk region. The museum illustrates the mineralogical wealth of the Ilmen Preserve which contains almost all known minerals.

   (xii) *Stradinia History of Medicine Museum* in Riga. The museum illustrates the history of medicine and modern developments.

   (xiii) *Fisheries Museum of the Pacific Scientific Research Institute of Fisheries and Oceanography* in Vladivostok. The museum contains more than 11,000 exhibits of fauna and flora of Pacific Ocean.

(B) Technology

   (i) *Popov Central Museum of Communications* in Leningrad. The museum has more than 4,000,000 exhibits including the State postage stamp collection.

   (ii) *Industrial Exhibition at the House for Dissemination of Scientific and Technical Propaganda* in Leningrad. The museum features the machine building and instrument manufacturing industries in Leningrad.

   (iii) *Leningrad Museum of Railway Transport*. It traces the history of railways in Russia and has a collection of miniature models of engines and carriages.

   (iv) *All-Union Permanent Exhibition of Labour Protection* in Moscow. It exhibits the latest safety techniques in industry and transport.

   (v) *Museum of History of Microscopy* in Moscow.

   (vi) *Pharmaceutical Museum of the Central Drug Research Institute* in Moscow.

   (vii) *Polytechnical Museum* in Moscow. The museum was founded in 1872 and has about 25,000 exhibits on the history and latest developments in science and technology.

(viii) Zhukovsky Memorial Museum in Moscow. It shows the work of N.E Zhukovsky and other Soviet contributions to aviation and astronautics.

(ix) Tsiolkovsky State Museum of the History of Astronautics in Moscow. It contains Tsiolkovsky's works together with exhibits on astronautics, aviation and rocket techniques, including the first experimental rockets launched in 1933, the Sputniks and Luniks, and Vostock.

In addition to these museums there are 10 botanical gardens at Leningrad, Riga, Kishinev, Moscow, Tashkent, Minsk and Kiev, and eight zoological gardens (containing more than 1,000 specimens) at Yerevan, Kiev, Leningrad, Moscow, Nikolaev, Tallin, Tashkent and Tbilisi.

20 The Regions

I Russian Soviet Federal Socialist Republic

The Russian Soviet Federal Socialist Republic is the largest of the Union Republics both in territory, population and economic strength. It occupies the eastern region of the European part of the Soviet Union and the northern region of the Asian part covering 17,100,000 sq. km. (or three quarters of the total area of the USSR). In 1970 the population of the Russian Federal Republic was 130,079,000, 54 per cent of the total population of the USSR,(241,000,000). The administrative capital is Moscow.

(A) Natural Resources

The Russian Federal Republic is very rich in mineral deposits, and its fuel and energy resources are particularly important to the Soviet economy as a whole. It contains nearly 70 per cent of the country's coal deposits; the main basins are the Kuzbas, the eastern wing of the Donbas, the Moscow basin, the Pechorsky basin, the Urals, Eastern Siberia and the Far East. Vast deposits of natural gas and oil have been discovered, the main ones in the Volga-Ural region and more recently around Tyumen in Western Siberia and in the Northern Caucasus. The Russian Federal Republic has 92 per cent of the peat resources of the USSR. There are large deposits of bituminous shale in the Volga region and the western part of the Leningrad region. Its iron ore reserves are among the largest in the country, the main deposits opened up since the revolution being in the Kursk Magnetic Anomaly, the Urals, the North-West, West and East of Siberia The Urals and Eastern Siberia are rich in non-ferrous and precious metals. The Republic has good deposits of non-metal minerals such as potassium salts, apatites, phosphorites and so on. A large diamond field has now been discovered in Yakutia. Timber is an important asset, the Russian Federal Republic having 96 per cent of the total resources in the country.

(B) Economy

The economy of the Russian Federal Republic has altered radically since the revolution. Production of heavy industry has increased some 162 times since 1913, whereas the production of consumer goods has only grown 22 times. In 1966, 56.7 per cent of the republic's income came from industry, 17.6 per cent from agriculture, 8.8 per cent from construction, 6.1 per cent from transport and communications, and 10.8 per cent from other sources.

(C) Industry

The Russian Federal Republic accounts for nearly 75 per cent of the total coal production in the USSR. The Kuzbas is the most important coal basin, followed by the eastern part of the Donbas, the Pechorsky basin, and the Moscow basin for brown coal. The republic accounts for nearly 80 per cent of total oil production in the USSR. At present the main oil field lies between the Volga and Urals and this produces 64 per cent of the country's oil production. Other worked oil fields are in the Tatar and Bashkir Autonomous Republics and the Kubyshev region; the giant fields round Tyumen are being developed rapidly. Saratov is the main area where natural gas is obtained. Gas is also found in the north, the Urals, Far East, and again now in the Tyumen region. Shale is used as a fuel and for the chemical industry. The republic accounts for 25 per cent of total USSR production in the shale industry. The industry is concentrated in the north-west Leningrad region and in the Volga basin. The Republic has large reserves of peat, which is used as fuel for power stations, as fertiliser in agriculture, as a raw material in the chemical industry, and in the construction industry. The fuel balance of the republic has altered radically over the past twenty years; oil has taken the lead over coal, and the production of natural gas is growing fast. The Russian Federal Republic produces some 67 per cent of total electric power output in the country. Hydroelectric power stations are responsible for 18 per cent of this output.

Since the revolution the republic has developed a powerful metallurgical industry. In 1966 it was responsible for 46 per cent of the iron, 54 per cent of the steel and 54 per cent of rolled stock production of the whole country. The Urals is the main centre of the ferrous metallurgy industry, with the large plants at Magnetogorsk, Chelyabinsk, Nizhny Tagil and Novo-Troitska. The Urals has good iron ore reserves of its own, ores are also brought in from Kazakhstan, and its power comes from the Kuzbas and Karaganda regions. The second important metallurgical centre is based around the Kuznetsk metallurgical plant in Western Siberia. The central regions of the republic, around Moscow, Tula, Lipetsk and Gorky occupy third place. They use iron ore from the Kursk Magnetic Anomaly.

The non-ferrous metallurgical industries are well developed. Production of copper is centred in the Urals, zinc and lead in the northern Caucasus, lead in the Primorsky region, zinc in the Urals and Western Siberia, nickel in the Kola Peninsula and Urals, nickel together with copper in the Krasnoyarsk region. Aluminium production is developing in the Krasnoyarsk-Irkutsk region based on the cheap supplies of electric power and large reserve of nepheline.

The Russian Federal Republic has an important engineering industry. In 1966 it produced, as a percentage of the USSR's total production, 54 per cent of the metal cutting tools; 50 per cent of metallurgical equipment; 65 per cent of chemical equipment; 71 per cent of generators and turbines; 55 per cent of freight wagons; 71 per cent of passenger carriages; 84 per cent of cars; 42 per cent of tractors; and 100 per cent of combine harvesters.

The chemical industry of the republic has grown up on the basis of large reserves of raw materials and the side products of various industries, such as the timber, ferrous and non-ferrous. There are large fertiliser-producing plants in the Moscow, Leningrad and Sverdlovsk districts.

The timber and woodworking industry is very important. It is centred in the North-West, the Urals and Siberia (particularly the East). The republic accounts for 91 per cent of the cellulose and 83 per cent of the paper production of the USSR.

The Republic has a well developed light industry, in particular textiles. Here production is based in the central regions round Moscow, and in the North-West round Leningrad. In the food industry there has been increased production particularly of granulated sugar, vegetable oils, conserves and fish.

(D) Agriculture

The Republic is important for its grain production (mainly wheat, rye and barley). Potato and vegetable production is growing particularly in connection with the development of new industrial centres. Flax and sugar beet are other important crops. Animal farming includes beef and dairy cattle, pigs, sheep and poultry.

(E) Development

The Russian Federal Republic is a very vast area, covering 17,100,000 sq. km., but much of its territory is as yet undeveloped. The Far North, Western and Eastern Siberia, and the Far East all have large reserves of power and mineral resources, but development has been hindered by the very severe climate, poor communications, difficult living conditions and inadequate population. However, all these areas have tremendous potential and great efforts are being made to develop them.

(i) Far North

The Far North and equivalent regions include the Murmansk Oblast, two-thirds of Arkhangelsk Oblast, almost all the Tyumen Oblast and the Krasnoyarsk Krai, two-thirds of the Komi Autonomous Republic, the Irkutsk Oblast, the Khabarovsk Krai, the Yakut Autonomous Republic, one quarter of the Amur Oblast, and the Magadan and Kamchatka Oblasts. They cover altogeher some 10,000,000 sq.km., or nearly half the entire area of the USSR.

The region is well provided with natural resources: over three-quarters of the country's coal reserves, a considerable proportion of its non-ferrous metal (including gold) reserves and over 12,000,000,000 tons of iron ore. There are huge diamond and chemical deposits in the Yakut Autonomous Republic. High grade petroleum has been found both on and off shore. The region's forests account for three-quarters of the USSR's timber resources, and the hydroelectric power resources are 800,000,000,000 kWh per year, or more than half the country's total.

During Stalin's time the Far North was developed on a highly localised and specialised basis of industrial complexes. There are petroleum works at Ukhta, coal at Pechora, non-ferrous metals at Norilsk, timber at Igarka and Maklakovo-Yeniseisk, gold and complex ores at Mama-Chuya, gold and non-ferrous metals at Kolyma and Chukotsky, gold at Aldam, complex ores at Yana-Inidgirka and diamond complexes at Mirny.

Twenty cities have developed in the area during the Soviet period. The biggest is Norilsk with a population of 124,000 followed by Magadan with 81,000 and Vorkuta with 63,000.

Development of the area is hindered by the severe climate which can drop to as low as $-75^{\circ}C$ at Yakutsk, very poor communications and transport facilities except for aeroplanes, inadequate buildings, and construction work, and the generally difficult living conditions which make it hard to encourage people to settle. In fact there appear to be two opinions on the question of settlement of the area. At a conference held at Magadan in 1965 views were divided between those who thought that a permanent population was not required, qualified workers being brought in for limited periods of time, and those who thought that permanent settlement was necessary. Whichever view prevails it is clear that people working in these areas will have to be offered the best possible living and working conditions and other material incentives.*

(ii) Western Siberia

During the 1960s the world's largest oil and gas deposits were discovered in Western Siberia, and this event has changed the whole character of this region. The Tyumen gas bearing province contains the three largest gas deposits in the world - the Urengoi on Pur River, the Zapolyarnoye and the Medvezhe. The joint capacity was estimated in 1969 to be more than 9,000,000,000,000 cubic metres. Altogether 32 gas deposits have been discovered in the Western Siberian plain.

---

*Source: Vredenskii, G.A.: The Development of the Soviet Far North, *Bulletin of the Institute for the Study of the USSR*, vol XIV, no 2, P 27.

Forty oil-bearing deposits have been found. Most of them are located near the surface and have a low sulphur content. One of the largest is at Samotlor on the Ob River. At the moment the West Siberian oil industry's output is relatively small compared with the country's as a whole - 21,300,000 tons in 1969 out of a total output of 328,000,000 tons. However, the rate of growth is highest in Western Siberia and within five years this new region should overtake the country's oldest oil-bearing area around Baku. Experts have estimated that the Tyumen Oblast will eventually be able to produce up to 500,000,000 tons of oil a year.

The difficulties of developing these regions are again the climate and more particularly the boggy terrain and lack of communications or any close industrial centres. At present 60,000 people are employed in the region. New towns are emerging, pipelines are being laid, roads are being built and power lines are being set up. In Tyumen educational facilities are being developed to train the specialist personnel needed for the area. An *Institute of Engineering and Construction* carrying out both training and research was opened in 1970; a new industrial institute is turning out engineers for geological organisations and the oil and gas industries.

Another problem is the development of associated industries such as oil refining and petrochemicals, and electricity generation. None of these exists at the present.

(iii) Eastern Siberia

Development of Eastern Siberia is centred round the Irkutsk-Bratsk-Krasnoyarsk industrial region. There is a wealth of power, both hydroelectric from the Angara, Yenisei and Lena rivers, and thermal based on the rich coalfields in the area. Iron, steel and aluminium industries are being developed on the basis of local raw materials - rich magnetic iron ore at the Rudnaya (Ore) Mountain some 100 miles north east of Bratsk; and nepheline from Goryachegorsk, south west of Krasnoyarsk. It is claimed that the world's largest timber industry is being developed in the Yenisei-Baikal area using electricity from the Bratsk power station. Products will be timber, viscose, cellulose, cardboard, plywood, alcohol, yeasts, and so on. Current plants allow for the development of the heavy engineering and chemical industries. The heavy engineering is needed to provide the machinery for industrial expansion, and the chemical is needed to provide fertiliser for the not-very-fertile soils of the region. It is hoped that the area will be able to produce at least part of the food needed by the population of the developing industrial towns.

(iv) The Far East

The Soviet Far East has had a slower economic growth rate, both before and after the Second World War, than other regions of the USSR. Notable developments since the war include the establishment of a modern coalmining industry on the Bureya coalfield and the enlargement of the Komsomolsk metallurgical works. However, it seems likely that major industrial developments in the area will have to wait until regions further west - Western and Eastern Siberia - have become more firmly established as industrial centres.

Industrial development of the Far East can stray into the ticklish area of politics. The region shares the basin of its main river - the Amur - with China, and so plans for hydroelectric power stations and flood control will have to be worked out on a joint basis. *

(F) Education

The Russian Federal Republic has some of the largest higher educational establishments in the country. In 1969/70 there were 454 VUZy of which twenty were universities, 128 were industrial, polytechnic and construction institutes, 107 were pedagogical institutes, forty six medical, seven drama institutes, seven conservatoires, and so on. These VUZy

---

* Source: Gregory: *Russian Land, Soviet People*, PP 663, 674, Harrap, London, 1968.

catered for some 2,655,800 students, with just under half attending full time day courses.

In the republic as a whole there were 7,700,000 specialists working, of whom 3,000,000 had higher education and 4,700,000 secondary education.

(G) Scientific Research*

The Russian Federal Republic plays a leading role in scientific research. The Academy of Sciences of the USSR is centred in Moscow; there is its Siberian branch in Novosibirsk and also four filials in Dagestan, Kolskii, Komi and Ural districts. Other all-union academies for agriculturs, medicine, pedagogy and so on are also situated in the republic. In 1966 there were 2,527 science establishments altogether, with 488,700 scientific workers working in them. These included 11,600 doctors of science and 100,400 candidates of science. Apart from Moscow, the main science centres are Leningrad, Novosibirsk, Tomsk, Sverdlovsk, Gorky, Kazan, Voronezh, Saratov, and Irkutsk.

II The Ukraine

The Ukrainian Soviet Socialist Republic was formed in December 1917. It is situated in the south-western part of the European region of the Soviet Union. It is bordered by Poland and Czechoslovakia on its west, by Hungary and Romania on its south-west, and by the Black and Azov Seas on its south. To the east is the Russian Federal Republic, and to its north the Byelorussian Republic. The Ukraine is 601,000 sq.km. in area and has a population of just under 47,126,000. Just over half (some 52 per cent in 1966) live in the towns. This compares with the overwhelming 80 per cent rural population that existed in 1913. The capital of the Ukraine is Kiev.

(A) Natural Resources

The Ukraine has abundant deposits of fuel, metal and iron ores, mineral, chemical raw materials, and construction materials. There are big coal basins in the Donbas and at Lvovsko-Volynskii; oil and natural gas are found in the Pred-Carpathians and Dneprovski-Donets regions. The largest iron ore deposits in the Soviet Union are found in the Ukraine at the Krivorog, Kremenchug and Kerchenskii basins and the Belozerskoye deposit; the republic also has very rich manganese ores at Nikopol and Bolshoi Tokmak. The Black and Azov Seas provide the basis for the development of sea transport, docks, fishing and health resorts. Half the area of the Ukraine is covered with the rich Chernozem soil (black soil), and the rest of the republic is fairly fertile as well.

(B) The Economy

The Ukraine is a highly industrialised republic. In overall volume of production it lies second only to the Russian Federal Republic. It contributes around 20 per cent of the gross national income of the USSR. 49.8 per cent of the Ukraine's income comes from industry, 25.8 per cent from agriculture, 9 per cent from construction, 4.6 per cent from transport and communications, and 10.8 per cent from other sources.

The Ukraine occupies an important place in the overall production of certain industrial and agricultural goods. In 1965 it produced as a percentage of the Soviet Union's

---

* The term 'scientific', denoting establishment and worker, used in this and following sections covers all academic disciplines from the humanities to the applied sciences.

overall production; 49.2 per cent of the pig iron; 40.6 per cent of steel; 42.4 per cent of finished rolling stock; 54.7 per cent of iron ore; 33.6 per cent of coal; 51.9 per cent of coke; 18.7 per cent of electrical energy; 30.4 per cent of turbines; 98.6 per cent of diesel engines; 33.4 per cent of tractors; 17 per cent of cement; 23.4 per cent of artificial fertilisers; 60.6 per cent of sugar; and 26.2 per cent of animal fats.

(C) Industry

Together with the development of the old branches of industry - coal, metallurgy and food - a large number of new ones have been created during the period of Soviet power. These include the aviation, automobile, tractor, instrument-making, radio-electronic, gas, chemical fibre, and textile industries.

Coal forms the basic ingredient of the Ukraine's power base. The Donbas is the main mining region, though the Lvovsko-Volynskii and Pridneprovskii basins are being developed. The role of gas and oil in the fuel balance is growing. In 1965 they accounted for 36 per cent of all fuels used. The gas reserves are estimated to be in the region of 3,600,000,000,000 cubic metres, and production in 1965 was 39,362,000,000 cubic metres, or 31 per cent of total production. Pipelines are being constructed to carry the gas to the industrial centres of the republic as well as to Moldavia, Byelorussia, the Baltic States and the central regions of the Russian Federal Republic. The main oil fields are situated in the Ivano-Frankovskaya, Lvov, Chernigov, Sumskaya and Poltava regions.

The bulk of electric power is produced by large thermal and hydroelectric power stations. These include the Pridneprovskaya, which is the largest thermal station in the world. The Ukrainian power network forms part of the unified system that is being built up in the European part of the Soviet Union. It is also connected with the electricity systems of Hungary, Czechoslovakia and Poland.

Two of the most important resources of the Ukraine, as far as the country as a whole is concerned, are its iron and manganese ores, and the metal production which is based on these two materials. The Republic produces about 10 per cent of the world's pig iron and 8 per cent of its steel. The main centres of ferrous metallurgy are Krivoi Rog, Dneprodzerzhinsk, Zaporozhe, Zhdanov, Makeevka, Donetsk and Dnepropetrovsk. Non-ferrous metallurgy is also an important industry. The Ukraine was the first region in the Soviet Union to master the production processes of titanium and zirconium.

Machine building is another important industry in the republic, employing more than 32 per cent of the labour force. The chemical industry is being developed on the basis of the coke and natural gases in the Donbas, Pridneprovie and Cherkass regions.

The food industry constitutes 25 per cent of the republic's gross industrial production. It also makes an important contribution to the whole country; in 1965 the Ukraine produced 60 per cent of sugar, 23 per cent of meat, 26 per cent of animal and 31 per cent of vegetable fats, and 27 per cent of the wine produced in the Soviet Union.

Light industry is also growing in the republic. In 1965 output had increased three times compared with that in 1940. However, it still occupies a relatively small place in production figures. Consumer goods only took up 27 per cent of total production in 1965 compared to 73 per cent for heavy industry.

(D) Agriculture

The Agriculture of the Ukraine is very varied. Its main sections are grain, industrial crops and cattle breeding.

(E) Education

In 1969/70 there were 138 institutes of higher education with 804,000 students studying at them. Of these 444,000 were doing evening and other part-time courses. These VUZy included eight universities, five polytechnical, six construction, fifteen medical, 32 pedagogical and 17 agriculture institutes. The leading universities are those situated in Kiev, Kharkov, Odessa and Lvov; the leading polytechnics are in Donetsk, Kiev, Lvov, Odessa and Kharkov.

(F) Scientific Research*

The *Ukrainian Academy of Sciences* was formed in February 1919. It was one of the first of the republican academies of science. At first it was very small and concentrated mainly on the social sciences. However, the first five-year plan in the 1930s saw the gradual development of research into the physical, mathematical, technological, chemical and biological sciences. At first all the Academy's research institutes were centred exclusively at Kiev; but then as the economy developed, the Academy attempted to develop scientific institutions in conjunction with the most important, industrial, agricultural and cultural centres. Research institutes exist in Kharkov, Lvov, Donetsk and Sevastopol.

In 1965 the Ukrainian Academy had 215 full and corresponding members and employed 7,720 scientific workers. There were sixty scientific establishments within its administration, including institutes of mathematics, cybernetics, physics, electrodynamics, engineering of thermal physics, the physical engineering institute of low temperatures, radiophysics and electronics, semiconductors, problems of materials, the Pisarzhevsk institute of physical chemistry, chemistry of high molecular compounds, the gas institute, organic chemistry, geology and geochemistry, combustible materials, botany, hydrobiology, zoology, the Bogomolts Institute of Physiology, the Paton Institute of Electrical Welding and the institute of foundry problems. The Academy has seven scientific departments of mathematics, mechanics and cybernetics; physics; physical engineering problems of materials; earth and space problems; chemistry and chemical technology; general biology; and biochemistry, biophysics and physiology.

In the Ukraine as a whole there were 740 scientific establishments in 1965 employing 93,984 scientific workers. These included 1,885 doctors and 19,291 candidates of science. Important research institutes, not under the administration of the Ukrainian Academy of Sciences, include the *All-Union Research Institute of the Sugar Industry* in Kiev, the *All-Union Research Institute of Monocrystals* in Kharkov, the *Research Institute of Basic Chemistry* in Kharkov, the *Ukrainian Research Institute of Synthetical Superhard Materials* in Kiev, the *All-Union Maize Research Institute* in Dnepropetrovsk, the *All-Union Research Institute of Selection and Genetics* in Odessa, and the *Institute of Gerontology* in Kiev.

III Byelorussia

The Byelorussian Soviet Socialist Republic was formed on 1st January 1919. It is situated in the western regions of the European part of the Soviet Union, bordering with Poland. It has the Baltic Republics to its north and the Ukraine to its south. The total area of Byelorussia is 207,600 sq.km., and it had a population of 9,002,000 in 1970. The capital is Minsk.

(A) Natural Resources

Good quality peat is one of the chief resources of Byelorussia. Deposits have been

---

*See footnote, page 220.

estimated at 4,800,000,000 tons dry weight. In the south of the republic reserves of high quality oil and coal and lignite have been found. In the Soligorskii region there is the very rich Starobinsk deposit of potassium and common salts, which has probable reserves of 10,000,000,000 tons. There are also good deposits throughout the republic, though concentrated in the north, of construction materials such as dolomite limestone, and refractory clays.

(B) Economy

During the period of Soviet government, Byelorussia has developed from a rather backward region into one of the most important economic regions of the USSR. The republic's economy combines highly developed industrial production with an intensive agriculture.

In 1965 it produced, as a percentage of total production in the Soviet Union, 19.8 per cent of tractors, 12.9 per cent of metal cutting machine tools, 6.1 per cent of lorries, 7.9 per cent of televisions, 8 per cent of radios, 11.8 per cent of linen, 6 per cent of milk, 5 per cent of meat, 14 per cent of potatoes, and 24 per cent of flax.

(C) Industry

Peat is the main local fuel of the republic. Nearly all the power stations run on it. Coal is brought in from the Ukraine.

Machine building and metal working are the two main branches of Byelorussian industry. The basic products that they turn out are tip-up and goods lorries, tractors, assembly lines, motor cycles, bicycles, pendulum bearings, various types of machine tools, agricultural machinery, and radioelectronic parts. There is also a growing chemical industry which specialises in the production of fertilisers, synthetic fibres, and lacquer-decorated objects. The Starobinsky deposit forms the basis for one of the largest enterprises producing potassium fertilisers both in the USSR and Europe.

The building materials industry is also developing. It produces the prefabricated large panels used in much of the domestic building. On the basis of local materials, the republic produces cement, bricks, lime, large silicate blocks, and drain pipes. Byelorussia is one of Russia's largest producers of window glass and plastic panes. The timber production has grown from a cottage trade into a large mechanised branch of industry. The wood is used to produce furniture, matches, paper and veneer.

Textiles and wool are the most developed branches of light industry. They rely on homegrown flax, but wool, cotton and silk have to be brought in from other republics. The food industry accounts for 25 per cent of all industrial production of the Byelorussian Republic.

(D) Agriculture

Agriculture in Byelorussia is varied. Dairy cattle and pigs are the main livestock and wheat, corn, potatoes, flax and vegetables the main crops.

(E) Education

At the time of the revolution there were no institutes of higher education in Byelorussia at all. In 1921 the first university was opened in Minsk. By 1970 there were 28 VUZy with some 137,300 students.

Among the most important of these are the university, and polytechnical institute in Minsk, the agricultural academy and institute in Grodno, the veterinary institute in Vitebsk, Gomel, and Mogilev. In addition to the VUZy, there are some 126 specialist secondary schools training 55,200 students.

(F) Scientific Research *

The first scientific institutes organised in the republic in the 1920s were mainly concerned with agricultural questions. The Institute of Byelorussian Culture became the focus point of scientific activity in 1922 and in 1929 it was reorganised to form the Byelorussian Academy of Sciences. The Academy now administers twenty-eight scientific establishments. These include institutes of physics, technological physics, solid state and semiconductor physics, physical and organic chemistry, genetics and cytology, experimental botany and microbiology, nuclear power, and peat. In 1965, the membership of the Academy consisted of eighty-three full academicians and corresponding members. In the republic as a whole there were 185 scientific establishments with 14,668 scientific workers engaged in them. These included 255 doctors, and 3,093 candidates of science. Important research institutes, not under the administration of the Byelorussian Academy of Sciences, include the *Institute of Geological Sciences*, and the *Byelorussian Polytechnic Institute*, both in Minsk, and the *Byelorussian Agricultural Academy* in Gorky.

IV   Uzbekistan

The Uzbekistan Soviet Socialist Republic was formed in October 1924. It lies in the middle of the Central Asian region of the Soviet Union between the two large rivers, the Amu Darya and the Syr Darya. To the south the Amu Darya forms the border with Afghanistan. Uzbekistan has an area of 449,600 sq.km., and had a population of 11,960,000 in 1970. The capital is Tashkent.

(A) Natural Resources

Uzbekistan has a wide variety of natural resources. There are large natural gas fields. Coal deposits lie in the south and a large gold seam has been found near Muruntau. Copper, zinc, and lead ores are found in the Almalykskii region. The Republic's deposits of tungsten and other rare metals are important for the Soviet Union. Other natural resources include large reserves of porcelain clay, and salt, natural sulphut, phosphorites, fluorite, barites, and building stone, sand, gravel, and marble.

(B) Economy

Uzbekistan has the strongest economy of the Central Asian Republics. It occupies 35 per cent of the territory, has 59.7 per cent of the population and is responsible for about 60 per cent of the industrial and agricultural production of the area. Cotton is the most important crop vis a vis the Soviet Union as a whole, followed by silk, but at the same time attempts are being made to build up heavy industry in the republic.

(C) Industry

The fuel resources have been expanded considerably over the past 20 years or so to provide a base for the development of heavy industry. Machine building forms the core of Uzbekistan's industrial development. Agricultural machinery, cotton picking machines, tractors, cultivators, bulldozers, and so on - most of the machinery produced is connected with the main crops of the region. Ferrous and non-ferrous metallurgy has grown up parallel with the machine building industry.

---

* See footnote, page 220.

The chemical industry's main product is fertiliser for the cotton crop. The building materials industry manufactures cement, bricks, reinforced concrete, slate, and large panels for housing. Light industry is nearly all connected with the processing of cotton, silk and bast. Tashkent has one of the largest textile mills in the whole country. The main products of the food industry are oil from the cotton seed, milk, meat, conserves, biscuits, cakes and flour grinding.

(D) Agriculture

With the extremely dry climate, irrigation is an important part of the agricultural economy. Uzbekistan has 27 per cent of all the irrigated lands in the USSR. Its irrigation system of some 800 canals, stretches for about 160,000 km.

Cotton is the most important crop and silk is the oldest. Other products include vegetables, fruit, grapes and wheat.

(E) Education

The Soviet government had to start from scratch in the education field as far as literacy and formal education in Uzbekistan were concerned. Teaching in the schools has to be provided in seven languages - Uzbek, Karakalpak, Tadzhik, Kirgiz, Turkmen and Russian. Literacy is now around 90 per cent compared with 1.8 per cent in the First World War. The higher educational system has developed since the civil war. In 1920 Lenin authorised the setting up of a university in Tashkent. There are now 38 VUZy in the Republic with 231,900 students. The VUZy include the Tashkent and Samarkand universities, the polytechnical, railway, textile, medical and agricultural institutes, and the institute of cotton production.

(F) Science *

The Uzbek Filial of the USSR Academy of Sciences was formed in 1940 and this formed the basis of the *Uzbek Academy of Sciences* created three years later in 1943. This has since become the scientific centre of the Republic. It administers twenty-six scientific establishments, including the following institutes - applied physics, nuclear physics, mathematics, chemistry, plant chemistry, geology and geophysics, experimental biology, industrial and grain crops, electronics, seismology, botany, zoology and parasitology. There is also a botanical garden and astronomical observatory. The membership of the Academy consists of two honorary academicians and 74 full and corresponding members.

The number of scientific establishments in the Republic as a whole was 159 in 1966 with around 18,000 scientific workers attached to them. These included 323 doctors and 4,221 candidates of science. There were 6,282 Uzbeks working in Soviet research organisation in 1964. Important research institutes, not under the administration of the Uzbek Academy of Sciences, include the *Central Asian Research Institute of Natural Gas* in Tashkent, the *Central Asian Research Institute of Silk Industry* at Dzhar-Aryk, and the *All-Union Research Institute for Astrakhan Raising*, in Samarkand, and the *Tashkent Agricultural Institute*.

V Kazakhstan

The Kazakh Soviet Socialist Republic was formed in December 1939. It had previously

---

* See footnote, page 220.

been an autonomous republic. It is situated in the Central Asian region of the Soviet Union, lying between the Caspian Sea and the China border, and with the Russian Federal Republic to its north and the Uzbek Republic to its south. It occupies an area of 2,715,100 sq. km., and is the second largest republic in the USSR. It had a population of 12,849,000 in 1970. The capital is Alma Ata.

(A) Natural Resources

Kazakhstan is very rich in mineral deposits, power resources, and in the acreage under cultivation and used for pasturage. It occupies first place in the USSR for known deposits of chromite ores, copper, lead, zinc, silver, tungsten, phosphorites, barites, and second place for its deposits of asbestos, cadmium and molybdenum. It also has significant reserves of iron ore, manganese, nickel, gold, antimony, cobalt, bismuth, titanium, bauxite, coal, oil and natural gas.

(B) Economy

During the period of Soviet government, Kazakhstan has developed from an underdeveloped region with a predominantly omadic form of life to a Republic with a highly developed industry and mechanised agriculture. The mining of non-ferrous and rare metals, the smelting of lead, zinc and copper, and the production of superphosphates and salts are the branches of industry that Kazakhstan is most famous for. In 1965, the Republic produced 9 per cent of the total iron ore mined in the USSR, 8 per cent of its coal and 10 per cent of its sulphuric acid. On the agricultural side Kazakhstan has some 66 per cent of all virgin lands in the country, 13 per cent of all irrigated land and 15 per cent of all land under cultivation.

(C) Industry

Kazakhstan's industry is based mainly on the Republic's own local resources. It is one of the fastest growing industries in the country. Heavy industry, apart from machine building, is mainly situated in the desert and mountain regions of the area, west and central Kazakhstan and Rudnii Altai, where most of the natural resources are to be found. Machine building, and the light and food industries are situated in the north and central Kazakhstan where the climate and living conditions are easier.

Most of Kazakhstan's power needs are met from its own resources, though it does import some coal from the Kuznets field and from Uzbekistan. The mining industry is centred in Karagand (with an output of 32,000,000 tons in 1966) and Ekibastuz. The oil industry is centred on the Embenskii basin but exploitation and development has begun of the vast reserves found in Mangyshlak. There are hydroelectric power stations on the Irtysh and its tributaries and on the Bol'shaya and Malaya Almaatinka. There are also several large thermal power stations.

As mentioned previously, the Rudnii Altai, central and southern regions of Kazakhstan are the main centres for the mining, enriching, smelting and processing of non-ferrous and precious metals.

The machine building industry, producing machines for agriculture, transport and mining, grew up during the Second World War on the basis of factories evacuated from the Ukraine and Russia. It is centred in the main towns of the Republic. There is a quickly growing chemical industry producing phosphates, superphosphates, sulphuric acid, fertilisers, synthetic fibre, and so on. The main branches of light industry are concerned with textiles (cotton, wool, etc) and leather (shoes, furs, etc). The food industry produces meat, oil, cheese, flour, sugar, fruit and vegetable conserves, wine tobacco and confectionery. There is also a quite well developed fishing industry in the Caspian and Aral Seas.

(D) Agriculture

There are two main branches to Kazakh agriculture - highly mechanised agriculture on the virgin and irrigated lands the majority of which is carried out by big grain state farms; and large-scale sheep and cattle farming (for meat and wool). The virgin lands lie in the north and mountainous regions of the Republic. Wheat is the main crop; other crops include millet, oats, barley, rice, sunflowers, flax, sugar beet, cotton and tobacco.

(E) Education

In 1917 there was a 2 per cent literacy rate, twelve secondary schools and no institutes of higher education. In the 1969/70 academic year there were 43 VUZy in Kazakhstan, with 195,700 students. The largest of the VUZy are the Kazakhstan University, the polytechnical, agricultural, veterinary, pedagogical, medical, and the Tselinograd Engineering-Construction Institutes. In the same year there were also 179 specialised secondary schools with a student population of 193,400. Altogether in 1966 there were 210,900 specialists with higher education working in the republic and 333,600 with secondary education.

(F) Science *

A filial of the Academy of Sciences of the USSR was organised in Kazakhstan in 1938. Seven years later in 1945 this was transformed into the Kazakh Academy of Sciences and became the centre of scientific activity in the Republic. It administers 34 scientific establishments, including institutes of geology, mining, chemical metallurgy, chemistry, petrochemistry and mineral salts, nuclear physics, astrophysics, botany, zoology, soil science, physiology, and experimental biology. The Academy has a membership of one honorary academician and 97 full and corresponding members.

In 1966 there were 147 scientific establishments in the Republic as a whole, employing a total of 19,467 scientific workers. These included 247 doctors and 3,670 candidates of sciences. Important research institutes, not under the administration of the Kazakh Academy of Sciences, include the *All-Union Research Institute of Non-Ferrous Metals* in Ust-Kamenogorsk, and the *Kazakh State Agricultural Institute* in Alma Ata.

VI Georgia

The Georgian Soviet Socialist Republic was formed in February 1921. It is situated in the central and western part of the Caucasus, bordering with Turkey, and the Armenian and Azerbaijan Republics to its south, the Russian Federal Republic to its north, and the Black Sea to its west. It occupies an area of 69,700 sq. km., and had a population of 4,686,000 in 1970. The capital is Tbilisi.

(A) Natural Resources

Georgia has deposits of coal, brown coal and oil. Its metal resources include an important deposit of manganese at Chiatursk. Its non-ferrous metals include copper and semi-metals. The region is rich in non-metallic deposits - barites, talc, diatomite, quartz, dolomite and refractory clays. The Republic has a large number of mineral and hot springs, and the great asset of a sub-tropical climate.

(B) Economy

Georgia is the leading sub-tropical economy in the USSR. It produces 98 per cent of the

---

* See footnote, page 220.

country's tea, nearly 100 per cent of its citrus fruits, 20 per cent of the high quality tobacco, and 20 per cent of the high quality table wines.

(C) Industry

The main branches of industry are the consumer goods - the food industry contributes up to 40 per cent of gross industrial production and the light industry up to 22 per cent.

The power industry of the Republic is mainly based on coal and hydroelectric power. There is a small oil industry and natural gas is brought in from Azerbaijan and the northern Caucasus. A large electro-metallurgical industry has developed on the basis of the Chiatursk manganese deposit and the availability of cheap power. The chemical industry is new but developing. Machine building and metal processing are important branches of industry. Particular lines include electric trains, cars, machine tools, and agricultural machinery. Construction materials is a growing branch of industry. Timber and wood processing industries are based on the Svanetiya forests.

The main branches of light industry are silk, wool, cotton, hosiery, shoes, and carpets. Tea is one of the main products of the food industry; there are some seventy processing factories lying in the western regions of the Republic. Wine is another well-established product. Mineral waters, tobacco, meat and dairy produce and oils are also produced.

(D) Agriculture

Irrigation and drainage schemes have increased the area of cultivation from 95,000 hectares in 1921 to 355,000 hectares in 1965. The crops are chiefly sub-tropical - tea, citrus fruits, vineyards, the tung tree, bamboo and laurel. Animal breeding also contributes significantly to the Republic's agricultural output - representing about 30 per cent to the gross turnover in agriculture.

(E) Education

The only higher education institute at the time of the Revolution was Tbilisi Polytechnical Institute. There are now 18 VUZy with more than 90,100 students. The largest of these are Tbilisi University, the polytechnical and agricultural institutes in Tbilisi, and the institute of sub-tropical agriculture in Sukhumi. In 1965 there were 138,000 specialists with higher education and 110,400 with secondary education working in the Republic.

(F) Science *

The Georgian Filial of the USSR Academy of Sciences was organised in 1932. In 1941 it was formed into the Georgian Academy of Sciences and became the scientific centre of the Republic. It administers 35 scientific establishments, including institutes of mathematics, physics, electronics, automation and telemechanics, cybernetics, geology, geophysics, geography, botany, physiology, zoology, experimental morphology, mining engineering and working of deposits, inorganic chemistry, electrochemistry, physico-organic chemistry, and psychology. A computing centre and the Abastuman Astrophysical Observatory also come under the administration of the Academy.

The number of scientific establishments in the Republic as a whole was 188 in 1965. They had some 14,200 scientific workers attached to them, of whom 647 were doctors, and 4,090 were candidates of science. Important research institutes, not under the administration of the Georgian Academy of Sciences, include the *All-Union Research Institute of Tea and Subtropical Plants,* at Makharadze, the *Institute of Stable Isotopes,*

---

*See footnote, page 220.

and the *Georgian Ageicultural Institute*, both in Tbilisi.

VII Azerbaijan

The Azerbaijan Soviet Socialist Republic was formed in April 1920. It lies in the western part of the Caucasus, bordering with Turkey and Iran in the south, with Armenia in the west, Georgia and the Russian Federal Republic to the north, and the Caspian Sea to the east. It has an area of 86,600 sq.km., and had a population of 5,117,000 in 1970. The capital is Baku.

(A) Natural Resources

Azerbaijan is well known for its oil and gas resources. The rich field centred round Baku and stretching out under the Caspian Sea is the oldest worked oil deposit in the country. In the mountainous region in the west of the Republic there are large deposits of iron ore, alunite, copper, cobalt, barite and pyrites. Non-ferrous metals have also been found.

(B) Economy

The emphasis of the Azerbaijan economy is on power. The Republic's significance on an All-Union scale is its supply of 8.9 per cent of the country's oil, 4.8 per cent of the natural gas, and 2 per cent of the electricity; it also contributes 5.9 per cent of the raw cotton. (These figures apply to 1965; presumably with the more recent development of the large oil and gas deposits in Western Siberia and Kazakhstan, Azerbaijan's share in the country's total output will become relatively smaller).

(C) Industry

The power industry, and oil in particular, accounted for 53.8 per cent of all industry in 1964. More than half the oil is obtained from wells drilled through the sea bottom. The relative output of Azerbaijan in the total for the whole country has dropped from 71 per cent in 1940 to 8.9 per cent in 1965, absolute production having remained fairly steady at around 20,000,000 tons a year. There are oil processing plants in Baku. The largest gas deposit is at Karadagskoye; the gas is used in the chemical industry. There are also a number of hydroelectric power stations. A small metallurgical industry has developed in Azerbaijan since the Second World War. The main products are aluminium, and pipes for the gas and oil industries. Machine building is also growing. The chemical industry is well developed and has a good raw material base in the ready supplies of oil and gas. Products include sulphuric acid, iodine and bromine, synthetic rubber, caustic soda and carbon black.

The light industry is centered on the raw materials of cotton, wool and silk. The Republic occupies first place in the Caucasus for the production of carpets. In the food industry, fish, tinned fruit and vegetables, meat and wine are the main products. The bulk of the fish comes from the Caspian and includes both the sturgeon and its roe - the famous caviar (now becoming rather a rarity).

(D) Agriculture

Cotton is one of the most important crops. Grain crops include wheat, corn, barley and rice. The vineyards are important. Sericulture (silk) has developed. Animal breeding is an important branch of the Republic's agriculture, and Azerbaijan mountain merinos and large horned cattle are specially bred.

(E) Education

In the 1966/67 academic year there were 12 VUZy in Azerbaijan with 99,200 students studying them. The most important of these include the Azerbaijan University, the Polytechnical Institute, the Institute of Oil and Chemistry, and the Pedagogical Institute, In the same year there were 78 specialised secondary schools and technical colleges with a student population of 65,000. All in all, in 1965, there were 95,000 specialists with higher education and 109,600 with secondary education working in the Republic.

(F) Science *

The *Azerbaijan Academy of Sciences* was founded in 1945. It administers twenty-two scientific establishments: these include the institutes of petrochemistry, chemistry, physics, mathematics and mechanics, power, exploitation of oil and gas deposits, geology, botany, and zoology; the Shemakhinskaya Observatory and the commission for studying the problems of the Caspian Sea, also come under the Academy. The Academy had 65 full and corresponding members in 1965, and its staff included 2,440 scientific workers.

In the Republic as a whole there are 130 scientific establishments employing some 12,350 scientific workers (this compares with a figure of 60 establishments and 1,933 scientists in 1940), including 329 doctors and 2,999 candidates of science. In the Soviet Union as a whole, there are 8,642 Azerbaijan scientists compared with only 900 in 1940. Important research institutes, not under the administration of the Azerbaijan Academy of Sciences, include the *All-Union Research Institute of Oil Refineries* and the *Azerbaijan Azizbekov Institute of Oil and Chemistry*, both in Baku, and the *Azerbaijan Agricultural Institute in Kirovavad*.

VIII   Lithuania

The Lithuanian Soviet Socialist Republic was formed in July 1940. It is situated in the Western region of the USSR, bordering with Poland and the Byelorussian Republic to its south and east, the Baltic Sea to its west, and the Latvian Republic to its north. It has an area of 65,300 sq.km., and had a population of 3,128,000 in 1970 The capital is Vilnius.

(A) Natural Resources

The mineral deposits of Lithuania include gypsum and anhydrides, chalk, marl, limestone, dolomite, various clays, sands, peat, marsh iron ores, phosphorites and mineral waters.

(B) Economy

Lithuania has a mixed industrial-agricultural economy.

(C) Industry

Heavy industry - machine building, metallurgy and construction - together with power resources have developed fast since the foundation of the Republic in 1940. The power industry is based on hydroelectric power stations built on the Neman River and thermal

---

* See footnote, page 220.

power stations, presumably using fuel brought in from other parts of the country.

Vilnius and Kaunus are the main centres of the machine building and metallurgical industries, which specialise in instruments, machine tools, electric machines, turbines and ships. The chemical industry is developing. Its products include artificial fibres, sulphuric acid, superphosphates, nitrogen fertilisers and plastics. The construction materials industry has developed around the local resources. There is also a well-developed timber industry producing sawn timber, veneer, cellulose, paper cartons, matches and furniture.

Lithuania's light and food industries are quite significant, each accounting for a third of the Republic's gross industrial production. Artistic goods such as wood carvings, ceramics, and ornaments made from amber are an important part of the light industry. The main products of the food industry include meat and dairy produce, fish, flour, and sugar, The fish production is based on a newly-equipped fishing fleet which sails in the North Atlantic and Baltic Sea. Lithuania has the largest fishing catch of the Baltic States.

(D) Agriculture

Agriculture is a highly developed branch of the Lithuanian economy. The main products are meat and milk, followed by flax, sugar beet, potatoes and vegetables. The successful development of agriculture has depended on the draining of large areas of swampy land. Since 1946, some 1,250,000 hectares of land have been added to the agricultural land.

(E) Education

Vilna University was founded at the beginning of the nineteenth century when Lithuania was part of the Russian Empire. It is now one of the twelve VUZy in the Republic. The others include the agricultural and veterinary academies, and the polytechnical, medical and pedagogical institutes. Attached to the twelve VUZy are 55,700 students. In addition there are 82 technical colleges and specialised secondary schools with 60,900 students. All in all the Republic employed 135,000 specialists with higher and secondary education in 1965.

(F) Science *

The *Lithuanian Academy of Sciences*, which is the scientific centre of the Republic, was founded in 1941. It has eleven scientific establishments under its administration. These include the institutes of physics and mathematics, chemistry and chemical technology, power and electro-technology, botany and zoology and parasitology. There is also a botanical garden. The Academy had 36 full and corresponding members in 1965.

All in all there were 106 scientific establishments in Lithuania in 1965. These employed 6,415 scientific workers, 57 were doctors and 1,349 candidates of science. One important research institute, not under the administration of the Lithuanian Academy of Sciences, is the *Lithuanian Agricultural Academy*, in Kaunas.

IX Moldavia

The Moldavian Soviet Socialist Republic was formed in August 1940. Previously, since

---

* See footnote, page 220

1924, it had been an autonomous republic. It is situated in the south-west corner of the European region of the Soviet Union, bordering with Romania to the west, and otherwise surrounded by the Ukraine. The republic has an area of 33,700 sq.km., and had a population of 3,569,000 in 1970. The capital is Kishinev.

(A) Natural Resources

There are quite abundant deposits of non-metallic minerals - limestone, marl, chalk, gypsum, diatomite, clays, and sands - which are mainly used for building materials. Oil was discovered in 1957.

(B) Economy

The economy of Moldavia is predominantly agricultural, with agriculture providing a greater share (45.8 per cent) of the gross income than industry (35.2 per cent). This contrasts with most of the other republics of the USSR, where industry usually predominates. Moldavia occupies first place in the production of wine, wine products, natural fruit juices, and certain tobaccos. The Republic generally occupies about third or fourth place in the country for the production of fruit, wine, conserves, vegetable oil, sugar, grain, and corn.

(C) Industry

The volume of industrial production has grown by a factor of 17 since 1940, the result of the construction of many new factories and enterprises. The power base depends on hydroelectric and thermal power stations. There seems to be little exploitation of the oil discovered in 1957.

In heavy industry, the machine building, metallurgy and construction materials fields have developed. The biggest enterprises here are a large tractor factory, factories producing oscillographs, electric motors, refrigerators and washing machines.

Food production remains the main industrial activity, although its share in industrial production has dropped from 84.5 per cent in 1940 to 58.5 per cent in 1966. The chief products are wine, fruit and vegetable preserves, meat, milk and dairy produce, flour, sugar, bread, confectionery, and tobacco. Light industry is represented by textiles, furniture, the production of glass jars and bottles for the food industry and so on. There is a small chemical industry producing paints and varnishes, pharmaceutical goods and rubber products.

(D) Agriculture

Two-thirds of Moldavia's industrial production depends on raw materials from agriculture. The main branches are viticulture, fruit and vegetable culture, and industrial crops. Nearly three-quarters of the Republic's population is engaged in agriculture.

(E) Education

In the academic year 1969/70, there were eight VUZy in Moldavia with a student population of 45,500. The VUZy include the Kishinev University, and the polytechnical agricultural, medical and pedagogical institutes. In addition to the VUZy there were 44 specialised secondary schools with 39,800 students studying in them. All in all the Republic had 55,700 specialists with higher education and 78,000 with secondary education working in it.

(F) Science *

The Moldavian filial of the Academy of Sciences of the USSR was organised in 1949.

---

* See footnote, page 220.

It was transformed into the *Moldavian Academy of Sciences* in 1961. In 1965 this had fifteen scientific establishments under its administration. These included a mathematics institute with a computer centre, institutes of applied physics, chemistry, plant physiology and biochemistry, and zoology. It has a botanical garden and a seismic station. The Academy has 27 full and corresponding members.

In the Republic as a whole there were 61 scientific establishments in 1965, with 3,737 scientific workers attached to them. This can be compared with a figure of six establishments and 140 scientific workers in 1940. One important research institute not under the administration of the Moldavian Academy of Sciences, is the *Kishinev Frunze Agricultural Institute*.

X Latvia

The Latvian Soviet Socialist Republic was formed in July 1940. It is one of the three Baltic States, lying between Estonia and Lithuania, with the Baltic Sea on its west and the Russian Federal Republic and the Byelorussian Republic bordering it on its east. It has an area of 63,700 sq.km., and had a population of 2,364,000 in 1970. The capital is Riga.

(A) Natural Resources

Mineral resources are mainly construction materials such as clays, dolomite, limestone, gypsum, sands and gravels. There are significant deposits of peat (some 11,400,000,000 cu m), and a workable oil deposit has been found.

(B) Economy

Latvia is a highly industrialised Republic with a well-developed agricultural sector. Industry contributes 58 per cent to the total income, and agriculture 22 per cent. Latvia's industry is significant on an All-Union scale for its electrical engineering, radio-electronics, instrument building, transport machines, construction, wood processing and paper.

(C) Industry

Machine building, metallurgy and the light and food industries predominate. The relative contribution of the latter two is twice the normal rate for the USSR as a whole. The power supply is based on hydro-electric and thermal power stations. Gas is piped in from the Ukraine, and a number of thermal stations use the local peat. As already mentioned, Latvia is well known for its machine building and metallurgical industries which turn out products such as diesel powered buses, semiconductor apparatus and instruments, electrical instruments, motorbikes, and so on. The chemical industry is developing. It produces synthetic fibres, microbiological preparations, bacteriological fertilisers, and lacquers. Furniture production has been well established since at least 1940. There are also various wood processing plants producing paper, cellulose, and punch card tape for the computer industry. The construction materials industry is developing on the basis of local materials.

In light industry, the production of knitted wear occupies an important place; production is now being diversified to include non-woven and synthetic materials. The food industry has developed considerably since 1940, and this applies particularly to fish, meat and dairy produce. The fishing fleet works in both the Baltic and the Atlantic.

(D) Agriculture

Livestock - cattle, pigs and sheep - are the main interest of agriculture. More than half the crops grown are used for fodder.

(E) Education

There are ten VUZy in Latvia in the academic year 1969/70. These included the University of Latvia, the polytechnical, medical, civil aviation, and agricultural institutes, the two pedagogical institutes. The VUZy population numbered 40,400. There were also 54 specialised secondary schools with 41,000 students. In 1965 the Republic had a total of 56,900 specialists with higher education and 82,500 with secondary education working in its economy.

(F) Science *

The *Latvian Academy of Sciences* was created in 1946. In 1967 it had a total of 29 full and corresponding members. and 14 scientific research establishments within its administration, employing 1,143 scientists. These include the institutes of inorganic chemistry, organic synthesis, the chemistry of wood, biology, microbiology, physics, power physics, electronics and computer technology, mechanics, polymers, and economics. The Academy also has a botanical garden and an astrophysical laboratory.

In the Republic as a whole there were 121 scientific establishments in 1965 with 6,019 scientists working in them (compared with 36 and 1,128 respectively in 1940), including 78 doctors of science and 1,325 candidates. Research institutes not under the administration of the Latvian Academy of Sciences, include the *All-Union Research Institute of Marine Geology and Geophysics* and the *Latvian Agricultural Academy*.

XI Kirghizia

The Kirgiz Soviet Socialist Republic was formed in December 1936, having previously been an autonomous republic for ten years. It is situated in the north-eastern region of Central Asia, with China as a neighbour on its south-west border, the Tadzhik SSR to its south and the Kazakh SSR to its north. It occupies an area of 198,500 sq.km. and in 1970 had a population of 2,933,000 people. The capital is Frunze.

(A) Natural Resources

Kirgizia is rich in non-ferrous metals (chiefly lead), rare metals, coal and minerals. There are good deposits of mercury and antimony in the south of the Republic. The coal reserves are the largest among the Central Asian Republics - the largest basins being at Uzgensky in the south and in the Central Tian'Shan. Exploitation of the latter has not started yet because they are very difficult to get at.

(B) Economy

Since the revolution, Kirgizia's economy has changed from one based almost entirely on nomadic agriculture to one which is varied and has both an industrial and agricultural

---

* See footnote, page 220.

base. In 1965 industry contributed 39 per cent and agriculture nearly 36 per cent to the Republic's total income.

(C) Industry

Industrial growth in Kirgizia is generally higher than the average for the rest of the country. The main industrial region is centred around Frunze in the north of the Republic, although the fuel base lies in the south where there are coal, oil and gas resources. Nearly half Kirgizia's electric power is produced by hydroelectric power stations, the biggest of these being at Uchkurganskaya on the River Naryn with a power of 180,000 kilowatts.

The mining of mercury, antimony and lead ores is of national importance. A developing construction materials industry produces bricks, cement, wall panels and other concrete building parts, and iron and ceramic pipes. Machine building is developing rapidly and the most important products are farm machinery, electrical equipment, metal working equipment, and instruments. Others include bicycles, washing machines, and pumps.

Textiles form the most important branch of light industry. There is a timber industry producing sawn timber and furniture. The food industry produces meat, cheese, cream, sugar, flour and bread.

(D) Agriculture

The Kirgizian agriculture has diversified since the revolution and new areas have come under cultivation with the construction of various irrigation schemes. Livestock rearing is now more important than crops, and sheep raising is the most important activity. Kirgizia occupies fourth place in the whole country (after the Russian Federal Republic, Kazakhstan and the Ukraine) for head of sheep, and first place for the specific weight of treated fine and semi-fine wool.

The crops sown are: grain crops (the main one being wheat), 51.8 per cent; industrial crops, 12.7 per cent; potatoes and vegetables, 3.4 per cent; fodder, 32.1 per cent.

(E) Education

The first higher educational establishment opened in Kirgizia in 1932. It was the pedagogical institute in Frunze, which later formed the basis for the Kirgiz University which was organised in 1951. By 1966/67 academic year there was a total of nine VUZy in the Republic. These included the university, a medical and a polytechnical institute, and three pedagogical institutes. The total student population was 46,200. In addition there were 36 vocational and specialised secondary schools with 35,400 students attending them. In 1965 there were 45,000 specialists with higher education and 61,400 with secondary education working in the Republic.

(F) Science *

The Kirgiz filial of the Academy of Sciences of the USSR was set up in 1943. It formed the basis for the organisation of the *Kirgiz Academy of Sciences*, established in 1954. In 1967 the Academy had 50 full and corresponding members and one honorary member. It employed 927 scientists. Its sixteen scientific establishments included institutes of physics and mathematics, physics and the mechanics of mining, automation, geology, organic chemistry, inorganic and physical chemistry, biochemistry, and physiology, biology, and regional medicine.

There were 45 scientific establishments in all in the Republic in 1967 employing 4,000 scientific workers. These included 85 doctors of science and more than 1,000

---

* See footnote, page 220.

candidates of science. One research institute, not under the Kirgiz Academy of Sciences, is the *Kirgiz Agricultural Institute* in Frunze.

XII Tadjikistan

The Tadzhik Soviet Socialist Republic was formed in 1929, on the basis of the autonomous republic that had been created five years previously. It is situated in the south-east of Central Asia, bordering with Afghanistan in the south, China in the east, the Kirgiz Republic in the north, and the Uzbek Republic in the west. It has a total area of 143,100 sq.km., and had a population of 2,900,000 in 1970. The capital is Dushanbe.

(A) Natural Resources

Tadzhikistan is rich in many mineral deposits - zinc, lead, tungsten, bismuth, arsenic, tin, antimony, mercury and fluorspar. Nepheline shale, dolomite, boron and natural gas have been found. There are large reserves of rock salt.

(B) Economy

Tadzhikistan is a mixed industrial-agricultural region. In 1965 39.8 per cent of the overall income came from industry and 36.1 per cent from agriculture. On a national scale, however, the Republic is especially important for some of its agricultural products, such as cotton, wine, fruit, vegetables and silk.

(C) Industry

Many branches of industry are occupied with the processing of agricultural products. At the same time, the power industry, machine tools, mining, and construction materials industries are also developing. The basis is being laid for a chemical industry and ferrous metallurgy. However, the light and food industries are basic, accounting for two-thirds of industrial production.

The silk industry is centred around Leninabad in the north. A large amount of the cotton thread produced in the Republic is exported to other textile centres in the country. There is a large carpet factory at Kairak-kum. The food industry concentrates on fruit preserves, wine and oil. Automatic looms, power transformers, light technological equipment, refrigerators, electric cable, farm machinery, cars, and equipment for the oil, cotton cleaning and textile industries are the main products of the machine tool industry.

The construction materials industry has a number of factories producing cement, slates, binding materials, concrete building parts, bricks, etc. In the power industry one of the most recent constructions has been the Nurek Hydroelectric Power Station on the Vaksh River, where one single and extremely powerful explosive blast was used to dam the river.

(D) Agriculture

Cotton growing is the most important branch of agriculture. Tadzhikistan is one of the leading cotton producers in the USSR and it has the highest cotton plant yield in the country. Nearly 40 per cent of the raw cotton is of the fine thread type. The valleys, the majority of which are sown with cotton, give the highest yield in the world - 23 to 26 centners per hectare. Other crops include grapes, apricots, apples and pears, wheat and barley; rice, corn, beans and other vegetables; and fodder.

Livestock rearing generally takes a good second place to crop growing, providing only some 14 per cent of the collective farms' income. However, in some regions, particularly in the mountainous ones, it is fairly significant, Sericulture is practised in many parts of the Republic. More than 2,000 tons of cocoons are provided each year - some 7 per cent of the total for the Soviet Union.

(E) Education

There was no form of secondary or higher education in Tadzhikistan prior to the revolution. In 1966/67 academic year there were seven VUZy in the Republic with a total of 42,700 students. The main establishments included the university at Dushanbe, the agricultural, polytechnical and medical institutes, and three pedagogical institutes in Dushanbe, Leninabad and Kuliab. In addition to the VUZy, there were 32 technical vocation schools and specialised secondary schools with 27,200 students attending them. In 1965, 35,400 specialists with higher education and 46,500 with secondary education were employed in the Republic.

(F) Science *

The *Tadzhik Academy of Sciences* was founded in 1951 on the basis of the previously existing filial of the Academy of Sciences of the USSR. In 1965 it had 34 full and corresponding members, and employed 703 scientific workers. The Academy administers seventeen scientific establishments, including institutes of astrophysics, earthquake resistant construction and seismology, chemistry, botany, plant physiology and biophysics, and zoology and parasitology. It has a Pamir Base, a mountain botany station, two botanical gardens in Dushanbe, and Leninabad, and a computer centre (organised in 1964).

Within the Republic there are 60 scientific establishments, which employ 3,538 scientific workers. These included 47 doctors of science and 737 candidates. Some 1,600 of the scientific workers were native Tadzhiks. One research institute, not administered by the Tadzhik Academy of Sciences, is the *Tadzhik Agricultural Institute* in Dushanbe.

XIII Armenia

The Armenian Soviet Socialist Republic was formed in November 1920. It is situated in the south of Trans-Caucasia. It borders with Iran and Turkey to the south and west, with Azerbaijan to the east, and Georgia to the north. It covers an area of only 29,800 sq.km.making it the smallest Republic in the Union, and had a population of 2,492,000 in 1970. The capital is Yerevan.

(A) Natural Resources

Armenia is rich in mineral resources - copper, molybdenum, gold, rare metals, and trace elements, iron ores, aluminium and magnesium raw materials, rock salt, volcanic rocks such as building tufa of various colours and shades, granites and marbles.

(B) Economy

Armenia's economy has undergone a radical change since the revolution. It is now a

---

* See footnote, page 220.

highly industrialised region. Its industry contributes 60 per cent to its annual income, compared with agriculture's 14 per cent. It is important on an All-Union scale for its ferrous metallurgy, chemical and electrotechnical industries.

(C) Industry

The main branches of industry are electric power generation, the chemical industry, machine building and metal working, construction materials production, wood processing, and the light and food industries.

The power industry has relied on hydroelectric power stations and a cascade of six has been built on the River Razdan flowing out of Lake Sevan. The fuel balance is beginning to change, however, with the import of gas into the Republic through the gas pipeline from Azerbaijan.

Ferrous metallurgy is represented by the mining of copper ore and smelting of copper. Aluminium is produced and ferrous metals rolled; trace metals are extracted, and a gold industry is being developed. The chemical industry is a highly important branch of the Armenian economy. It first arose on the basis of the cheap sources of electric power and the deposits of limestone near Mount Ararat. Products now include synthetic rubber, latex, polyvinylacetate, polyvinyl, alcohol, varnishes, glues, films, paints, nitrogen fertilisers, and synthetic corundum. Imported natural gas is used to produce acetylene.

The machine building industry has developed particularly since the Second World War. Its products include electrical engineering machines, machine tools, instruments, and radio electronic and electronic equipment. Other important products are chemical equipment, car parts, stone cutting machines, and the production of cars themselves has recently been started. The construction materials industry relies on the local resources such as tufa, natural light aggregates, basalt, granite and marble.

The woollen, textile and show industries are the most important light industries. The food industry is well developed and based on the agricultural products of the Republic - brandy, wines, fruit preserves, tobacco, oils, cheese and so on.

(D) Agriculture

The agriculture is highly mechanised. The chief sectors are the vineyards, fruit and vegetable growing, and livestock rearing. The grapes, fruit and vegetables are mainly grown in the valleys and here irrigation is important because of the hot dry climate. The livestock rearing is centred on the higher ground - dairy cattle and sheep are the main animals. Industrial crops include tobacco and sugar beet.

(E) Education

There were no higher education establishments in Armenia before the revolution. There are now twelve VUZy with a total of 53,300 students. These include Yerevan University, and the polytechnical, medical, agricultural, and pedagogical institutes. In addition, there are 53 specialised secondary schools with 35,000 students attending them.

(F) Science *

The Armenian filial of the Academy of Sciences of the USSR was organised in 1935, and this formed the basis for the *Armenian Academy of Sciences*, founded in 1943. In 1966 it had 33 scientific establishments under its wing, including the *Byurakan Astrophysical Observatory*, institutes of mathematics and mechanics, radiophysics and electronics,

---

* See footnote, page 220.

organic chemistry, biochemistry, physical chemistry, geology, zoology, botany, archaeology, and ethnography; a computer centre, the *Sevanskii Hydrobiological and Seismic Station*, and a botanical garden.

There were 98 scientific establishments in the Republic as a whole, with 8,580 scientific workers attached to them. Of these 298 were doctors of science and 2,293 were candidates. Research institutes, not under the administration of the Armenian Academy of Sciences, include the *Research Institute of Mathematical Computing Machines*, and the *Armenian Agricultural Institute*, both in Yerevan.

XIV Turkmenia

The Turkmen Soviet Socialist Republic was formed in October 1924. It is situated in the south-west of Central Asia, bordering with Iran and Afghanistan in the south, the Uzbek and Kazakh Republics to the north, and it has the Caspian Sea to its west. The area is 488,100 sq. km., and the population was 2,159,000 in 1970. The capital is Ashkhabad.

(A) Natural Resources

Oil and natural gas are the Turkmen Republic's most important natural resources. The oil and gas bearing fields stretch from the Caspian to the Amu Darya. Other resources include natural sulphur, potassium and rock salts, Glauber salt or mirabilite (these deposits have a world-wide significance), iodine, bromine, benthonite and construction materials such as limestone, gypsum, chalk, and quartz sands.

(B) Economy

Oil refining, the treatment of chemical raw materials (sulphur and mirabilite), the production of fine thread cotton, raw silk and wool are the branches of the Turkmen economy significant for the Soviet Union as a whole. Four-fifths of the Republic is occupied by the Kara Kum desert and so one of the most important developments for the economy is the construction of the Kara Kum Canal. This runs from the Amu Darya River at the point near the Soviet-Afghan border across the desert to Ashkhabad. It is planned that the canal will eventually reach the Caspian.

(C) Industry

Oil is the main field of the power industry. Turkmen production is highest for the Central Asian Republics and third for the whole country (after the Russian Federal Republic and Azerbaijan). The production of natural gas is also growing and a number pipelines are being constructed. Thermal power stations provide 95 per cent of the electric power.

The chemical industry is developing on the basis of the local raw materials. Turkmenia produces more than 40 per cent of all the sodium sulphate produced in the country. Other products are sulphuric acid, mineral fertilisers, iodine and bromine. The metal working and machine building industries are growing on the basis of the production of farm machinery spare parts, industrial and transport equipment, oil pumps, ventilators and electric cables. The construction materials industry produces cement, slate, concrete parts, and glass.

The chief branches of light industry are cotton, silk, knitted goods, leather goods and footwear, and carpets. The main products of the food industry are meat, oil, wine, flour, fish-canning, vegetable preserves and confectionery.

(D) Agriculture.

Cotton, sheep and silk are the major agricultural products, followed by viticulture, horticulture and melon cultivation. Irrigation is very important for agriculture.

Turkmenia is one of the chief cotton producers in the USSR. It produces 10 per cent of the country's raw cotton crop. Astrakhan sheep form the bulk of the livestock. In numbers and production of hides, Turkmenia comes second only to Uzbekhistan. Sericulture is carried on in all the important oases, and the Republic contributes 10 per cent of all raw silk in the Soviet Union.

(E) Education

There were no higher educational establishments in Turkmenia before the revolution. By 1969/70 there were five with 29,200 students studying at them; namely, the Turkmen University founded in 1950, and the polytechnical, medical, agricultural and pedagogical institutes. In addition there were 28 secondary specialised schools with 25,300 students attending them. In 1966 there were 33,100 specialists with higher education and 45,600 with secondary education working in the Republic.

(F) Science *

There were no scientific establishments in Turkmenia before the revolution. The first ones were organised in the 1920s. The Turkmen filial of the Academy of Sciences of the USSR was founded in 1940 and formed the basis for the *Turkmen Academy of Sciences* set up in 1951. In 1965 this had 39 full and corresponding members and two honorary members, and employed 577 scientific workers. The Academy has fourteen scientific establishments under its wing - institutes of technical physics, chemistry, the desert, botany, zoology and parasitology, archaeology and ethnography, and a central botanical garden.

In the whole Republic there were 56 scientific establishments in 1965. These employed 2,607 scientific workers including 39 doctors of science and 2,607 candidates. 1,140 of the scientific workers were native Turkmen. One research institute, not under the administration of the Turkmen Academy of Sciences, is the *Turkmen Katinin Agricultural Institute*, in Ashkhabad.

XV Estonia

The Estonian Soviet Socialist Republic was formed in July 1940. It is situated in the north-west part of the USSR, on the shores of the Baltic sea; it is bordered to the east by the Russian Federal Republic and to the south by the Latvian Republic. It has an area of 45,100 sq km, and had a population of 1,356,000 in 1970. The capital is Tallinn.

(A) Natural Resources

Bituminous shale is the Estonian Republic's most important mineral resource. It has

---

* See footnote, page 220.

56.8 per cent of all Soviet reserves. There are also deposits of peat, phosphorites, limestone, dolomite, marl, and clay.

(B) Economy

Estonia is a highly industrialised Republic with a well-developed agriculture. In volume of production, it comes first in the country for mining of bituminous shales and production of shale oil and gas; third in the production of oil equipment and light roofing materials; fourth for the production of paper, and cotton cloth; and fifth for the production of mineral fertilisers and fish.

(C) Industry

Estonia's diverse industry developed on the basis of the local mineral deposits, but now a large amount of the raw materials has to be brought in from other republics. Since the war, eight new shale mines have been opened, a large peat-briquette factory built, and the biggest shale processing factory in the world constructed. A nitrogen fertiliser plant is to be attached to this.

Machine building and metal working are important industries, producing electrical and radio equipment, precision instruments, and equipment for the oil, mining and chemical industries. The chemical industry produces shale chemicals, mineral fertilisers, and pharmaceutical products. The building materials industry produces cement, concrete building parts, wall materials (including new ones made from silicates), and mineral wool.

The light and food industries are old established industries in Estonia. The light industries are dominated by enormous textile enterprises, and the food by the fishing industry which provides 29.6 per cent of the gross food production. Other products are meat, dairy, bread and confectionery.

(D) Agriculture

The Republic has one of the highest levels of mechanisation and use of fertiliser in the whole Soviet Union. Livestock rearing is the most important agricultural branch; the bulk of the crops are used for fodder. Dairy cattle come first and the average production of milk per animal is higher than in any other republic.

(E) Education

In the 1969/70 academic year there were six VUZy in Estonia, with 22,500 students studying in them. These included the Tartuskii University and agricultural institute in Tartu, and the polytechnical and pedagogical institutes in Tallinn. In addition there were 37 vocational colleges and specialised secondary schools with 27,100 students attending them. In 1965 there were 33,200 specialists with higher education and 50,700 with secondary education working in the Republic.

(F) Science [*]

The *Estonian Academy of Sciences* was founded in 1946. In 1965 it had 44 full and corresponding members, and employed 554 scientific workers. It administers fifteen scientific establishments, including institutes of physics and astronomy, cybernetics, thermophysics and electrophysics, chemistry, experimental biology, zoology and botany. In 1965 there were 75 scientific establishments in Estonia as a whole with 3,473 scientific workers attached to them. These included 79 doctors and 973 candidates of science. One research institute, not under the administration of the Estonian Academy of Sciences, is the *Estonian Agricultural Academy* in Tartu.

---

[*] See footnote, page 220.

Index of Establishments mentioned in the Text

Academies of Sciences, Council for the Co-ordination of Scientific Activities of the Republican	21 ff
Academies of Sciences of the Union Republics	46-48
Academy Mechanics Institute	75
Academy of Agricultural Sciences	48
Academy of Arts of the USSR	48
Academy of Construction and Architecture of the USSR	48,152
Academy of Construction of the Ukrainian Republic	48,152
Academy of Medical Sciences of the USSR	48,151,180-81
Academy of Municipal Economy of the RSFSR	49
Academy of Pedagogical Sciences of the USSR	48,67
Academy of Sciences, Armenian	238
Academy of Sciences, Azerbaijan	230
Academy of Sciences, Byelorussian	224
Academy of Sciences, Estonian	241
Academy of Sciences, Georgian	228
Academy of Sciences, Kazakh	227
Academy of Sciences, Kirgiz	235
Academy of Sciences, Latvian	234
Academy of Sciences, Lithuanian	231
Academy of Sciences, Moldavian	233
Academy of Sciences, Tadzhik	237
Academy of Sciences, Turkmen	240
Academy of Sciences, Ukrainian	222
Academy of Sciences, Uzbek	225
Academy of Sciences of the USSR	10,16ff,20,21ff,36-46
Academy of Sciences of the USSR: Historical Background	36-38
Academy of Sciences of the USSR: Membership	41
Academy of Sciences of the USSR: Objectives and Functions	38-39
Academy of Sciences of the USSR: Scientific Establishments	43
Academy of Sciences of the USSR: Structure	39-41
Acoustics, Institute of	79
Acoustics Institute, Moscow	77
Acoustics research	77
Aerial Surveying and Cartography, Central Scientific Institute of Geodesy and, Moscow	79
Aero- and Hydrodynamic Institute, Central	200
Aerodynamic Institute, Central	9,75,197
Aeroflot	199
Aeronautical Engineering Academy, Mozhaisky Military	200
Aeronautical Engineering Academy, Shukovsky	200
Aeronautics, Bureau of Commissars for Aviation and	197
Affiliates of the Academy of Sciences of the USSR	45-46
Agrarian Industrial Associations	164
Agricultural Academy, Byelorussian	224
Agricultural Academy, Estonian	241
Agricultural Academy, Latvian	234
Agricultural Academy, Lithuanian	231
Agricultural Construction, Ministry of	154
Agricultural Institute, Armenian	239
Agricultural Institute, Azerbaijan	230
Agricultural Institute, Georgian	229
Agricultural Institute, Kazakh State	227
Agricultural Institute, Kirgiz	236
Agricultural Institute, Kishinev Frunze	233
Agricultural Institute, Tadzhik	237
Agricultural Institute, Tashkent	225
Agricultural Institute, Turkmen Katinin	240
Agricultural Machinery Construction, Ministry of Tractor and	132
Agricultural Machinery to State and Collective Farms, All-Union Association for the Sale of	19
Agricultural Research Institute, Krasnodar	176
Agricultural Research Institute of the South-East	176

Agricultural Science Town	174
Agricultural Sciences, Academy of	84
Agricultural Sciences, All-Union Lenin Academy of	48,173
Agricultural Technical Associations	164
Agriculture, Ministry od	173
Agriculture, Republican Academies of	48
Air-Defence Command	97
Air Fleet, Civil	200
Air Fleet, Leningrad Institute of the Civil	200
Air Fleet, Kiev Institute of the Civil	200
Air Force	98
Air Forces, Scientific Testing Institute of the	200
Aircraft Experimental Design Bureaux	200
Akademgorodok	44-45
Alcohol, State Research Institute of Synthetic	82
All-Union Association for the Sale of Agricultural Machinery to State and Collective Farms	19
All-Union Bank for Capital Investment Finance	157
All-Union Capital Investment Plan for the Development of Science	29-30
All-Union Chemico-Pharmaceutical Research Institute	181
All-Union Electrotechnical Institute	9,11
All-Union Heat Engineering Institute	9
All-Union Institute for Scientific and Technical Information	210
All-Union Institute of Aviation Materials	200
All-Union Institute of Hydraulic Engineering, Leningrad	75
All-Union Institute of Mineral Raw Materials	80
All-Union Institute of Plant Breeding	84
All-Union Lenin Academy of Agricultural Sciences	48,173
All-Union Maize Research Institute	222
All-Union Research Institute for Astrakhan Raising, Samarkand	225
All-Union Research Institute for Metallurgical Machine Building	136
All-Union Research Institute of Marine Geology and Geophysics	234
All-Union Research Institute of Medical Instruments and Equipment	181
All-Union Research Institute of Medicinal Plants	181
All-Union Research Institute of Monocrystals	222
All-Union Research Institute of Non-ferrous Metals	132
All-Union Research Institute of Non-ferrous Metals, Ust-Kamenogorsk	227
All-Union Research Institute of Oil Refineries, Baku	230
All-Union Research Institute of Selection and Genetics, Odessa	222
All-Union Research Institute of Tea and Subtropical Plants, Makharadze	228
All-Union Research Institute of the Cellulose and Paper Making Industry	147
All-Union Research Institute of the Sugar Industry	222
All-Union Scientific Research Geological Institute	80
All-Union Scientific Research Geology Prospecting Institute	80
All-Union Scientific Research Institute for Technical Information, Classification and Codification	211
All-Union Scientific Research Institute of Electromechanics	105,116
All-Union Scientific Research Institute of Medical and Medico-Technical Information	211
All-Union Scientific Research Institute of Natural Gases	80
All-Union Society of Inventors and Rationalisers	207
All-Union Technical Patent Library	208
Alma-Ata Aviation Institute	200
AMO Works	190
Analytical chemistry	82
Analytical Chemistry im V.I.Vernadskii, Institute of Geochemistry and	80,82
Animal Ecology, Severtsov Institute of Evolutionary Morphology and	84
Animal farming	169-72
Antarctic Institute, Arctic and	81
Antarctic Institute, State Oceanographic, Arctic and	79
Applied Chemistry, Institute of	81
Applied Geophysics, Institute of	80
Arbuzov Chemical Institute, Kazan	82
Architectural Theory and History, Scientific Research Institute for	158
Architecture, Central Institute for Scientific Information on Construction and	158,211
Architecture, State Committee for Civil Construction and	147ff
Architecture of the USSR, Academy of Construction and	48,158
Arctic, Scientific Research Institute of the Geology of the	80
Arctic and Antarctic Institute	81
Arctic and Antarctic Institute, Oceanographic and, State	79
Armaments, Scientific Testing Institute of Aviation	200
Armaments, State Committee of	200
Armenian Academy of Sciences	238
Armenian Agricultural Institute	239
Armenian Soviet Socialist Republic	237-39
Army	94
Arts of the USSR, Academy of	48
Asbestos industry, rubber and	140-41

Association of Scientists and Engineers for Assistance in the Building of Socialism	56
Astrakhan Raising, All-Union Research Institute for, Samarkand	225
Astronomical Council of the Academy of Sciences of the USSR	78
Astronomical-Geodesical Society	57
Astronomical Institute, Shternberg State	78
Astronomical Observatory, Mountain	78
Astronomy	78
Astronomy, Institute of Theoretical, Leningrad	78
Astrophysical Laboratory, Latvian Republic	78
Astrophysical Institute, Kazakh Republic	78
Astrophysical Observatory, Byruakan	43, 238
Astrophysical Observatory, Crimean	78
Astrophysical Observatory, Estonian Academy of Sciences	78
Astrophysical Observatory, Shemekha	78
Astrophysics, Institute of, Tadzhik Republic	78
Atmosphere, Institute of Physics of the, Moscow	77, 79
Atomic and molecular physics research	77
Atomic Energy, Kurchatov Institute of, Moscow	76, 119
Atomic Energy, State Committee for the Use of	110
Atomic nucleus research	76
Atomic Power Stations	112
Automation and Control Systems, Ministry of Instrument Construction, Means of	104
Automobile Energetics Problems, Central Laboratory for Neutralisation and for	193
Automobile Industry, Ministry of the	190
Automobiles	190-94
Automobiles and Engines, Central Scientific Research Institute for	192-93
Aviation, Ministry of Civil	199
Aviation and Aeronautics, Bureau of Commissars for	197
Aviation Armaments, Scientific Testing Institute of	200
Aviation Engines, Central Institute of	200
Aviation Industry, Ministry of the	199
Aviation Industry, State Committee of the	200
Aviation Institute, Alma Ata	200
Aviation Institute, Kazan	200
Aviation Institute, Kharkov	200
Aviation Institute, Kuizyshev	200
Aviation Institute, Ordzhonikidze, Moscow	200
Aviation Institute, Ufa	200
Aviation Materials, All-Union Institute of	200
Aviation Materials, Institute of	132
Aviation Technological Institute, Moscow	200
Aviation Transport	197-201
Avtooperator System	103
Azerbaijan Academy of Sciences	230
Azerbaijan Agricultural Institute	230
Azerbaijan Azizbekov Institute of Oil and Chemistry	230
Azerbaijan Soviet Socialist Republic	229-30

Baikal Pulp Mills	145
Baikonur Cosmodrome	85
Baikov Institute of Metallurgy	131-33
Bakh Institute of Biochemistry	83, 181
Bakulev Institute of Cardiac and Vascular Surgery	181
Bank, State	19, 30
Bank for Construction	30
Bardin Central Scientific Research Institute for Ferrous Metallurgy	131
Bashkir Affiliate of the Academy of Sciences	46
Basic chemicals industry	138-39
Biochemistry, Bakh Institute of	83, 181
Biochemistry, Institute of Plant Physiology and, Irkutsk	45
Biochemistry, Pavlov Institute of Evolutionary Physiology and, Leningrad	83
Biochemistry, Sechenov Institute of Evolutionary Physiology and	181
Biochemistry and Physiology and Micro-organisms, Institute of, Moscow	83
Biology	83-84
Biology, Institute of Molecular	181
Biophysics Institute, Physics and, Moscow	76
Botanical Society	57
Botany, Genetics and Selection, Bureau for Applied	83
Botany Institute	81
Bratsk Timber Conversion Plant	146
Building Affairs, State Committee for (Gosstroi)	19
Building and Architecture, Central Institute for Scientific Information on	158
Building materials	155
Building Materials Industry, Ministry of the	154
Bureau for Applied Botany, Genetics and Selection	83
Bureau of Commissars for Aviation and Aeronautics	197
Buryat Affiliate of the Academy of Sciences	46
Byelorussian Academy of Sciences	224
Byelorussian Agricultural Academy	224
Byelorussian Motor Works	192

Byelorussian Polytechnical Institute	224
Byelorussian Soviet Socialist Republic	222-24
Byelorussian University	77
Byeloyarsk Atomic Power Station	113
Byrukan Astrophysical Observatory	43, 238
CINIS	211
CPSU	15, 16, 18
CSA	101
CSU	16
Calculating Machines, Scientific Research Institute of	103, 105
Calculating Machines Plant, Moscow	103
Cardiac and Vascular Surgery, Bakulev Institute of	181
Cardiology, Myashikov Institute of	181
Cartography, Central Scientific Institute of Geodesy, Aerial Surveying and, Moscow	79
Cartography, Chief Administration for Geodesy and	79
Catalysis, Institute of, Novosibirsk	45
Catalysis, Institute of, Siberia	143
Caucasus Institute of Mineral Raw Materials	80
Cellulose and Paper Making Industry, All-Union Research Institute of the	147
Central Aero and Hydrodynamic Institute	200
Central Aerodynamic Institute	9, 75, 197
Central Asian Research Institute of Natural Gas, Tashkent	225
Central Chemical Laboratory	9
Central Institute for Scientific Information on Building and Architecture	158
Central Institute of Aviation Engines	200
Central Laboratory for Neutralisation and for Automobile Energetics Problems	193
Central Paper Research Institute	147
Central Research Institute of Non-ferrous Metallurgy	132
Central Scientific Institute of Geodesy, Aerial Surveying and Cartography, Moscow	79
Central Scientific Research Institute for Automobiles and Engines	192-93
Central Scientific Research Institute for City Construction and Regional Planning	158
Central Scientific Research Institute for Design of Steel Construction	158
Central Scientific Research Institute for Experimental Planning of Housing	158
Central Scientific Research Institute for Organisation, Mechanisation and Technical Assistance	158
Central Scientific Research Institute of Patent Information and Technical-Economic Studies	208
Central Scientific Research Institute of the Maritime Fleet	195-96
Central Scientific Research Institute of the River Fleet	197
Central Statistical Administration	101
Central Statistical Authority	19
Central Statistical Board	16
Chamber of Commerce	207
Chelyabinsk Plant	132
Chemical and Oil Equipment Production, Ministry of	134, 137
Chemical fibres industry	139
Chemical Industry, Ministry of the	137
Chemical Institute, Arbuzov, Kazak	82
Chemical Kinetics and Combustion, Institute of, Novosibirsk	45
Chemical Laboratory, Central	9
Chemical Mining Industry	140
Chemical Physics, Institute of, Moscow	82
Chemical Pigments, Institute of	143
Chemical Technology, Moscow Institute of Fine	82
Chemicals Industry, basic	138-39
Chemicals Industry, organic	139-40
Chemico-Pharmaceutical Research Institute, All-Union	181
Chemistry	81-83
Chemistry, Azerbaijan Azizbekov Institute of Oil and	230
Chemistry, Institute of Applied	81
Chemistry, Institute of General and Inorganic	82
Chemistry, Institute of General and Organic	143
Chemistry, Institute of Inorganic	143
Chemistry, Institute of Organic, Irkutsk	45
Chemistry, Institute of Organic, Novosibirsk	45
Chemistry, Institute of Physical, Moscow	77
Chemistry, Research Institute of Basic	222
Chemistry im V.I.Vernadskii, Institute of Geochemistry and Analytical	80, 82
Chemistry of Natural Compounds, Institute of, Moscow	82
Chemistry of Plant Substances, Institute of, Tashkent	82
Chemistry of Silicates, Institute of, Leningrad	82
Chief Administration for Architecture and Planning	154
Chief Administration for Geodesy and Cartography	79
Chief Geophysical Observatory	81

Civil Air Fleet	200
Civil Construction and Architecture, State Committee for	153ff
City Construction and Regional Planning, Central Scientific Research Institute for	158
Co-operative Institute	65
Coal Industry, Ministry of the	110
Collective Farmers Congress	162-63
Collective Farmers Council	163
Collective Farms	161-62
Combustion, Institute of Chemical Kinetics and, Novosibirsk	45
Commerce, Chamber of	207
Committee for Inventions and Discoveries	206-207
Committee for State Security	19
Communications, Leningrad Electrical Engineering Institute of	204
Communications, Moscow Electrical Engineering Institute of	204
Communications, Novosibirsk Electrical Engineering Institute of	204
Communications, Scientific Research Institute of Urban and Rural Telephone	204
Communications Network, Orbita Space	203
Communist Party of the Soviet Union	15,16,18
Compounds, elementary	82
Compounds, High molecular	82
Compounds, Institute of Elementary Organic	82
Computers, Institute of Electronic Control	100,104
Computers, Severodonetsk Scientific Research Institute of Control	105
Computer Centre, USSR Academy of Sciences	105
Computer Centres, State Network of	101
Computer Engineering, Institute of Precise Mechanics and, Moscow	100,104
Computing Machines, Institute of, Armenia	100,105
Computing Machines, Research Institute of Mathematical	239
Concrete, Scientific Research Institute for Reinforced	158
Construction, Ministry of	154
Construction, Ministry of Agricultural	154
Construction, Ministry of Heavy Industrial	154
Construction, Ministry of Transport	154,185-87
Construction, Scientific Research Institute for Rural	158
Construction, Scientific Research Institute for the Economics of	158
Construction, State Committee for	153ff
Construction and Architecture, Central Institute for Scientific Information on	211
Construction and Architecture, State Committee for Civil	153
Construction and Architecture of The USSR, Academy of	48,158
Construction and Erection Trusts	110
Construction of the Ukrainian Republic, Academy of	48,158
Conrol Systems, Ministry of Instrument Construction, Means of Automation and	102
Council for the Co-ordination of Scientific Activities of the Republican Academies of Sciences	21ff
Council of Experts (Inventions and Discoveries)	209
Council of Ministers of the USSR	14ff,19ff
Council of Scientific and Technical Societies	57
Councils of Ministers, Republican	20
Cosmic Ray Research	76
Cosmos Programme	93
Crimean Astrophysical Observatory	78
Cryogenics Laboratory, Physical Technical Institute, Kharkov	77
Crystallochemistry of Rare Elements, Institute of Mineralogy, Geochemistry and	80
Crystallography research	77
Culture, Ministry of	148
Cybernetics, Institute of	43
Cybernetics, Institute of, Ukrainian Academy of Sciences	100,103,104
Cybernetics, Scientific Council on	100
Cytology and Genetics, Siberian Institute of	83
Daghestan Affiliate of the USSR Academy of Sciences	46
Dairy Industry, Scientific Research Institute of the	82
Defence, Ministry of	106ff,199
Defence Command, Air	97
Department for Science and Establishments of Higher Education	18
Dielectric Physics Research	77
Discoveries, Committee for Inventions and	206-207
Dneprospetssta1	132
Dyestuffs Industry	139
ENIMS	135
Earth Physics, Institute of the	79,80
Earth's Crust, Institute of the	80
Earth's Mantle, Institute for the Study of, Irkutsk	45
Economics, Institute of	81

Education	58-73
Education, Department for Science and Establishments of Higher	18
Education, Institutes of Higher (VUZy)	52-56,59ff
Education, Ministry of Higher and Specialised Secondary	52-53,59ff
Education, Republican Ministries of	53
Education, State Committee for Vocational and Technical	19
Electric Power Production	109-120
Electric Welding Institute, Paton	132
Electrical Engineering Industry, Ministry of the	134
Electrical Engineering Institute of Communications, Leningrad	204
Electrical Engineering Institute of Communications, Moscow	204
Electrical Engineering Institute of Communications, Novosibirsk	204
Electrical Machines Plant, Leningrad	103
Electrification, Ministry of Power and	109ff
Electrification of Russia, State Commission for the	9,11
Electro-Physical Apparatus, Institute of, Leningrad	76
Electro-Technical Industry, Ministry of the	110
Electromechanics, All-Union Scientific Research Institute of	105,116
Electronic Control Computers, Institute of	100,104
Electronics, Institute of Radio Engineering and, Moscow	77
Electronics Industry, Ministry of the	101
Electronics Institute, Radiophysics and, Ukrainian Academy of Sciences	78
Electronics research	77
Electrosila Works, Leningrad	110,116,117,134
Electrostal Plant	132
Electrotechnical Institute, All-Union	9,11
Electrotiazhmash Plant	116,117
Elementary Organic Compounds, Institute of	82
Elementary particle research	76
Engelhardt Observatory	78
Engineering, All-Union Institute of Hydraulic	75
Engineering Academy, Mozhaisky Military Aeronautical	200
Engineering Academy, Zhukovsky Aeronautical	200
Engineering Maritime School, Vladivostok Higher	197
Engineers, Odessa Institute of Marine	197
Engineers, Novosibirsk Institute of Marine	197
Engineers of the Merchant Fleet, Odessa Institute of	197
Engines, Central Institute of Aviation	200
Engines, Central Scientific Research Institute for Automobiles and	192-193
Entomological Society	57
Estonian Academy of Sciences	241
Estonian Agricultural Academy	241
Estonian Soviet Socialist Republic	240-41
Ethnography, Institute of	81
Evolutionary Morphology and Animal Ecology, Severtsov Institute of	84
Evolutionary Physiology and Biochemistry, Sechenov Institute of Leningrad	83,181
Experimental Design Bureaux (Aircraft)	200
Experimental Research Institute of Metal Cutting Machine Tools	135
Experimental Scientific Research Institute of Metal-cutting Lathes	101
Far East Affiliate of the Academy of Sciences of the USSR	46
Farmers Congress, Collective	162-63
Farmers Council, Collective	163
Farms, Collective	161-62
Farms, State	161
Farms, State Institute for Prototype Design of Rural	158
Ferrous Metallurgy, Bardin Central Scientific Research Institute for	131
Ferrous Metallurgy, Ministry of	128
Fertiliser Institute	81
Fertilisers and Insecto-Fungicides, Institute for, Moscow	143
Fibres, Institute for Synthetic, Kalinin	143
Field Crops	168-69
Finance of Science and Technology	28-31
Finance Plan	25
Flight Research Institute, Zhukovskaya	200
Food Additives	151-52
Food Industry, Ministry of the	148
Foreign Economic Relations, State Committee for	19
Forestry, Institute of	147
Forestry Committee, State	19
Fuel Production	120-27
Fuel Resources, Institute for the Geology and Exploitation of	80
Fund for the Assimilation of New Technology	31
Fund for Consumer Goods	31

Fund for Financing Scientific Research, Special	31
Fund for the Development of Production	30
Funds, Special	31
GOELRO	9,11,109
GUGMS	79
Gas, Central Asian Research Institute of Natural, Tashkent	225
Gas Industry, Ministry of the	110
Gas Production	121
Gases, All-Union Scientific Research Institute of Natural	80
Genetics, All-Union Research Institute of Selection and, Odessa	222
Genetics, Institute of General, Moscow	83
Genetics, Siberian Institute of Cytology and	83
Genetics and Selection, Bureau for Applied Botany and	83
Geochemical Institute, Irkutsk	45
Geochemistry, Institute of Geology of Ore Mine Deposits, Petrography, Mineralogy and	80
Geochemistry and Analytical Chemistry, im V.I.Vernadskii, Institute of	80,82
Geochemistry and Crystallochemistry of Rare Elements, Institute of Mineralogy and	80
Geodesy	79
Geodesy, Aerial Surveying and Cartography, Central Scientific Institute of, Moscow	79
Geodesy and Cartography, Chief Administration for	79
Geographical Society	57,81
Geography, Institute of, Academy of Sciences	81
Geography, Institute of Siberian and Far Eastern, Irkutsk	45
Geography of Siberia and the Far East, Institute of	81
Geological Institute, Academy of Sciences	80
Geological Institute, All-Union Scientific Research	80
Geological Prospecting Oil Institute, Ukrainian Scientific Research	80
Geological Sciences, Institute of, Minsk	224
Geology	79-80
Geology, Ministry of	80
Geology and Exploitation of Fuel Resources, Institute for the	80
Geology and Geophysics, All-Union Research Institute of Marine	234

Geology and Geophysics, Institute of, Siberia	79,80
Geology and Mineral Raw Materials, Siberian Scientific Research Institute of	80
Geology of the Arctic, Scientific Research Institute of the	80
Geology Prospecting Institute, All-Union Scientific Research	80
Geophysical Observatory, Chief	81
Geophysics	79
Geophysics, All-Union Research Institute of Marine Geology and	234
Geophysics, Institute of Applied	79,80
Geophysics, Institute of Geology and, Siberia	79,80
Georgian Academy of Sciences	228
Georgian Agricultural Institute	229
Georgian Soviet Socialist Republic	227-29
Gerontology, Institute of, Kiev	222
Gipromez	131
Giproplast	143
Gorky Vehicle Plant	190
Gosgrazhdanstroi	153ff
Gosplan USSR	16,19,20,21ff
Gosstroi	153ff
Health, Ministry of	148,151,178ff
Heat Engineering Institute, All-Union	9
Helminthology, Skryabin Institute of	181
Heterocyclic compounds	82
High molecular compounds	82
High Pressure Institute	81
High temperature plasma physics research	76
Housing, Central Scientific Research Institute of Experimental Planning of	158
Housing construction	154
Hydraulic Engineering, All-Union Institute of	75
Hydrobiological and Seismic Station, Sevanskii	239
Hydrodynamic Institute, Central Aero- and	200
Hydrodynamics, Institute of, Novosibirsk	200
Hydroelectric Power Stations	111-112
Hydrophysical (Sea) Institute, Ukrainian Academy of Sciences	79
Hydrological Institute, State	81
IGEM	80
IPMCE	100
Information, All-Union Institute for Scientific and Tec-nical	210

Information, All-Union Scientific Research Institute of Medical and Medico-Technical	211
Information, Classification and Codification, All-Union Scientific Research Institute for	211
Information on Construction and Architecture, Central Institute for Scientific	211
Information Services	210-12
Inorganic chemistry	82
Inorganic Chemistry, Institute of	143
Inorganic Chemistry, Institute of General and	82
Insecto-Fungicides, Institute for Fertilisers and, Moscow	143
Institute for Fertilisers and Insecto-Fungicides, Moscow	143
Institute for Machine Sciences	136
Institute for Polymerised Plastics, Leningrad	143
Institute for Synthetic Fibres, Kalinin	143
Institute for the Study of Properties of Materials	43
Institute for the Study of the Earth's Mantle, Irkutsk	45
Institute of Acoustics	79
Institute of Applied Chemistry	81
Institute of Applied Geophysics	79,80
Institute of Applied Mathematics	75
Institute of Astrophysics of the Tadzhik Republic	78
Institute of Automation and Remote Control	104
Institute of Aviation Materials	132
Institute of Biochemistry and Physiology of Micro-organisms, Moscow	83
Institute of Catalysis, Novosibirsk	45
Institute of Catlysis, Siberia	143
Institute of Chemical Kinetics and Combustion, Novosibirsk	45
Institute of Chemical Physics, Moscow	82
Institute of Chemical Pigments	143
Institute of Chemistry of Natural Compounds, Moscow	82
Institute of Chemistry of Plant Substances, Tashkent	82
Institute of Chemistry of Silicates, Leningrad	82
Institute of Computing Machines, Armenia	100,105
Institute of Cybernetics	43
Institute of Cybernetics, Ukrainian Academy of Sciences	100,103-104
Institute of Earth Physics	79,80
Institute of Economics	81
Institute of Electro-Physical Apparatus, Leningrad	76
Institute of Electronic Control Computers	100,104
Institute of Elementary Organic Compounds	82
Institute of Ethnography	81
Institute of Experimental and Clinical Oncology	181
Institute of Forestry	147
Institute of General and Inorganic Chemistry	82,143
Institute of General Genetics, Moscow	83
Institute of Geochemistry and Analytical Chemistry im V.I. Vernadskii	80,82
Institute of Geography, Academy of Sciences of the USSR	81
Institute of Geography of Siberia and the Far East	81
Institute of Geological Sciences, Minsk	224
Institute of Geology and Geophysics, Siberia	79,80
Institute of Geology of Ore Mine Deposits, Petrography, Mineralogy and Geochemistry	80
Institute of Gerontology, Kiev	222
Institute of Higher Nervous Activity and Neurophysiology, Moscow	83
Institute of Hydrodynamics, Novosibirsk	200
Institute of Inorganic Chemistry	143
Institute of Mathematics	69
Institute of Mechanics and Instrument Design	100,105
Institute of Microbiology, Moscow	83
Institute of Mineralogy, Geochemistry and Crystallochemistry of Rare Elements	80
Institute of Molecular Biology	181
Institute of Nuclear Physics, Novosibirsk	45,77
Institute of Nuclear Research, Dubna	76
Institute of Oceanology	79,84
Institute of Organic Chemistry, Irkutsk	45
Institute of Organic Chemistry, Novosibirsk	45
Institute of Petrochemical Synthesis, Moscow	82
Institute of Physical Chemistry, Moscow	77
Institute of Physical Problems, Moscow	77
Institute of Physics, Byelorussia	43
Institute of Physics, Kiev	77
Institute of Physics and Mathematics, Byelorussian Academy of Sciences	100
Institute of Physics and Mathematics, Minsk	77
Institute of Physics of Metals, Sverdlovsk	77
Institute of Physics of the Atmosphere	77,79

Institute of Plant Physiology and Biochemistry, Irkutsk	45
Institute of Poliomyelitis and Virus Diseases, Moscow	181
Institute of Precise Mechanics and Computer Engineering, Moscow	100,104
Institute of Proteins, Akademgorodok	83
Institute of Radio Engineering and Electronics, Moscow	77
Institute of Radio Reception and Acoustics, Leningrad	77
Institute of Semiconductors, Kiev	77
Institute of Semiconductors, Leningrad	77
Institute of Siberian and Far Eastern Geography, Irkutsk	45
Institute of Stable Isotopes, Tbilisi	228
Institute of Steel and Alloys	132
Institute of Surgery, Moscow	181
Institute of Synthetic Rubber, Leningrad	143
Institute of Terrestrial Magnetism, Irkutsk	45
Institute of Terrestrial Magnetism, Radio Research and the Ionosphere	77
Institute of the Earth's Crust	80
Institute of the Geology and Exploitation of Fuel Resources	80
Institute of the Nitrogen Industry	143
Institute of Theoretical and Applied Mechanics, Novosibirsk	45
Institute of Theoretical Astronomy, Leningrad	78
Institute of Theoretical Physics, Moscow and Kiev	76
Institute of Virology, Moscow	181
Institutes of Higher Education (VUZy)	52-56,59ff
Instrument Construction, Means of Automation and Control Systems, Ministry of	102
Instrument Construction Plant, Severodonetsk	103
Instrument Design, Institute of Mechanics and	100,105
Instrument Industry, Ministry of the Machine Construction and	101
Intercosmos Programme	95
Inventions and Discoveries, Committee for	206-207
Inventors and Rationalisers, All-Union Society of	207
Ioffe Physico-Technical Institute	9,11,77
Ionosphere and Radio Waves, Siberian Institute of Terrestrial Magnetism, and the	77
Irkutsk Centre of the Academy of Sciences	45
Iron and Steel Plants, State Institute for the Design of	131
Isotopes, Institute of Stable, Tbilisi	228
Joint Institute for Nuclear Research	76
KazakhIMS	80
KIMS	80
Kama Lorry Factory	192
Kapustin Yar launching site	87
Karelian Affiliate of the Academy of Sciences	46
Karpov Institute	143
Karpov Physical-Chemical Institute	81,82
Kaunas Polytechnical Institute	65
Kazakh Academy of Sciences	227
Kazakh Institute of Mineral Raw Materials	80
Kazakh Soviet Socialist Republic	225-27
Kazakh State Agricultural Institute	227
Kazan Aviation Institute	200
Kazan University	82
Kharkov Polytechnical Institute	65
Kiev Institute of the Civil Air Fleet	200
Kinetics and Combustion, Institute of Chemical, Novosibirsk	45
Kirgiz Academy of Sciences	235
Kirgiz Agricultural Institute	236
Kirgiz Soviet Socialist Republic	234-36
Kishinev Frunze Agricultural Institute	233
Kola Affiliate of the Academy of Sciences	46
Kolkhozy	161-62
Komarov Affiliate of the Academy of Sciences	46
Komi Affiliate of the Academy of Sciences	46
Komsomolsk Mill	146
Kotlas Mill	146
Kramatorskii Machine-building Plant	134
Krasnodar Agricultural Research Institute	176
Krasnoyarsk Mill	146
Krasny Oktyabr Plant	132
Kuibyshev Aviation Institute	200
Kurchatov Institute of Atomic Energy, Moscow	76,119

LEIS	204
Labour and Wages, State Committee for	19
Land Reclamation and Water Conservancy, Ministry of	161
Latvian Academy of Sciences	234
Latvian Agricultural Academy	234
Latvian Soviet Socialist Republic	233-34
Launching sites	85-89
Launching vehicles	95-96
Lebedev Physics Institute	76-78
Lenin Komsomol Works, Moscow	192
Leningrad Electrical Engineering Institute of Communications	204
Leningrad Electrical Machines Plant	103
Leningrad Institute of Precision Mechanics and Optics	77
Leningrad Institute of the Civil Air Fleet	200
Leningrad Metallurgical Plant	134
Leningrad Polytechnical Institute	65,75,201
Leningrad Technological Institute	82
Leningrad University	76,77
Libraries	212-13
Licensintorg	208
Light Industry, Ministry of	148
Likhachev Automobile Works	190
Limnological Institute, Irkutsk	45
Lithuanian Academy of Sciences	231
Lithuanian Agricultural Academy	231
Lithuanian Soviet Socialist Republic	230-31
Low temperature physics research	77
Luganskii Diesel Engine Plant	134
Lunar Programme	87-91
Lvov TV Plant	103
MEIS	204
Machine and Hand Tools Industry, Ministry of the	132ff
Machine Building, All-Union Research Institute for Metallurgical	136
Machine Building, Ministry of General	132ff
Machine Building, Ministry of Heavy Power and Transport	134
Machine Building, Ministry of Medium	132ff
Machine Building for the Light, Food and Household Equipment Industries, Ministry of	132,148
Machine Construction and Instrument Industry, Ministry of the	101
Machine Science, Institute for	136
Machine Tools, Experimental Research Institute for Metal-Cutting	135
Machinery, Ministry of Communal Building and Road Building Machinery	132
Magnetism research	77
Maize Research Institute, All-Union	222
Manned-flight Programme	90ff
Manpower (science and technology)	31-35
Marine Engineers, Novosibirsk Institute of	197
Marine Engineers, Odessa Institute of	197
Marine Geology and Geophysics, All-Union Research Institute of	234
Maritime Fleet, Central Scientific Research Institute of the	195-96
Maritime School, Vladivostok Higher Engineering	197
Materials, All-Union Institute of Aviation	200
Materials, Institute for the Study of Properties of	43
Materials, Ukrainian Research Institute of Synthetic Superhard	222
Mathematical Computing Machines, Research Institute of	239
Mathematics	75
Mathematics, Institute of	69
Mathematics, Institute of Applied	75
Mathematics, Institute of Physics and, Byelorussian Academy of Sciences	100
Mathematics, Institute of Physics and, Minsk	77
Mathematics Institute, Siberian	75
Mathematics Institute, Steklov	75
Meat and Milk Industry, Ministry of the	148
Mechanics	75-76
Mechanics, Institute of Theoretical and Applied, Novosibirsk	45
Mechanics and Computer Engineering, Institute of Precise, Moscow	100
Mechanics and Instrument Deisgn, Institute of	100,105
Mechanics and Optics, Leningrad Institute of Precision	77
Mechanics Institute, Academy	75
Medical and Medico-Technical Information, All-Union Scientific Research Institute of	211
Medical Instruments and Equipment, All-Union Research Institute of	181
Medical Sciences of the USSR, Academy of	48,151,180-81
Medical Services	179-80
Medical Supplies Industry	180
Medicinal Plants, All-Union Research Institute of	181
Merchant Fleet, Odessa Institute of Engineers of the	197
Merchant Marine, Ministry of the	194

Metal-cutting Lathes, Experimental Scientific Research Institute of	101
Metal Cutting Machine Tools, Experimental Research Institute of	135
Metallurgical Machine Building, All-Union Research Institute for	136
Metallurgy, Baikov Institute of	131
Metallurgy, Bardin Central Scientific Research Institute for Ferrous	131
Metallurgy, Central Research Institute of Non-ferrous	132
Metallurgy, Ministry of Ferrous	128
Metallurgy, Ministry of Non-ferrous	128
Metals, All-Union Research Institute of Non-ferrous	132
Metals, All-Union Research Institute of Non-ferrous, Ust-Kamenogorsk	227
Metals, Institute of Physics of, Sverdlovsk	77
Meteor Programme	93
Micro-organisms, Institute of Biochemistry and Physiology of, Moscow	83
Microbiology, Institute of, Moscow	83
Milk Industry, Ministry of the Meat and	148
Mineral Raw Materials, All-Union Institute of	80
Mineral Raw Materials, Caucasus Institute of	80
Mineral Raw Materials, Kazakh Institute of	80
Mineral Raw Materials, Siberian Scientific Research Institute of Geology and	80
Mineralogical Society	57
Mineralogy, Geochemistry and Crystallochemistry of Rare Elements, Institute of	80
Mineralogy and Geochemistry, Institute of Geology of Ore Mine Deposits, Petrography and	80
Mining, Moscow Academy of	9,80
Ministerial System of the USSR	49-52
Ministries of the USSR (organisation)	50-51
Ministry of Agricultural Construction	154
Ministry of Agriculture	173
Ministry of Chemical and Oil Equipment Production	134,137
Ministry of Civil Aviation	199
Ministry of Communal Building and Road Building Machinery	132
Ministry of Construction	154
Ministry of Culture	148
Ministry of Defence	96ff,199
Ministry of Education	59ff
Ministry of Ferrous Metallurgy	128
Ministry of General Machine Building	128ff
Ministry of Geology	80
Ministry of Health	148,151,168ff
Ministry of Heavy Industrial Construction	154
Ministry of Heavy Power and Transport Engineering	110
Ministry of Heavy Power and Transport Machine Building	134
Ministry of Higher and Specialised Secondary Education	52-53,59ff
Ministry of Instrument Construction, Means of Automation and Control Systems	102,132
Ministry of Land Reclamation and Water Conservancy	161
Ministry of Light Industry	148
Ministry of Machine Building for the Light, Food and Household Equipment Industries	132,148
Ministry of Medium Machine Building	132ff
Ministry of Non-ferrous Metallurgy	128
Ministry of Oil-Refining and Petrochemical Industry	137
Ministry of Postal Services and Telecommunications	202
Ministry of Power and Electrification	109ff
Ministry of Railways	184-85
Ministry of the Automobile Industry	190
Ministry of the Aviation Industry	199
Ministry of the Building Materials Industry	154
Ministry of the Chemical Industry	137
Ministry of the Coal Industry	110
Ministry of the Electrical Engineering Industry	134
Ministry of the Electro-Technical Industry	110
Ministry of the Electronics Industry	101
Ministry of the Food Industry	148
Ministry of the Gas Industry	110
Ministry of the Machine and Hand Tools Industry	132ff
Ministry of the Machine Construction and Instrument Industry	101
Ministry of the Meat and Milk Industry	148
Ministry of the Merchant Marine	194
Ministry of the Oil-Extracting Industry	110
Ministry of the Pulp and Paper Industry	146
Ministry of the Radio Industry	100,101,202
Ministry of the River Fleet	194
Ministry of the Shipbuilding Industry	194
Ministry of the Timber Industry	146
Ministry of Tractor and Agricultural Machinery Construction	132

Ministry of Transport Construction	81,154,185
Minsk Ordzhonikodze Plant	103,105
Mir Power Grid	115
Mironovsky Institute for Wheat Selection and Seed Production	176
Moldavian Academy of Sciences	233
Moldavian Soviet Socialist Republic	231-33
Molecular Biology, Institute of	181
Molecular physics research	77
Molniya Programme	93
Monocrystals, All-Union Research Institute of	222
Morphology and Animal Ecology, Severtsov Institute of Evolutionary	84
Moscow Academy of Mining	9,80
Moscow Aviation Technological Institute	200
Moscow Calculating Machines Plant	103
Moscow Electrical Engineering Institute of Communications	204
Moscow Higher Technical College	201
Moscow Institute of Fine Chemical Technology	82
Moscow Physical-Technical Institute	204
Moscow SAM Plant	102
Moscow Society of Naturalists	57
Moscow State University	76,77,82,84,100,201
Mountain Astronomical Observatory	78
Mozhaisky Military Aeronautical Engineering Academy, Leningrad	200
Museums	213-15
Myashikov Institute of Cardiology	181
NIIAV	200
NIIGA	80
NIIPlastmass	143
NIIVVS	200
Natural Compounds, Institute of Chemistry of, Moscow	82
Natural Gas, Central Asian Research Institute of, Tashkent	225
Natural Gases, All-Union Scientific Research Institute of	80
Natural Sciences	74-84
Naturalists, Moscow Society of	57
Navy	97-98
Neurophysiology, Institute of Higher Nervous Activity and, Moscow	83
Neutralisation and for Automobile Energetics Problems, Central Laboratory for	193
Nitrogen Industry, Institute of the	143
Nizhegorod Radio Laboratory	9
Non-ferrous Metallurgy, Central Research Institute of	132
Non-ferrous Metallurgy, Ministry of	128
Non-ferrous Metals, All-Union Research Institute of	132
Non-ferrous Metals, All-Union Research Institute of, Ust-Kamenogorsk	227
North Pole I Drift Station	81
Novosibirsk Centre at Akademgorodok	44-45
Novosibirsk Electrical Engineering Institute of Communications	204
Novosibirsk Institute of Marine Engineers	197
Novovoronezh Atomic Power Station	113
Nuclear Physics, Institute of, Novosibirsk	45,77
Nuclear Research, Institute of, Dubna	76
Nuclear Research, Joint Institute for	76
Observatories	78
Observatory, Byruakan Astrophysical	43
Oceanographic, Arctic and Antarctic Institute, State	79
Oceanographical Institute, State	81
Oceanology, Institute of	79,81,84
Odessa Institute of Engineers of the Merchant Fleet	197
Odessa Institute of Marine Engineers	197
Oil and Chemistry, Azerbaijan Azizbekov Institute of	230
Oil Equipment Production, Ministry of Chemical and	134,137
Oil-Extracting Industry, Ministry of the	110
Oil Production	121
Oil-Refining and Petrochemical Industry, Ministry of	137
Oil Institute, Ukrainian Scientific Research Geological Prospecting	80
Oil Refineries, All-Union Research Institute of, Baku	230
Oncology, Institute of Experimental and Clinical	181
Optical Institute, State	9,76,77
Optics, Leningrad Institute of Precision Mechanics and	77
Optics research	77
Orbita Space Communications Network	94,203
Ordzhonikidze Moscow Aviation Institute	200
Ore Mine Deposits, Petrography, Mineralogy and Geochemistry, Institute of Geology and	80

Organic Chemicals Industry	139-40
Organic Chemistry	82
Organic Chemistry, Institute of, Irkutsk	45
Organic Chemistry, Institute of, Novosibirsk	45
Organic Chemistry, Institute of General and	143
Organic Compounds, Institute of Elementary	82
PVO	97
Paints and lacquers industry	140
Palaeontological Institute, Moscow	84
Paper Industry, Ministry of the Pulp and	146
Paper Making Industry, All-Union Research Institute of the Cellulose and	147
Paper Research Institute, Central	147
Patent	207
Patent Examination Research Institute of the USSR	208
Patent Information and Technical-Economic Studies, Central Scientific Research Institute of	208
Patent Library, All-Union Technical	208
Patent Services	206-10
Paton Electric Welding Institute	132
Patrice Lumumba Friendship University	65
Pavlov Institute of Physiology	83,181
Pedagogical Sciences of the USSR, Academy of	48,67
Penza SAM Plant	105
People's Control Committee	19
Petrochemical Synthesis, Institute of, Moscow	82
Petrography, Mineralogy and Geochemistry, Institute of Geology of Ore Mine Deposits and	80
Pharmaceutical Research Institute, All-Union Chemico-	181
Physical-Chemical Institute, Karpov	81,82
Physical chemistry	81
Physical Chemistry, Institute of, Moscow	77
Physical Institute, Lebedev	78
Physical Problems, Institute of, Moscow	77
Physical Technical Institute of Tashkent	82
Physico-Technical Institute, Ioffe	76
Physico-Technical Institute, Kharkov	77
Physico-Technical Institute, Siberian	76
Physico-Technical Institute, Sukhumi	77
Physico-Technical Institute, Ukrainian	76
Physics	76-78
Physics, Institute of, Byelorussia	43
Physics, Institute of, Kiev	77
Physics, Institute of Chemical, Moscow	82
Physics, Institute of Earth	79
Physics, Institute of Nuclear, Novosibirsk	45
Physics and Biophysics Institute, Moscow	76
Physics and Mathematics, Institute of, Minsk	77
Physics Institute, Lebedev	76,77
Physics of Metals, Institute of, Sverdlovsk	77
Physics of the Atmosphere, Institute of, Moscow	77,79
Physiology, Timiyazev Institute of Plant	83
Physiology and Biochemistry, Institute of Plant, Irkutsk	45
Physiology and Biochemistry, Pavlov Institute of Evolutionary, Leningrad	83,181
Physiology and Biochemistry, Sechenov Institute of Evolutionary	181
Physiology of Micro-organisms, Institute of Biochemistry and, Moscow	83
Pigments, Institute of Chemical	143
Plan, Finance	25
Plan for Capital Investments	25
Plan for Experimental Design Work	24
Plan for Introducing the Achievements of Science and Technology	24
Plan for Scientific Research	24
Plan for Training Scientific Personnel	25
Plan Providing for the Material and Technical Execution of Scientific Research and Experimental Work	24
Planetary Probes	93
Planning Committee, State (Gosplan)	16,19,20,21ff
Planning of Science and Technology	23-28
Plant Breeding, All-Union Institute of	84
Plant Physiology, Timiyazev Institute of	83
Plant Physiology and Biochemistry, Institute of, Irkutsk	45
Plant Substances, Institute of Chemistry of, Tashkent	82
Plasma physics research, high-temperature	76

Plastics, Institute for Polymerised, Leningrad	143
Plastics Industry, Resins and	139
Plesetsk Launching Site	87
Poliomyelitis and Virus Diseases, Institute of, Moscow	181
Polymerised Plastics, Institute of, Leningrad	143
Polytechnical Institutes and Colleges	65
Postal Services	204
Postal Services and Telecommunications, Ministry of	202
Power, Siberian Institute of	45
Power and Electrification, Ministry of	109ff
Precise Mechanics and Computer Engineering, Institute of, Moscow	100, 104
Precision Mechanics and Optics, Leningrad Institute of	77
Presidium of the Council of Ministers of the USSR	14, 20
Pridneprovskaya Thermal Power Station	111
Proteins, Institute of, Akademgorodok	83
Prototype Design and Technical Research, State Institute for	158
Pulkovo Observatory	78
Pulp and Paper Industry, Ministry of the	146
Purchasing Committee, State	19
Radio and Television	203
Radio Engineering and Electronics, Institute of, Moscow	77
Radio Industry, Ministry of the	100, 101, 202
Radio Laboratory, Mizhegorod	9
Radio Reception and Acoustics, Institute of, Leningrad	77
Radio Research and the Ionosphere, Institute of Terrestrial Magnetism,	77
Radio Waves, Siberian Institute of Terrestrial Magnetism, the Ionosphere and	77
Radioelectronics, State Committee on	101
Radiophysics and Electronics Institute, Ukrainian Academy of Sciences	78
Radiophysics Institute, Gorky	77
Radiophysics research	77
Radium Institute, Leningrad	82
Railways	183-89
Railways, Ministry of	184-85
Railways Network	185-87
Railways Research Institute	188
Rare Elements, Institute of Mineralogy, Geochemistry and Crystallochemistry of	80
Raw Materials, All-Union Institute of Mineral	80
Raw Materials, Caucasus Institute of Mineral	80
Raw Materials, Kazakh Institute of Mineral	80
Raw Materials, Siberian Scientific Research Institute of Geology and Mineral	80
Regional Planning, Central Scientific Research Institute for City Construction and	158
Reinforced Concrete, Scientific Research Institute for	158
Republican Councils of Ministers	20
Republican Ministries of Education	53
Research Institute of Basic Chemistry	222
Research Institute of Mathematical Computing Machines	239
River and Sea Transport	194-97
River Fleet, Central Scientific Research Institute of the	197
River Fleet, Ministry of the	194
Romashka Experimental Unit	119
Rostselmach Agricultural Machinery Plant	134
Rubber, Institute of Synthetic, Leningrad	143
Rubber and Asbestos Industey	140-41
Rural construction	155-56
Rural Construction, Scientific Research Institute for	158
Russian Soviet Federal Socialist Republic	216-20
SNCC	101
SNIIGIMS	80
Saratov State University	201
Science and Establishments of Higher Education, Department for	18
Science and Technology, State Committee for	16, 19, 20ff
Science Councils	22
Scientific Council on Cybernetics	100
Scientific Establishments (structure)	51-52
Scientific Research Institute for Architectural Theory and History	158
Scientific Research Institute for Reinforced Concrete	158
Scientific Research Institute for Rural Construction	158
Scientific Research Institute for the Economics of Construction	158
Scientific Research Institute of Calculating Machines	103, 105

Scientific Research Institute of the Dairy Industry	82
Scientific Research Institute of the Geology of the Arctic	80
Scientific Research Institute of Urban and Rural Telephone Communications	204
Scientific Societies	56-57
Scientific-Technical Council, Ministry of Education	53
Scientific Testing Institute of Aviation Armaments	200
Scientific Testing Institute of the Air Forces	200
Sea and River Transport	194-97
Sea Hydrophysical Institute, Ukrainian Academy of Sciences	79
Sechenov Institute of Evolutionary Physiology and Biochemistry, Leningrad	83,181
Security, Committee for State	19
Seed Production, Mironovsky Institute for Wheat Selection and	176
Seismic Station, Sevanskii Hydrobiological and	239
Selkhoztekhnika	19
Semiconductor Institute, Academy of Sciences	76
Semiconductor physics research	77
Semiconductors, Institute of, Kiev	77
Semiconductors, Institute of, Leningrad	77
Serpukhov Station, Lebedev Physical Institute	78
Sevanskii Hydrobiological and Seismic Station	239
Severodonetsk Instrument Construction Plant	103
Severodonetsk Scientific Research Institute of Control Computers	105
Severtsov Institute of Evolutionary Morphology and Animal Ecology	84
Shemekha Astrophysical Observatory	78
Shipbuilding Industry, Ministry of the	194
Shipbuilding Institute im Admiral S.O.Makarov	197
Shternberg State Astronomical Institute	78
Siberian Branch of the Academy of Sciences	44
Siberian Institute of Cytology and Genetics	83
Siberian Institute of Power	45
Siberian Institute of Terrestrial Magnetism, the Ionosphere and Radio Waves	77
Siberian Mathematics Institute	75
Siberian Physico-Technical Institute	76
Siberian Scientific Research Institute of Geology and Mineral Raw Materials	80
Sigma Association of Lithuania	105
Silicates, Institute of Chemistry of, Leningrad	82
Skryabin Institute of Helminthology	181
Societies, Scientific	56-57
Soil Institute	81
Solid State Physics research	77
Soviet of Nationalities	14
Soviet of the Union	14
Soviet of Workers' Deputies	14
Sovkhozy	161
Soyuzselkhozkhimia	164
Soyuzselkhoztekhnika	164
Space Programme	87-94
Spectroscopy research	77
Sputniks	87
Stanko-konstruktsiya	135
State Bank	19,30
State Commission for the Electrification of Russia	9,11
State Committee for Building Affairs (Gosstroi)	19
State Committee for Civil Construction and Architecture	153ff
State Committee for Construction	153ff
State Committee for Foreign Economic Relations	19
State Committee for Labour and Wages	19
State Committee for Material and Technical Supply	19
State Committee for Science and Technology	16,19,20ff
State Committee for the Use of Atomic Energy	110
State Committee for Vocational and Technical Education	19
State Committee of Armaments	200
State Committee of the Aviation Industry	200
State Committee on Popular Control	18
State Committee on Radioelectronics	101
State Farms	161
State Forestry Committee	19
State Hydrological Institute	81
State Institute for Prototype Design and Technical Research	158
State Institute for the Design of Iron and Steel Plants	131
State Network of Computer Centres	101
State Oceanogr aphic, Arctic and Antarctic Institute	79
State Oceanographical Institute	81
State Optical Institute	9,76,77
State Planning Committee (Gosplan USSR)	16,19,20,21ff
State Procurement Committee	163
State Purchasing Committee	19
State Research Institute of Synthetic Alcohol	82
Statistical Administration, Central	101
Statistcal Authority, Central	19
Statistical Board, Central	16
Steel and Alloys, Institute of	132
Steel Construction, Central Scientific Research Institute for Design of	158

Steel Plants, State Institute for the Design of Iron and	131
Steklov Mathematics Institute	75
Stroibank	30,157
Subtropical Plants, All-Union Research Institute for Tea and, Makharadze	228
Sugar Industry, All-Union Research Institute of the	222
Supply, State Committee for Material and Technical	19
Supply Associations	164
Supreme Attestation Commission, Ministry of Education	53
Supreme Soviet of the USSR	14,18
Surgery, Bakulev Institute of Cardiac and Vascular	181
Surgery, Institute of, Moscow	181
Surveying (Aerial) and Cartography, Central Scientific Institute of Geodesy and, Moscow	79
Syktyvkar Timber Conversion Plant	146
Synthetic Fibres, Institute of, Kalinin	143
Synthetic Rubber, Institute of, Leningrad	143
TsAGI	197,200
TsIAM	200
Tadzhik Academy of Sciences	237
Tadzhik Agricultural Institute	237
Tadzhik Soviet Socialist Republic	236-37
Tashkent Agricultural Institute	226
Tea and Subtropical Plants, All-Union Institute of, Makharadze	228
Teachers and Teacher Training	63
Technology, State Committee for Science and	16,19,20ff
Telecommunications, Ministry of Postal Services and	202
Telephone and Telegraph Services	202-203
Telephone Communications, Scientific Research Institute of Urban and Rural	204
Television and Radio	203
Teploelectro Project Institute	116
Terrestrial Magnetism, Institute of, Irkutsk	45
Terrestrial Magnetism, Radio Research and the Ionosphere, Institute of	77
Terrestrial Magnetism, the Ionosphere and Radio Waves, Siberian Institute of	77
Theoretical Physics, Institute of, Moscow and Kiev	76
Thermal Power Stations	111
Thermal-electric power	116-17
Timber Conversion Plant, Bratsk	146
Timber Industry, Ministry of the	146
Timiryazev Agricultural Academy	84
Timiyazev Institute of Plant Physiology	83
Tomsk Polytechnical Institute	205
Tools Industry, Ministry of the Machine and Hand	132ff
Tractor and Agricultural Machinery Construction, Ministry of	132
Trade Union School	65
Trans-Siberian Railway	10
Transport, Sea and River	194-97
Transport Construction, Ministry of	154,185-87
Turkmen Academy of Sciences	240
Turkmen Katinin Agricultural Institute	240
Turkmen Soviet Socialist Republic	239-40
Tyuratam launching site	85ff
UKRNIGRI	80
Ufa Aviation Institute	200
Ukrainian Academy of Sciences	222
Ukrainian Physico-Technical Institute	76
Ukrainian Research Institute of Synthetic Superhard Materials	222
Ukrainian Scientific Research Geological Prospecting Oil Institute	80
Ukrainian Soviet Socialist Republic	220-22
Universities	65
University, Patrice Lumumba Friendship	65
University system	52-56
Ural Affiliate of the Academy of Sciences of the USSR	46
Uralmash	135
Urals Polytechnical Institute	65
Uzbek Academy of Sciences	225
Uzbekistan Soviet Socialist Republic	224-25
VIAM	200
VIMS	80
VMF	97-98
VNIGNI	80
VNIGRI	80
VNIIGaz	80
VNIINefthim	143
VINKI	211
VNITI	210
VSEGEI	80
VTPB	208
VUZy	52-56,59ff

VVS	98
Vascular Surgery, Bakulev Institute of Cardiac and	181
Vavolov Institute of Physical Problems	76
Virology, Institute of, Moscow	181
Virus Diseases, Institute of Poliomyelitis and, Moscow	181
Vladivostok Higher Engineering Maritime School	197
Voronezh State University	76, 82

Water Conservancy, Ministry of Land Reclamation and	161
Wave thoery and radiophysics research	77
Welding Institute, Paton Electric	132
Wheat Selection and Seed Production, Mironovsky Institute for	176

Yakutsk Affiliate of the Academy of Sciences of the USSR	46

Zaporozhy Transformer and Switchgear Factories	110
Zaporozhya Works	192
Zavolzhsk Motor Works	193
Zhukovsky Aeronautical Engineering Academy, Moscow	200
Zlatoust Plant	132
Zond Programme	94

English Cover to Cover Translations of Soviet Journals

Listed in this Appendix are the cover-to-cover translations, in English, of Russian journals. Arrangement is alphabetically by transliterated Russian title; there is also an alphabetical list of publishers.

Many guides to translations have been issued and a most useful summary guide is that by J.P. Chillag, "Translations and their guides", *NLL Review 1*, 1971, P 46-53. The European Translations Centre (101 Doelenstraat, Delft, Netherlands) publishes the *World Index of Scientific Translations* which once a year incorporates a section entitled "Translations Journals. List of periodicals translated cover-to-cover, abstracted publications and periodicals containing selected articles".

Abstract Journal. Referativnyi Zhurnal. Informatics
    Publisher: USSR State Committee for Science and Technology, Institute of
          Scientific Information, Baltyskaya ulitza 14, Moscow, A-219, USSR.

Akusticheskiĭ Zhurnal (Soviet Physics - Acoustics)
    Publisher: American Institute of Physics, 335 East 45th Street, New York, N.Y.
          10017, USA

Algebra i Logika (Algebra and Logic)
    Publisher: Plenum Publishing Corporation, 227 West 17th Street, New York, N.Y.
          10011, USA

Astrofizika (Astrophysics)
    Publisher: Faraday Press, 84 Fifth Avenue, New York, N.Y. 10011, USA

Astronomicheskiĭ Vestnik (Solar System Research)
    Publisher: Plenum Publishing Corporation, 227 West 17th Street, New York, N.Y.
          10011, USA

Astronomicheskiĭ Zhurnal (Soviet Astronomy)
    Publisher: American Institute of Physics, 335 East 45th Street, New York,
          N.Y. 10017, USA

Atomnaya Energiya (Soviet Atomic Energy)
    Publisher: Plenum Publishing Corporation, 227 West 17th Street, New York, N.Y. 10011, USA

Avtomaticheskaya Svarka (Automatic Welding)
    Publisher: The Welding Institute, Abington Hall, Abington, Cambridge CB1, 6AL, UK

Avtomatika (Soviet Automatic Control)
    Publisher: Scripta Publishing Corporation, 1511 K Street N.W., Washington DC 20005, USA

Avtomatika i Telemekhanika (Automation and Remote Control)
    Publisher: Plenum Publishing Corproation, 227 West 17th Street, New York, N.Y. 10011, USA

Avtomatika Telemekhanika i Svyaz (Railway Automation Telemechanics)
    Publisher: Railway Research Index Division, Railroad Engineering Index Institute, P.O. Box 4045, Amsterdam, Netherlands.

Avtomatika i Vychislitel'naya Tekhnika (Automatic Control)
    Publisher: Faraday Press, 84 Fifth Avenue, New York, N.Y. 10011, USA

Avtometriya (Automatic Monitoring and Measuring)
    Publisher: Plenum Publishing Corporation, 227 West 17th Street, New York, N.Y. 10011, USA

Biofizika (Biophysics)
    Publisher: Pergamon Press, Headington Hill Hall, Oxford OX3 OBW, UK

Biokhimiya (Biochemistry)
    Publisher: Plenum Publishing Corporation, 227 West 17th Street, New York, N.Y. 10011, USA

Byulleten Eksperimental'noi Biologii i Meditsiny (Bulletin of Experimental Biology and Medicine)
    Publisher: Plenum Publishing Corporation, 227 West 17th Street, New York, N.Y. 10011, USA

Byulleten Stantsii Opticheskogo Nablyudeniya Iskusstvennykh Sputnikov Zemli (Bulletin of Optical Artificial Earth Satellite Tracking Stations)
    Publisher: National Technical Information Service, 5285 Port Royal Road, Springfield Virginia, 22151, USA

Defektoskopiya (The Soviet Journal of Destructive Testing)
    Publisher: Plenum Publishing Corporation, 227 West 17th Street, New York, N.Y. 10011, USA

Differentsial'nye Uravneniya (Differential Equations)
    Publisher: Faraday Press, 84 Fifth Avenue, New York, N.Y. 10011 USA

Doklady Akademiya Nauk SSSR
   Translations are issued in the following subject sections:

   Biochemistry, Biological Sciences, Biophysics, Botanical Sciences,
   Chemical Technology, Chemistry, Physical Chemistry.
   Publisher: Plenum Publishing Corporation, 227 West 17th Street,
       New York, N.Y. 10011, USA

   Earth Science -
   Publisher: American Geological Institute, 2201 M Street, NW.,
       Washington DC 20037, USA

   Soviet Mathematics -
   Publisher: American Mathematical Society, P.O. Box 6248, Providence,
       Rhode Island, 02904, USA

   Soviet Physics -
   Publisher: American Institute of Physics, 335 East 45th Street,
       New York, N.Y. 10017, USA

Ekologiya (Ecology)
   Publisher: Plenum Publishing Corporation, 227 West 17th Street, New York,
       N.Y. 10011, USA

Elektrichestvo (Electric Technology USSR)
   Publisher: Pergamon Press, Headington Hill Hall, Oxford, OX3 OBW, UK

Elektrokhimiya (Soviet Electrochemistry)
   Publisher: Plenum Publishing Corporation, 227 West 17th Street, New York,
       N.Y. 10011, U.S.A.

Elektrokhimiya Rasplavlennykh Solevykh i Tverdykh Elektrolitov (Electrochemistry of
Molten and Solid Electrolytes)
   Publisher: Plenum Publishing Corporation, 227 West 17th Street, New York,
       N.Y. 10011, USA

Elektronnaya Obrabotka Materialov (Applied Electrical Phenomena)
   Publisher: Plenum Publishing Corporation, 227 West 17th Street, New York,
       N.Y. 10011, USA

Elektrosvyaz i Radiotekhnika (Telecommunications and Radio Engineering)
   Publisher: Scripta Publishing Corporation, 1511 K Street NW., Washington DC.,
       20005, USA

Elektrotekhnika (Soviet Electrical Engineering)
   Publisher: Faraday Press, 84 Fifth Avenue, New York, N.Y. 10011, USA

Entomologicheskoe Obozrenie (Entomological Review)
   Publisher: Scripta Publishing Corporation, 1511 K Street, NW., Washington DC.,
       20005, USA

Farmakologiya i Toksikologiya (Russian Pharmacology and Toxicology)
   Publisher: Euromet Publications, 97 Moore Park Road, London, SW6, UK

Farmatsiya (Pharmaceutics)
    Publisher: National Technical Information Service, 5285 Port Royal Road,
        Springfield, Virginia 22151, USA

Fizika Goveniya i Vzrȳva (Combustion Explosion and Shock Waves)
    Publisher: Faraday Press, 84 Fifth Avenue, New York N.Y. 10011, USA

Fizika Metallov i Metallovedenie (Physics of Metals and Metallography)
    Publisher: Pergamon Press, Headington Hill Hall, Oxford OX3 0BW, UK

Fizika i Tekhnika Poluprovodnikov (Soviet Physics - Semiconductors)
    Publisher: American Institute of Physics, 335 East 45th Street,
        New York, N.Y. 10017, USA

Fizika Tverdogo Tela (Soviet Physics - Solid State)
    Publisher: American Institute of Physics, 335 East 45th Street, New York,
        N.Y. 10017, USA

Fiziko-Khimicheskaya Mekhanika Materialov (Soviet Materials Science)
    Publisher: Faraday Press, 84th Fifth Avenue, New York, N.Y. 10011
        USA

Fizikotekhnicheskie Problemȳ Razrabotki Poleznȳkh Iskopaemȳkh (Soviet Mining Science)
    Publisher: Plenum Publishing Corporation, 227 West 17th Street, New York,
        N.Y. 10011, USA

Fiziologiya i Biokhimiya Kulturnȳkh Rasteniĭ (Physiology and Biochemistry of Cultivated Plants).
    Publisher: Plenum Publishing Corporation, 227 West 17th Street, New York, N.Y.
        10011, USA

Fiziologiya Rasteniĭ (Soviet Plant Physiology)
    Publisher: Plenum Publishing Corporation, 227 West 17th Street, New York, N.Y.
        10011, USA

Funktsional'nyi analiz i ego prilozheniya (Functional analysis and its application)
    Publisher: Plenum Publishing Corporation, 227 West 17th Street, New York,
        N.Y. 10011, USA

Geoliotekhnika (Applied Solar Energy)
    Publisher: Faraday Press, 84 Fifth Avenue, New York, N.Y. 10011,
        USA

Genetika (Soviet Genetics)
    Publisher: Faraday Press, 84 Fifth Avenue, New York, N.Y. 10011,
        USA

Geogizicheskiĭ Byulleten (Geophysical Bulletin)
    Publisher: National Technical Information Service, 5285 Port Royal Road,
        Springfield, Virginia 22151, USA

Geofizicheskiĭ Byulleten (Geophysical Bulletin)
    Publisher: National Technical Information Service, 5285 Port Royal Road, Springfield, Virginia 22151, USA

Geologiya i Geofizika (Geology and Geophysics)
    Publisher: Aztec School of Languages Inc., P.O.Box 323, West Acton, Mass. 01780, USA

Geologiya Nefti i Gaza (Petroleum Geology)
    Publisher: Petroleum Geology, Box 171, McLean, Virginia, USA

Geologiya Rudnȳkh Mestorozhdeniĭ (Economic Geology USSR)
    Publisher: Pergamon Press, Headington Hill Hall, Oxford, OX3 0BW, UK

Geomagnetizm i Aeronomiya (Geomagnetism and Aeronomy)
    Publisher: American Geophysical Union, Suite, 435, 2100 Pennsylvania Avenue NW., Washington DC., 20037, USA

Geomorfologiya (Geomorphology)
    Publisher: Plenum Publishing Corporation, 227 West 17th Street, New York, N.Y. 10011, USA

Geotektonika (Geotectonics)
    Publisher: American Geophysical Union, Suite 435, 2100 Pennsylvania Avenue NW., Washington DC., 20037, USA

Gidrobiolugicheskiĭ Zhurnal (Hydrobiological Journal)
    Publisher: Scripta Publishing Corporation, 1511 K Street NW., Washington DC 20005, USA

Gidrotekhnicheskoe Stroitel'stvo (Hydrotechnical Construction)
    Publisher: American Society of Civil Engineering, 345 East 47th Street, New York, N.Y. 10017, USA

Gigiena i Sanitariya (Hygiene and Sanitation)
    Publisher: National Technical Information Service, 5285 Port Royal Road, Springfield, Virginia 22151, USA

Gigiena Truda i Professionalnȳe Zaboleviya (Labour hygiene and occupational diseases)
    Publisher: National Technical Information Service, 5285 Port Royal Road, Springfield, Virginia 22151, USA

Informatsionnȳĭ Byulleten Sovetskaya Antarkticheskaya Ekspeditsiya (Information Bulletin Soviet Antarctic Expedition)
    Publisher: American Geophysical Union, Suite 435, 2100 Pennsylvania Avenue, NW., Washington DC., 20037, USA

Inzhenerno-Fizicheskiĭ Zhurnal (Journal of Engineering Physics)
    Publisher: Faraday Press, 84 Fifth Avenue, New York, N.Y.10011, USA

Inzhenernyĭ Zhurnal Mekhanika Tverdogo Tela (Mechanics of Solids)
    Publisher: Faraday Press, 84 Fifth Avenue, New York, N.Y. 10011, USA

Issledovaniya v Oblasti Poverkhnostnȳkh Sil (Research in Surface forces)
    Publisher: Plenum Publishing Corporation, 227 West 17th Street, New York. N.Y.10011, USA

Itogi Nauki i Seriya Matematika (Progress in Mathematics)
    Publisher: Plenum Publishing Corporation, 227 West 17th Street, New York, N.Y.10011, USA

Izmeritel'naya Tekhnika (Measurement Techniques)
    Publisher: Plenum Publishing Corporation, 227 West 17th Street, New York, N.Y.10011, USA

Izobreteniya, Promyshlennȳe Obraztsȳ, Tovarnȳe Znaki (Soviet Inventions Illustrated)
    Publisher: Derwent Information Service, Theobalds Road, London, WC1, UK

Izvestiya. Akademiya Nauk SSSR. Fizicheskaya (Bulletin of the Academy of Sciences USSR. Physical Series)
    Publisher: Columbia Technical Translations, 5 Vermont Avenue, White Plains, N.Y. 10606, USA

Izvestiya. Akademiya Nauk SSSR. Fizika Atmosferȳ i Okeana (Izvestiya. Atmospheric and Ocean Physics)
    Publisher: American Geophysical Union, Suite 435, 2100 Pennsylvania Avenue NW., Washington DC 20037, USA

Izvestiya. Akademiya Nauk SSSR. Fizika Zemli. (Izvestiya. Physics of the Solid Earth)
    Publisher: American Geophysical Union, Suite 435, 2100 Pennsylvania Avenue NW., Washington DC 20037, USA

Izvestiya. Akademiya Nauk SSSR. Khimicheskaya. (Bulletin of the Academy of Sciences of the USSR. Division of Chemical Science)
    Publisher: Plenum Publishing Corporation, 227 West 17th Street, New York, N.Y.10011, USA

Izvestiya. Akademiya Nauk SSSR. Matematicheskaya (Mathematics of the USSR - Izvestiya)
    Publisher: American Mathematical Society, P.O. Box 6248, Providence, Rhode Island 02904, USA

Izvestiya-Akademiya Nauk SSSR. Mekhanika Zhidkosti i Gaza (Fluid Dynamics)
    Publisher: Faraday Press, 84 Fifth Avenue, New York, N.Y.10011, USA

Izvestiya. Akademiya Nauk SSSR. Metallȳ. (Russian Metallurgy and Mining)
    Publisher: Scientific Information Consultants, 661 Finchley Road, London NW2 2HN, UK

Izvestiya. Akademiya Nauk SSSR. Neorganicheskie Materialy. (Inorganic Materials)
    Publisher: Plenum Publishing Corporation, 227 West 17th Street, New York,
        N.Y.10011, USA

Izvestiya. Akademiya Nauk SSSR. Sibirskoe Otdelenie. Khimicheskikh (Siberian Chemistry Journal of the Academy of Sciences of the USSR)
    Publisher: Plenum Publishing Corporation, 227 West 17th Street, New York,
        N.Y.10011, USA

Izvestiya Vysshikh Uchebnykh Zavedenii. Aviatsionnaya Tekhnika. (Soviet Aeronautics)
    Publisher: Faraday Press, 84 Fifth Avenue, New York, N.Y.10011, USA

Izvestiya Vysshikh Uchebnykh Zavedenii Fizika (Soviet Physics Journal)
    Publisher: Scientific Information Consultants, 661 Finchley Road,
        London NW2 2HN, UK

Izvestiya Vysshikh Uchebnykh Zavedenii. Geodeziya i Aerofotosemka. (Geodesy and Aerophotography)
    Publisher: American Geophysical Union, Suite 435, 2100 Pennsylvania Avenue NW.,
        Washington DC., 20037, USA

Izvestiya Vysshikh Uchebnykh Zavedenii. Radiofizika (Radiophysics and Quantum Electronics)
    Publisher: Faraday Press, 84 Fifth Avenue, New York, N.Y.10011, USA

Izvestiya Vysshikh Uchebnykh Zavedenii. Tekhnologiya Tekstil'noi Promyshlennosti (Technology of the Textile Industry USSR)
    Publisher: Textile Institute, 10 Blackfriars Street, Manchester 3, UK

Kauchuk i Rezina (Soviet Rubber Technology)
    Publisher: MacLaren and Sons Ltd., P.O. Box 109, Davis House, 69-77 High Street,
        Croydon, Surrey, UK

Khimicheskaya Promyshlennost' (Soviet Chemical Industry)
    Publisher: Ralph McElroy Company Inc., 504 West 24th Street, Austin, Texas
        78705, USA

Khimicheskie Volokna (Fibre Chemistry)
    Publisher: Plenum Publishing Corporation, 227 West 17th Street, New York,
        N.Y.10011, USA

Khimicheskoe i Neftyanoe Mashinostroenie (Chemical and Petroleum Engineering)
    Publisher: Plenum Publishing Corporation, 227 West 17th Street, New York,
        N.Y.10011, USA

Khimiko-Farmatsevticheskii Zhurnal (Pharmaceutical Chemistry Journal)
    Publisher: Plenum Publishing Corporation, 227 West 17th Street, New York,
        N.Y.10011, USA

Khimiya Geterotsiklicheskikh Soedinenii (Chemistry of Heterocyclic Compounds)
    Publisher: Faraday Press, 84 Fifth Avenue, New York, N.Y.10011, USA

Khimiya Prirodnykh Soedinenii (Chemistry of Natural Compounds)
    Publisher: Faraday Press: 84 Fifth Avenue, New York, N.Y.10011, USA

Khimiya i Tekhnologiya Topliv i Masel (Chemistry and Technology of Fuels and Oils)
    Publisher: Plenum Publishing Corporation, 227 West 17th Street, New York,
        N.Y.10011, USA

Khimiya Vysokikh Energii (High-Energy Chemistry)
    Publisher: Plenum Publishing Corporation, 227 West 17th Street, New York,
        N.Y.10011, USA

Kibernetika (Cybernetics)
    Publisher: Faraday Press, 84 Fifth Avenue, New York, n.Y.10011, USA

Kinetika i Kataliz (Kinetics and Catalysis)
    Publisher: Plenum Publishing Corporation, 227 West 17th Street, New York,
        N.Y.10011, USA

Koks i Khimiya (Coke and Chemistry USSR)
    Publisher: The Coal Tar Research Association, Oxford Road, Comersal, Cleckheaton,
        Yorkshire, UK

Kolloidnyi Zhurnal (Colloid Journal of the USSR)
    Publisher: Plenum Publishing Corporation, 227 West 17th Street, New York,
        N.Y.10011, USA

Kosmicheskaya Biologiya i Meditsina (Environmental Space Sciences)
    Publisher: Plenum Publishing Corporation, 227 West 17th Street, New York,
        N.Y.10011, USA

Kosmicheskie Issledovaniya (Cosmic Research)
    Publisher: Plenum Publishing Corporation, 227 West 17th Street, New York,
        N.Y.10011, USA

Kristallografiya (Soviet Physics - Crystallography)
    Publisher: American Institute of Physics, 335 East 45th Street, New York,
        N.Y.10017, USA

Liteinoe Proizvodstvo (Russian Castings Production)
    Publisher: The British Cast Iron Research Association, Bordesley Hall,
        Alvechurch, Birmingham B48 7QB, UK

Litologiya i Poleznye Iskopaemye (Lithology and Mineral Resources)
    Publisher: Plenum Publishing Corporation, 227 West 17th Street, New York,
        N.Y.10011, USA

Magnitnaya Gidrodinamika (Magnetohydrodynamics)
    Publisher: Faraday Press, 84 Fifth Avenue, New York, N.Y.10011, USA

Matematicheskie Zametki (Mathematical Notes)
    Publisher: Plenum Publishing Corporation, 227 West 17th Street, New York,
        N.Y.10011, USA

Matematicheskiĭ Sborniki (Mathematics of the USSR - Sbornik)
    Publisher: American Mathematical Society, P.O. Box 6248, Providence, Rhode Island
        02904, USA

Meditsinskaya Teknika (Biomedical Engineering)
    Publisher: Plenum Publishing Corporation, 227 West 17th Street, New York,
        N.Y.10011, USA

Mekhanika Polimerov (Polymer Mechanics)
    Publisher: Faraday Press, 84 Fifth Avenue, New York, N.Y.10011, USA

Metallovedenie i Termicheskaya Obrabotka Metallov (Metal Science and Heat Treatment)
    Publisher: Plenum Publishing Corporation, 227 West 17th Street, New York,
        N.Y.10011, USA

Metallurg (Metallurgist)
    Publisher: Plenum Publishing Corporation, 227 West 17th Street, New York,
        N.Y.10011, USA

Meteoritika (Meteoritica)
    Publisher: Scripta Publishing Corporation, 1511 K Street NW., Washington DC
        20005, USA

Meteorologiya i Gidrologiya (Meteorology and Hydrology)
    Publisher: National Technical Information Service, 5285 Port Royal Road,
        Springfield, Virginia 22151, USA

Mikrobiologiya (Microbiology)
    Publisher: Plenum Publishing Corporation, 227 West 17th Street, New York,
        N.Y.10011, USA

Molekulyarnaya Biologiya (Molecular Biology)
    Publisher: Plenum Publishing Corporation, 227 West 17th Street, New York,
        N.Y.10011, USA

Nauchno-Tekhnicheskaya Informatsiya (Automatic Documentation and Mathematical Linguistics)
    Publisher: Faraday Press, 84 Fifth Avenue, New York, N.Y.10011, USA

Neftekhimiya (Petroleum Chemistry USSR)
    Publisher: Pergamon Press, Headington Hill Hall, Oxford OX3 OBW, UK

Neirofiziologiya (Neurophysiology)
    Publisher: Plenum Publishing Corporation, 227 West 17th Street, New York,
N.Y.10011, USA

Nukleonika (Nucleonics)
    Publisher: National Technical Information Service, 5285 Port Royal Road, Springfield, Virginia 22151, USA

Ogneupory̅ (Refractories)
    Publisher: Plenum Publishing Corporation, 227 West 17th Street, New York, N.Y.10011, USA

Okeanologiya (Oceanology)
    Publisher: American Geophysical Union, Suite 435, 2100 Pennsylvania Avenue NW., Washington DC, 20037, USA

Ontogenez (Soviet Journal of Developmental Biology)
    Publisher: Plenum Publishing Corporation, 227 West 17th Street, New York, N.Y.10011, USA

Optika i Spektroskopiya (Optics and Spectroscopy)
    Publisher: American Institute of Physics, 335 East 45th Street, New York, N.Y.10017, USA

Optiko. Mekhanicheskaya Promy̅shlennost' (Soviet Journal of Optical Technology)
    Publisher: American Institute of Physics, 335 East 45th Street, New York, N.Y.10017, USA

Osnovaniya, Fundamenty̅ i Mekhanika Gruntov (Soil Mechanics and Foundation Engineering)
    Publisher: Plenum Publishing Corporation, 227 West 17th Street, New York, N.Y.10011, USA

Paleontologicheskiĭ Zhurnal (Paleontological Journal)
    Publisher: American Geological Institute, 2201 M Street, NW., Washington DC., 20037, USA

Plasticheskie Massy̅ (Soviet Plastics)
    Publisher: Rubber & Technical Press Ltd., Tenterden, Kent, UK

Poroshkovaya Metallurgiya (Soviet Powder Metallurgy and Metal Ceramics)
    Publisher: Plenum Publishing Corporation, 227 West 17th Street, New York, N.Y.10011, USA

Pribory̅ i Sistemy̅ Upravleniya (Soviet Journal of Instrumentation and Control)
    Publisher: Scripta Publishing Corporation, 1511 K Street NW., Washington DC., 20005, USA

Pribory̅ i Tekhnika Eksperimenta (Instruments and Experimental Techniques)
    Publisher: Plenum Publishing Corporation, 227 West 17th Street, New York N.Y.10011 USA

Prikladnaya Biokhimiya i Mikrobiologiya (Applied Biochemistry and Microbiology)
    Publisher: Faraday Press, 84 Fifth Avenue, New York, N.Y.10011, USA

Prikladnaya Geofizika (Exploration Geophysics)
    Publisher: Plenum Publishing Corporation, 227 West 17th Street, New York,
        N.Y.10011, USA

Prikladnaya Matematika i Mekhanika (Journal of Applied Mathematics and Mechanics)
    Publisher: Pergamon Press, Headington Hill Hall, Oxford OX3 OBW, UK

Prikladnaya Mekhanika (Soviet Applied Mechanics)
    Publisher: Faraday Press, 84 Fifth Avenue, New York, N.Y.10011, USA

Problemy Matematicheskoĭ Fiziki (Topics in Mathematical Physics)
    Publisher: Plenum Publishing Corporation, 227 West 17th Street, New York
        N.Y.10011, USA

Problemy Peredachi Informatsii (Problems of Information Transmission)
    Publisher: Faraday Press, 84 Fifth Avenue, New York, N.Y.10011 USA

Problemy Prochnosti (Strength of Materials)
    Publisher: Plenum Publishing Corporation, 227 West 17th Street, New York,
        N.Y.10011 USA

Problemy Severa (Problems of the North)
    Publisher: National Research Council, Ottawa 7, Canada

Put i Putevoe Khozyaĭstvo (Railway Research and Engineering News)
    Publisher: Railway Research Index Division, Railroad Engineering Index
        Institute, P.O. Box 4045, Amsterdam, Netherlands

Radiobiologiya (Radiobiology)
    Publisher: National Technical Information Service, 5288 Port Royal Road,
        Springfield, Virginia 22151, USA

Radiokhimiya (Soviet Radiochemistry)
    Publisher: Plenum Publishing Corporation, 227 West 17th Street, New York,
        N.Y.10011, USA

Radiotekhnika i Elektronika (Radio Engineering and Electronic Physics)
    Publisher: Scripta Publishing Corporation, 1511 K Street NW., Washington DC.,
        20005, USA

Reaktsionnaya Sposobnost'Organicheskikh Soedineniĭ (Organic Reactivity)
    Publisher: Plenum Publishing Corporation, 227 West 17th Street, New York,
        N.Y.10011, USA

Rost Kristallov (Growth of Crystals)
    Publisher: Plenum Publishing Corporation, 227 West 17th Street, New York,
        N.Y.10011, USA

Savremena Polioprivreda (Contemporary Agriculture)
    Publisher: National Technical Information Service, 5285 Port Royal Road,
            Springfield, Virginia 22151, USA

Sibirskiĭ Matematicheskiĭ Zhurnal (Siberian Mathematical Journal of the Academy of Sciences of the USSR, Novosibirsk)
    Publisher: Plenum Publishing Corporation, 227 West 17th Street, New York,
            N.Y. 10011, USA

Sovetskoe Zdravookhranenie (Soviet Public Health)
    Publisher: National Technical Information Service, 5285 Port Royal Road,
            Springfield, Virginia 22151, USA

Stanki i Instrument (Machines and Tooling)
    Publisher: Production Engineering Research Association, Melton Mowbray,
            Leicestershire, UK

Steklo i Keramika (Glass and Ceramics)
    Publisher: Plenum Publishing Corporation, 227 West 17th Street, New York,
            N.Y. 10011, USA

Stekloobraznoe Sostoyanie (Structure of Glass)
    Publisher: Plenum Publishing Corporation, 227 West 17th Street, New York,
            N.Y. 10011, USA

Svarochnoe Proizvodstvo (Welding Production)
    Publisher: Welding Institute, Abington Hall, Abington, Cambridge CB1 6AL, UK

Teoreticheskaya i Eksperiment-alnaya Khimiya (Theoretical and Experimental Chemistry)
    Publisher: Faraday Press, 84 Fifth Avenue, New York, N.Y. 10011, USA

Teoreticheskaya i Matematicheskaya Fizika (Theoretical and Mathematical Physics)
    Publisher: Plenum Publishing Corporation, 227 West 17th Street, New York,
            N.Y. 10011, USA

Teoreticheskie Osnovy Khimicheskoĭ Tekhnologii (Theoretical Foundations of Chemical Engineering)
    Publisher: Plenum Publishing Corporation, 227 West 17th Street, New York,
            N.Y. 10011, USA

Teoriya Veroyatnosteĭ i Ee Primeneniye (Theory of Probability and its Application)
    Publisher: Society for Industrial and Applied Mathematics (SIAM), Box 7541,
            Philadelphia, Pennsylvania 19100, USA

Teploenergetika (Thermal Engineering)
    Publisher: Pergamon Press, Headington Hill Hall, Oxford, OX3 OBW, UK

Teplofizika Vȳsokikh Temperatur (High Temperature)
    Publisher: Plenum Publishing Corporation, 227 West 17th Street, New York,
            N.Y. 10011 USA

Trenie i Iznos v Mashinakh (Friction and wear in Machinery)
    Publisher: American Society of Mechanical Engineers, United Engineering Center,
        345 East 47th Street, New York, N.Y.10017, USA

Trudy. Matematicheskii Institut. Akademiya Nauk SSSR (Proceedings of the Steklov Institute of Mathematics)
    Publisher: American Mathematical Society, P.O. Box 6248, Providence,
        Rhode Island 02904, USA

Trudy. Moskovskoe Matematicheskoe Obshchestvo (Translations of the Moscow Mathematical Society)
    Publisher: American Mathematical Society, P.O. Box 6248, Providence,
        Rhode Island 02904, USA

Tsvetnye Metally (Soviet Journal of Non-Ferrous Metals)
    Publisher: Primary Sources, 11 Bleecker Street, New York, N.Y.10002, USA

Ukrayinskyĭ Biokimicheskyĭ Zhurnal (Ukrainian Biochemistry Journal)
    Publisher: National Technical Information Service, 5285 Port Royal Road,
        Springfield, Virginia 2215, USA

Ukrayinskyĭ Khimicheskyĭ Zhurnal (Soviet Progress in Chemistry)
    Publisher: Faraday Press, 84 Fifth Avenue, New York, N.Y.10011, USA

Ukrayinskyĭ Fizychnyi Zhurnal (Ukrainian Physics Journal)
    Publisher: National Technical Information Service, 5285 Port Royal Road,
        Springfield, Virginia 22151, USA

Ukrayinskyĭ Matematicheskyĭ Zhurnal (Ukrainian Mathematical Journal)
    Publisher: Plenum Publishing Corporation, 227 West 17th Street, New York,
        N.Y.10011, USA

Uspekhi Fizicheskikh Nauk (Soviet Physics - Uspekhi)
    Publisher: American Institute of Physics, 335 East 45th Street, New York,
        N.Y.10017, USA

Uspekhi Fiziologicheskikh Nauk (Progress in Physiological Sciences)
    Publisher: Plenum Publishing Corporation, 227 West 17th Street, New York,
        N.Y.10011, USA

Uspekhi Khimii (Russian Chemical Reviews)
    Publisher: The Chemical Society, Burlington House, London W1V OBN, UK

Uspekhi Matematicheskikh Nauk (Russian Mathematical Surveys)
    Publisher: Macmillan & Company, 4 Little Essex Street, London WC2, UK

Vestnik Akademiya Meditsinskikh Nauk SSSR. (Herald of the Academy of Medical Sciences)
    Publisher: National Technical Information Service, 5285 Port Royal Road,
        Springfield, Virginia 22151, USA

Vestnik.Akademiya Nauk SSSR (Herald of the USSR Academy of Sciences)
    Publisher: National Technical Information Service, 5285 Port Royal Road, Springfield, Virginia 22151, USA

Vestnik Mashinostroeniya (Russian Engineering Journal)
    Publisher: Production Engineering Research Association, Melton Mowbray, Leicestershire, UK

Vestnik Moskovskii Universitet Seriya.Fizika, Astronomiya (Moscow University Physics Bulletin)
    Publisher: Faraday Press, 84 Fifth Avenue, New York, N.Y.10011, USA

Vestnik. Moskovskii Universitet.Seriya.Khimiya (Moscow University Chemistry Bulletin)
    Publisher: Faraday Press, 84 Fifth Avenue, New York, N.Y.10011, USA

Vestnik. Moskovskii Universitet. Seriya. Matematika Mekhanika (Moscow University Mathematics Bulletin)
    Publisher: Faraday Press, 84 Fifth Avenue, New York, N.Y.10011, USA

Vestnik Protivovozdushnoĭ Oborony (Anti-Aircraft Defense Herald)
    Publisher: National Technical Information Service, 5285 Port Royal Road, Springfield, Virginia 22151, USA

Vestnik Svyazi (Herald of Communications)
    Publisher: National Technical Information Service, 5285 Port Royal Road, Springfield, Virginia 22151, USA

Voennyĭ Vestnik (Military Herald)
    Publisher: National Technical Information Service, 5285 Port Royal Road, Springfield, Virginia 22151, USA

Voprosy Ikhtiologii (Journal of Ichthology)
    Publisher: Scripta Publishing Corporation, 1511 K Street NW., Washington DC., 20005, USA

Voprosy Radiobiologii (Problems of Radiobiology)
    Publisher: National Technical Information Service, 5285 Port Royal Road, Springfield, Virginia 22151, USA

Voprosy Teorii Plazmy (Reviews of Plasma Physics)
    Publisher: Plenum Publishing Corporation, 227 West 17th Street, New York, N.Y.10011, USA

Vsesoyuznoe Khimicheskoe Obshchestvo Zhurnal (Mendeleev Chemistry Journal)
    Publisher: Faraday Press, 84 Fifth Avenue, New York, N.Y.10011, USA

Vysokomolekulyarnye Soedineniya (Polymer Science USSR)
    Publisher: Pergamon Press, Headington Hill Hall, Oxford OX3 OBW, UK

Yadernaya Fizika (Soviet Journal of Nuclear Physics)
    Publisher: American Institute of Physics, 335 East 45th Street, New York
          N.Y.10017, USA

Zashchita Metallov (Protection of Metals)
    Publisher: Plenum Publishing Corporation, 227 West 17th Street, New York,
          N.Y.10011, USA

Zavodskaya Laboratoriya (Industrial Laboratory)
    Publisher: Plenum Publishing Corporation, 227 West 17th Street, New York,
          N.Y.10011, USA

Zhurnal Analiticheskoĭ Khimii (Journal of Analytical Chemistry of the USSR)
    Publisher: Plenum Publishing Corporation, 227 West 17th Street, New York,
          N.Y.10011, USA

Zhurnal Eksperimental'noi i Teoreticheskoĭ Fiziki (Soviet Physics - JETP)
    Publisher: American Institute of Physics, 335 East 45th Street, New York,
          N.Y.10017, USA

Zhurnal Eksperimental'noi i Teoreticheskoĭ Fiziki. Pisma v Redaktsiya (JETP Letters)
    Publisher: American Institute of Physics, 335 East 45th Street, New York,
          N.Y.10017, USA

Zhurnal Evolyutsionnoĭ Biokhimii i Fiziologii (Journal of Evolutionary Biochemistry and Physiology)
    Publisher: Plenum Publishing Corporation, 227 West 17th Street, New York,
          N.Y.10011, USA

Zhurnal Fizicheskoĭ Khimii (Russian Journal of Physical Chemistry)
    Publisher: The Chemical Society, Burlington House, London WIV OBN, UK

Zhurnal Neorganicheskoĭ Khimii (Russian Journal of Inorganic Chemistry)
    Publisher: The Chemical Society, Burlington House, London WIV OBN, UK

Zhurnal Obshcheĭ Khimii (Journal of General Chemistry of the USSR)
    Publisher: Plenum Publishing Corporation, 227 West 17th Street, New York,
          N.Y.10011, USA

Zhurnal Organicheskoĭ Khimii (Journal of Organic Chemistry of the USSR)
    Publisher: Plenum Publishing Corporation, 227 West 17th Street, New York,
          N.Y.10011, USA

Zhurnal Prikladnoĭ Khimii (Journal of Applied Chemistry of The USSR)
    Publisher: Plenum Publishing Corporation, 227 West 17th Street, New York,
          N.Y.10011, USA

Zhurnal Prikladnoĭ Mekhaniki i Tekhnicheskoĭ Fiziki (Journal of Applied Mechanics and Technical Physics)
    Publisher: Faraday Press, 84 Fifth Avenue, New York, N.Y.10011, USA

Zhurnal Prikladnoĭ Spektroskopii (Journal of Applied Spectroscopy)
    Publisher: Faraday Press, 84 Fifth Avenue, New York, N.Y.10011, USA

Zhurnal Strukturnoĭ Khimii (Journal of Structural Chemistry)
    Publisher: Plenum Publishing Corporation, 227 West 17th Street, New York,
        N.Y.10011, USA

Zhurnal Tekhnicheskoĭ Fiziki (Soviet Physics – Technical Physics)
    Publisher: American Institute of Physics, 335 East 45th Street, New York,
        N.Y.10017, USA

Zhurnal Vychislitelnoĭ Matematiki i Matematicheskoĭ Fiziki (USSR Computational Mathematics and Mathematical Physics)
    Publisher: Pergamon Press, Headington Hill Hall, Oxford OX3 OBW, UK

Publishers of English Cover to Cover Translations of Soviet Journals

AMERICAN GEOLOGICAL INSTITUTE
   Address: 2201 M Street NW., Washington DC, 20037, USA

   Doklady Akademiya Nauk SSSR (Earth Science)
   Paleontologicheskii Zhurnal (Paleontological Journal)

AMERICAN GEOPHYSICAL UNION
   Address: Suite 435, 2100 Pennsylvania Avenue NW., Washington DC., 20037, USA

   Geomagnetizm i Aeronomiya (Geomagnetism and Aeronomy)
   Geotektonika (Geotectonics)
   Informatsionnyi Byulleten. Sovetskaya Antarkticheskaya Ekspeditsiya
      (Information Bulletin Soviet Antarctic Expedition)
   Izvestiya. Akademiya Nauk SSSR. Fizika Atmosferyi Okeana (Izvestiya.
      Atmospheric and Ocean Physics)
   Izvestiya. Akademiya Nauk SSSR. Fizika Zemli (Izvestiya. Physics of
      the Solid Earth)
   Izvestiya Vysshikh Uchebnykh Zavedenii. Geodeziya i Aerofotosemka.
      (Geodesy and Aerophotography)
   Okeanologiya (Oceanology)

AMERICAN INSTITUTE OF PHYSICS
   Address: 335 East 45th Street, New York, N.Y. 10017, USA

   Akusticheskii Zhurnal (Soviet Physics - Acoustics)
   Astronomicheskii Zhurnal (Soviet Astronomy)
   Doklady Akademiya Nauk SSSR (Soviet Physics)
   Fizika i Tekhnika Poluprovodnikov (Soviet Physics - Semiconductors)
   Fizika Tverdogo Tela (Soviet Physics - Solid State)
   Kristallografiya (Soviet Physics - Crystallography)
   Optika i Spektroskopiya (Optics and Spectroscopy)
   Optiko Mekhanicheskaya Promyshlennost' (Soviet Journal of Optical
      Technology)
   Uspekhi Fizicheskikh Nauk (Soviet Physics - Uspekhi)
   Yadernaya Fizika (Soviet Journal of Nuclear Physics)
   Zhurnal Eksperimental'noi i Teoreticheskoi Fiziki (Soviet Physics - JETP)
   Zhurnal Eksperimental'noi i Teoreticheskoi Fiziki. Pisma v Redaktsiya
      (JETP Letters)
   Zhurnal Tekhnickeskoi Fiziki (Soviet Physics - Technical Physics)

AMERICAN MATHEMATICAL SOCIETY
   Address: P.O. Box 6248 Providence, Rhode Island, 029L4, USA

       Doklady Akademiya Nauk SSSR (Soviet Mathematics)
       Izvestiya. Akademiya Nauk SSSR. Matematicheskaya (Mathematics of the USSR - Izvestiya)
       Matematicheskiĭ Sbornik (Mathematics of the USSR. Sbornik)
       Trudy. Matematicheskii Institut. Akademiya Nauk SSSR (Proceedings of the Steklov Institute of Mathematics)
       Trudy. Moskovskoe Matematicheskoe Obshchestvo (Transactions of the Moscow Mathematical Society)

AMERICAN SOCIETY OF CIVIL ENGINEERS
   Address: 345 East 47th Street, New York, N.Y. 10017, USA

       Gidrotekhnicheskoe Stroitel'stvo (Hydrotechnical Construction)

AMERICAN SOCIETY OF MECHANICAL ENGINEERS
   Address: United Engineering Center, 345 East 47th Street, New York, N.Y. 10017, USA

       Trenie i Iznos v Mashinakh (Friction and wear in Machinery)

AZTEC SCHOOL OF LANGUAGES INCORPORATED
   Address: P.O. Box 323, West Acton, Mass. 01780, USA

       Geologiya i Geofizika (Geology and Geophysics)

THE BRITISH CAST IRON RESEARCH ASSOCIATION
   Address: Bordesley Hall, Alvechurch, Birmingham B48 7QB, UK

       Liteĭnoe Proizvodstvo (Russian Castings Production)

THE CHEMICAL SOCIETY
   Address: Burlington House, London W1V OBN, UK

       Uspekhi Khimii (Russian Chemical Reviews)
       Zhurnal Fizicheskoĭ Khimii (Russian Journal of Physical Chemistry)
       Zhurnal Neorganicheskoĭ Khimii (Russian Journal of Inorganic Chemistry)

THE COAL TAR RESEARCH ASSOCIATION
   Address: Oxford Road, Comersal, Cleckheaton, Yorkshire, UK

       Koks i Khimiya (Coke and Chemistry)

COLUMBIA TECHNICAL TRANSLATIONS
   Address: 5 Vermont Avenue, White Plains, N.Y. 10606, USA

       Izvestiya. Akademiya Nauk SSSR. Fizicheskaya (Bulletin of the Academy of Sciences USSR - Physical Series)

DERWENT INFORMATION SERVICE
   Address: Theobalds Road, London WC1, UK

       Izobreteniya, Promyshlennye Obrazsty, Tovarnye Znaki (Soviet Inventions Illustrated)

EUROMED PUBLICATIONS
  Address: 97 Moore Park Road, London SW6, UK

   Farmakologiya i Toksikologiya (Russian Pharmacology and Toxicology)

FARADAY PRESS
  Address: 84 Fifth Avenue, New York, N.Y. 10011, USA

   Astrogizika (Astrophysics)
   Avtomatika i Vychislitel'naya Tekhnika (Automatic Control)
   Differentsial'nye Uravneniya (Differential Equations)
   Elektrotekhnika (Soviet Electrical Engineering)
   Fizika Goveniya i Vzryva (Combustion Explosion and Shock Waves)
   Fiziko-Khimicheskaya Mekhanika Materialov (Soviet Materials Science)
   Geliotekhnika (Applied Solar Energy)
   Genetika (Soviet Genetics)
   Inzhenerno-Fizicheskii Zhurnal (Journal of Engineering Physics)
   Inzhenernyi Zhurnal. Mekhanika Tverdogo Tela (Mechanics of Solids)
   Izvestiya. Akademiya Nauk SSSR. Mekhanika Zhidkosti i Gaza (Fluid Dynamics)
   Izvestiya Vysshikh Uchebnykh Zavedenii. Aviatsionnaya Tekhnika. (Soviet Aeronautics)
   Izvestiya Vysshikh Uchebnykh Zavedenii. Radiofizika (Radiophysics and Quantum Electronics)
   Khimiya Geterotsiklicheskikh Soedinenii (Chemistry of Heterocyclic Compounds)
   Khimiya Prirodnykh Soedinenii (Chemistry of Natural Compounds)
   Kibernetika (Cybernetics)
   Magnitnaya Gidrodinamika (Magnetohydrodynamics)
   Mekhanika Polimerov (Polymer Mechanics)
   Nauchno-Tekhnicheskaya Informatsiya (Automatic Documentation and Mathematical Linguistics)
   Prikladnaya Biokhimiya i Mikrobiologiya (Applied Biochemistry and Microbiology)
   Prikladnaya Mekhanika (Soviet Applied Mechanics)
   Problemy Peredachi Informatsii (Problems of Information Transmission)
   Teoreticheskaya i Eksperimentalnaya Khimiya (Theoretical and Experimental Chemistry)
   Ukrayinskyi Khimicheskyi Zhurnal (Soviet Progress in Chemistry)
   Vestnik Moskovskii Universitet. Seriya Fizika, Astronomiya. (Moscow Univeristy Physics Bulletin): *Khimiya* (Moscow University Chemistry Bulletin): *Matamatika.Mekhanika* (Moscow University Mathematics Bulletin)
   Vsesoyuznoe Khimicheskoe Obshchestvo Zhurnal (Mendeleev Chemistry Journal)
   Zhurnal Prikladnoi Mekhaniki i Tekhnicheskoi Fiziki (Journal of Applied Mechanics and Technical Physics)
   Zhurnal Prikladnoi Spektroskopii (Journal of Applied Spectroscopy)

RALPH McELRAY COMPANY INCORPORATED
  Address: 504 West 24th Street, Austin, Texas 78705, USA

   Khimicheskaya Promyshlennost' (Soviet Chemical Industry)

MacLAREN & SONS LTD.
  Address: P.O. Box 109, Davis House, 69-77 High Street, Croydon, Surrey, UK

   Kauchuk i Rezina (Soviet Rubber Technology)

MacMILLAN & COMPANY
   Address: 4 Little Essex Street, London WC2, UK

      Uspekhi Matematicheskikh Nauk (Russian Mathematical Surveys)

NATIONAL RESEARCH COUNCIL
   Address: Ottawa 7, Canada

      Problemy Severa (Problems of the North)

NATIONAL TECHNICAL INFORMATION SERVICE
   Address: 5285 Port Royal Road, Springfield, Virginia 22151, USA

      Byulleten Stantsii Opticheskogo Nablyudeniya Isk usstvennykh Sputnikov Zemli (Bulletin of Optical Artificial Earth Satellite Tracking Stations)
      Farmatsiya (Pharmaceutics)
      Geofizicheskiĭ Byulleten (Geophysical Bulletin)
      Gigiena i Sanitariya (Hygiene and Sanitation)
      Gigiena Truda i Professionalnye Zaboleviya (Labour Hygiene and Occupational Diseases)
      Meteorologiya i Gidrologiya (Meteorology and Hydrology)
      Nukleonika (Nucleonics)
      Radiobiologiya (Radiobiology)
      Savremena Polioprivreda (Contemporary Agriculture)
      Sovetskoe Zdravookhranenie (Soviet Public Health)
      Ukrayinskyĭ Biokimicheskyĭ Zhurnal (Ukrainian Biochemistry Journal)
      Ukrayinskyĭ Fizychnyi Zhurnal (Ukrainian Physics Journal)
      Vestnik. Akademiya Meditsinskikh Nauk SSSR (Herald of the Academy of Medical Sciences)
      Vestnik. Akademiya Nauk SSSR (Herald of the USSR Academy of Sciences)
      Vestnik Protivovozdushnoĭ Oborony (Anti-Aircraft Defense Herald)
      Vestnik Svyazi (Herald of Communications)
      Voennyĭ Vestnik (Military Herald)
      Voprosy Radiobiologiĭ (Problems of Radiobiology)

PERGAMON PRESS
   Address: Headington Hill Hall, Oxford, OX3 OBW, UK

      Biofizika (Biophysics)
      Elektrichestvo (Electric Technology USSR)
      Fizika Metallov i Metallovedenie (Physics of Metals and Metallography)
      Geologiya Rudnykh Mestorozhdeniĭ (Economic Geology USSR)
      Neftekhimiya (Petroleum Chemistry USSR)
      Prikladnaya Matematika i Mekhanika (Journal of Applied Mathematics and Mechanics)
      Teploenergetika (Thermal Engineering)
      Vysokomolekulyarnye Soedineniya (Polymer Science, USSR)
      Zhurnal Vychislitelnoĭ Matematiki i Matematicheskoĭ Fiziki (USSR Computational Mathematics and Mathematical Physics)

PETROLEUM GEOLOGY
   Address: Box 171, McLean, Virginia, USA

      Geologiya Nefti i Gaza (Petroleum Geology)

PLENUM PUBLISHING CORPORATION
Address: 227 West 17th Street, New York, N.Y.10011, USA
Davis House, 8 Scrubs Lane, London NW10 6SE, UK

Algebra i Logika (Algebra and Logic)
Astronomicheskiĭ Vestnik (Solar System Research)
Antomnaya Energiya (Soviet Atomic Energy)
Avtomatika i Telemekhanika (Automation and Remote Control)
Avtometriya (Automatic Monitoring and Measuring)
Biokhimiya (Biochemistry)
Byulleten Eksperimental'noi Biologii i Meditsinȳ (Bulletin of Experimental Biology and Medicine)
Defektoskopiya (The Soviet Journal of Non-Destructive Testing)
Doklady Akademiya Nauk SSSR. Sections: Biochemistry, Biological Sciences, Biophysics, Botanical Sciences, Chemical Technology, Chemistry, Physical Chemistry.
Ekologiya (Econogy)
Elektrokhimiya (Soviet Electrochemistry)
Elektrokhimiya Rasplavlennȳkh Solevȳkh i Tverdȳkh Elektrolitov (Electrochemistry of Molten and Solid Electrolytes)
Elektronnaya Obrabotka Materialov (Applied Electrical Phenomena)
Fizikotekhnickeskie Problemȳ Razrabotki Poleznȳkh Iskopaemȳkh (Soviet Mining Science)
Fiziologiya i Biokhimiya Kulturnȳkh Rasteniĭ (Physiology and Biochemistry of Cultivated Plants)
Fiziologiya Rasteniĭ (Soviet Plant Physiology)
Funktsional'nyi analiz i ego prilozheniya (Functional analysis and its Application)
Geomorfologiya (Geomorphology)
Issledovaniya v Oblasti Poverkhnostnȳkh Sil (Research in Surface Forces)
Itogi Nauki i Seriya Matematika (Progress in Mathematics)
Izmeritel'naya Tekhnika (Measurement Techniques)
Izvestiya. Akademiya Nauk SSSR. Khimicheskaya (Bulletin of the Academy of Sciences of the USSR. Division of Chemical Science)
Izvestiya. Akademiya Nauk SSSR. Neorganicheskie Materialȳ (Inorganic Materials)
Izvestiya. Academiya Nauk SSSR. Sibirskoe Otdelenie. Khimicheskikh (Siberian Chemistry Journal of the Academy of Sciences of the USSR)
Khimicheskie Volokna (Fibre Chemistry)
Khimicheskoe i Neftyanoe Mashinostroenie (Chemical and Petroleum Engineering)
Khimino-Farmatsevticheskiĭ Zhurnal (Pharmaceutical Chemistry Journal)
Khimiya i Tekhologiya Topliv i Masel (Chemistry and Technology of Fuels and Oils)
Khimiya Vȳsokikh Energiĭ (High-Energy Chemistry)
Kinetika i Kataliz (Kinetics and Catalysis)
Kolloidnȳĭ Zhurnal (Colloid Journal of the USSR)
Kosmicheskaya Biologiya i Meditsina (Environmental Space Sciences)
Kosmicheskie Issledovaniya (Cosmic Research)
Litologiya i Poleznȳe Iskopaemȳe (Lithology and Mineral Resources)
Matematicheskie Zametki (Mathematical Notes)
Meditsinskaya Teknika (Biomedical Engineering)
Metallovedenie i Termicheskaya Obrabotka Metallov (Metal Science and Heat Treatment)
Metallurg (Metallurgist)
Mikrobiologiya (Microbiology)
Molekulyarnaya Biologiya (Molecular Biology)
Neirofiziologiya (Neurophysiology)
Ogneuporȳ (Refractories)
Ontogenez (Soviet Journal of Developmental Biology)
Osnovaniya, Fundamentȳi Mekhanika Gruntov (Soviet Powder Metallurgy and Metal Ceramics)

PLENUM PUBLISHING CORPORATION (Continued)

Pribory i Tekhnika Eksperimenta (Instruments and Experimental Techniques)
Prikladnaya Geofizika (Exploration Geophysics)
Problemy Matematicheskoĭ Fiziki (Topics in Mathematical Physics)
Problemy Prochnosti (Strength of Materials)
Radiokhimiya (Soviet Radiochemistry)
Reaktsionnaya Sposobnost'Organicheskikh Soedineniĭ (Organic Reactivity)
Rost Kristallov (Growth of Crystals)
Sibirskiĭ Matematicheskiĭ Zhurnal (Siberian Mathematical Journal of the Academy of Sciences of the USSR, Novosibirsk)
Steklo i Keramika (Glass and Ceramics)
Stekloobraznoe Sostoyanie (Structure of Glass)
Teoreticheskaya i Matematicheskaya Fizika (Theoretical and Mathematical Physics)
Teoretickeskie Osnovy Khimicheskoĭ Tekhnologii (Theoretical Foundations of Chemical Engineering)
Teplofizika Vysokikh Temperatur (High Temperature)
Ukrayinskĭ Matematicheskyĭ Zhurnal (Ukrainian Mathematical Journal)
Voprosy Teorii Plazmy (Reviews of Plasma Physics)
Zashchita Metallov (Protection of Metals)
Zavodskaya Laboratoriya (Industrial Laboratory)
Zhurnal Analiticheskoĭ Khimii (Journal of Analytical Chemistry of the USSR)
Zhurnal Evolyutsionnoĭ Biokhimii i Fiziologii (Journal of Evolutionary Biochemistry and Physiology)
Zhurnal Obshcheĭ Khimii (Journal of General Chemistry of the USSR)
Zhurnal Organicheskoĭ Khimii (Journal of Organic Chemistry of the USSR)
Zhurnal Prikladnoĭ Khimii (Journal of Applied Chemistry of the USSR)
Zhurnal Strukturnoĭ Khimii (Journal of Structural Chemistry)

PRIMARY SOURCES
Address: 11 Bleecker Street, New York, N.Y. 10002, USA

Tsvetnye Metally (Soviet Journal of Non-Ferrous Metals)

PRODUCTION ENGINEERING RESEARCH ASSOCIATION
Address: Melton Mowbray, Leicestershire, UK

Stanki i Instrument (Machines and Tooling)
Vestnik Mashinostroeniya (Russian Engineering Journal)

RAILROAD ENGINEERING INDEX INSTITUTE
Address: P.O. Box 4045, Amsterdam, Netherlands

Avtomatika Telemekhanika i Svyaz (Railway Automation Telemechanics)
Put i Putevoe Khozyaĭstvo (Railway Research and Engineering News)

RUBBER & TECHNICAL PRESS LIMITED
Address: Tenterden, Kent, UK

Plasticheskie Massy (Soviet Plastics)

SCIENTIFIC INFORMATION CONSULTANTS
Address: 661 Finchley Road, London NW2 2HN, UK

Izvestiya. Akademiya Nauk SSSR. Metally (Russian Metallurgy and Mining)
Izvestiya. Nysshikh Uchebnykh Zavedeniĭ Fizika (Soviet Physics Journal)

SCRIPTA PUBLISHING CORPORATION
  Address: 1511 K Street NW., Washington DC., 20005, USA
           Transcripta Journals Ltd., 30 Craven Street, Strand, London WC2, UK

   Avtomatika (Soviet Automatic Control)
   Elektrosvyaz i Radiotekhnika (Telecommunications and Radio Engineering)
   Entomologicheskoe Obozrenie (Entomological Review)
   Gidrobiologicheskiĭ Zhurnal (Hydrobiological Journal)
   Meteorikika (Meteoritica)
   Pribory i Sistemy Upravleniya (Soviet Journal of Instrumentation and Control)
   Radiotekhnika i Elektronika (Radio Engineering and Electronic Physics)
   Voprosy Ikhtiologii (Journal of Ichthology)

SOCIETY FOR INDUSTRIAL AND APPLIED MATHEMATICS (SIAM)
  Address: Box 7541, Philadelphia, Pennsylvania, 19100, USA

   Teoriya Veroyatnosteĭ i Ee Primeneniye (Theory of Probability and its Application)

TEXTILE INSTITUTE
  Address: 10 Blackfriars, Street, Manchester 3, UK

   Izvestiya Nysshikh Uchebnykh Zavedeniĭ. Tekhnologiya Tekstil'noi Promyshlennosti (Technology of the Textile Industry USSR)

USSR STATE COMMITTEE FOR SCIENCE AND TECHNOLOGY
  Address: Institute of Scientific Information, Baltyskaya ulitza 14, Moscow A-219, USSR

   Referativnyi Zhurnal. Informatics (Abstract Journal)

THE WELDING INSTITUTE
  Address: Abington Hall, Abington, Cambridge CB1 6AL, UK

   Avtomaticheskaya Svarka (Automatic Welding)
   Svarochnoe Proizvodstvo (Welding Production)

Academies of Sciences of the USSR

ACADEMY OF SCIENCES OF THE USSR
  Address: Leninsky Prospekt 14, Moscow
  V-71, USSR

  Section of Physical-Technical and
  Mathematical Sciences
    Address: Leninsky Prospekt 14, Moscow
      Electronic Microscopy Science Council
      Geophysical Joint Committee
      Problems of Sun and Earth Science
        Council
      Scientific and Technological
        Terminology Committee
      Soviet Physicists National Committee
      Transport Development Commission

  Mathematics Department
    Address: Leninsky Prospekt 14, Moscow
      Applied Mathematics Institute
        Address: Miuskaya Pl.4, Moscow
      Computing Centre
        Address: Ul.Vavilova 40, Moscow
          Computing Technology Commission
          Mathematical Education Commission
          Mathematics V.A.Steklov Institute
            Address: Vavilov 42, Moscow
          National Committee of Soviet
            Mathematicians

  General Physics and Astronomy Department
    Address: Leninsky Prospekt 14, Moscow
      Acoustics Council
        Chairman: Academician A.V.
          Rimsky-Korsakov
      Astronomical Council
        Chairman: E.R. Mustel
      Astronomical and Geodesical All-Union
        Society
        Chairman: D.Y. Martynov
      Astro-Physical Observatory
        Address: Stavropolsky Territory
      Central Astronomical Observatory
        Address: Leningrad, Pulkovo

ACADEMY OF SCIENCES OF THE USSR (Continued)
  Crystallography Institute
    Address: Leninsky Prospekt 49, Moscow
  Electron Physics Council
    Chairman: Academician S.A.Vekshinsky
  Exploration of the Sun Commission
  International Scientific Radio Union
    Soviet National Committee
    Chairman: Academician A.M. Prokhorov
  Low Temperature Physics Council
    Chairman: Academician P.L. Kapitza
  Physical Laboratory
    Director: Academician P.L. Kapitza
  Physical P.N. Lebedey Institute
    Address: Leninsky Prospekt 53, Moscow
  Physical Problems S.I.Vavilov Institute
    Address: Vorobyevskoe chaussee 2, Moscow
  Physical-Technical A.F.Ioffee Institute
    Address: Politekhnicheskaya ul. 2,
      Leningrad
  Physical-Technical Kazan Institute
  Physics and Chemistry of Semiconductors
    Council
    Chairman: Academician B.M. Vul
  Physics of High Pressure Institute
    Address: Leninsky Prospekt 31, Moscow
  Physics of Metals Institute
    Address: Ul.S.Kovalevskoy 13, Sverdlovsk
  Physics of Plasma Council
    Chairman: Academician L.A.Artsimovich
  Precision Mechanics and Computing
    Equipment Institute
    Address: Leninsky Prospekt 51, Moscow
  Radio Astronomy Council
    Chairman: Academician V.A.Kotelnikov
  Radio Engineering and Electronics
    Institute
    Address: Prospekt K.Marks 18, Moscow
  Radio Propagation Council
    Chairman: Academician A.N.Shchukin
  Semiconductors Institute
    Address: Naberezhnaya Kutuzova 10,
      Leningrad

ACADEMY OF SCIENCES OF THE USSR (Continued)
- Solid Physics Council
  - Chairman: Academician G.V. Kurdyumov
- Solid Physics Institute
  - Address: Ul.Radio 23/29, Moscow
- Soviet Astronomers National Committee
  - Chairman: E.R. Mustel
- Soviet Crystallographers National Committee
  - Chairman: Academician N.V. Belov
- Space Research Institute
  - Director: Academician G.I. Petrov
- Spectroscopy Commission
  - Chairman: S.L. Mandelshtam
- Spectroscopy Institute
  - Director: Professor S.L. Mandelshtam
- Terrestrial Magnetism, Radio Research and Ionosphere Institute
  - Address: Skademgorodok, Podolsky Region
- Theoretical Astronomy Institute
  - Address: Mendeleyevskaya Liniya 1, Leningrad
- Theoretical Physics Institute
  - Address: Pos.Chernogolovka, Moscow Region
- Ultrasonic Physics and Technology Council
  - Chairman: Academician T.G. Mikhailov
- Zvenigorod Experimental Station

Nuclear Physics Department
  - Address: Leninsky Prospekt 14, Moscow
    - Cosmic Rays Council
      - Chairman: Academician S.N. Vernov
    - Nuclear Spectroscopy Council
      - Chairman: B.S. Dzelepov
    - Physical Institute P.N. Lebedev
      - Address: Leninsky Prospekt 53, Moscow
    - Physical-Technical Institute A.F. Ioffee
      - Address: Politekhnicheskaya ul. 2, Leningrad

Physical and Technical Energy Problems Department
  - Address: Leninsky Prospekt 14, Moscow
    - Energy Council
      - Chairman: Academician M.A. Styrikovich
    - Gas Turbines Commission
      - Chairman: A.M. Lyulka
    - High Temperatures Institute
      - Address: Krasnokazarmennaya ul. 17-a, Moscow
    - High Temperature Thermophysics Council
      - Chairman: Academician V.A. Kirillin
    - Transformation of Thermal Energy into Electrical Energy Council
      - Chairman: Academician A.P. Alexandrov

Mechanics and Control Processes Department
  - Address: Leninsky Prospekt 14, Moscow
    - Automatic Control National Committee
      - Chairman: Academician V.A. Trapeznikov
    - Electrical Measurements and Measuring Systems Council
      - Chairman: B.S. Sotskov
    - Fluid and Gas Mechanics Council
      - Chairman: Academician L.I. Sedov

ACADEMY OF SCIENCES OF THE USSR (Continued)
- Friction and Greasing Council
  - Chairman: Academician A.Y. Ishlinsky
- Fundamentals of Strength and Plasticity Council
  - Chairman: A.A. Ilyushin
- Information Transmission Institute
  - Address: Aviomotornaya ul. 18, Moscow
- Navigation and Automatic Control Council
  - Chairman: Academician B.N. Petrov
- Problems of Mechanics Institute
  - Address: Leningradsky Prospekt 7, Moscow
- Reliability Council
  - Chairman: Academician N.G. Bruyevich
- Technical Cybernetics Council
  - Chairman: M.A. Gavrilov
- Theoretical and Applied Mechanics National Committee
  - Chairman: Academician N.I. Muskhelishvili
- Theory of Machines Council
  - Chairman: Academician I.I. Artobolevsky

Section of Earth Sciences
  - Address: Leninsky Prospekt 14, Moscow
    - Geographical National Society
      - President: S.V. Kalesnik
    - Geologists National Committee
      - Chairman: D.S. Korzhinsky
    - Meteorites Committee
      - Chairman: Academician V.G. Fesenkov
    - Mineralogical All-Union Society
      - President: P.M. Tatarinov
    - Quaternary Era Commission
      - Chairman: Academician G.Y. Goretsky
    - Soviet Co-ordinated Committee on Geophysics
      - Address: Molodezhnava ul. 3, Moscow
    - Soviet Geographers National Committee
      - Chairman: Academician I.P. Gerasimov

Geology, Geophysics and Geochemistry Department
  - Address: Leninsky Prospekt 14, Moscow
    - Earth's Crust and Upper Mantle Council
      - Chairman: V.V. Belousov
    - Earth Physics O.Y. Schmidt Institute
      - Address: B. Gruzinskaya ul.10, Moscow
    - Engineering Geology and Ground Studies Council
      - Chairman: E.M. Sergeyev
    - A.E. Fersman Mineralogical Museum
      - Director: Professor G.P. Barsanov
    - Geochemistry and Analytical Chemistry V.I. Vernadsky Institute
      - Address: Vorobyevskoye chaussee 47a, Moscow
    - Geology Institute
      - Address: Pyzhevsky per. 7, Moscow
    - Geology of Ore Deposits, Petrography, Mineralogy and Geochemistry Institute
      - Address: Staromonetny per. 35, Moscow

ACADEMY OF SCIENCES OF THE USSR(Continued)
  Geological Formations Commission
    Chairman: G.D. Afanasyev
  Geology and Precambrian Geochronology
    Institute
    Address: Nab. Makarova 2, Leningrad
  Geomagnetism Council
    Chairman: Professor V.A.Tropitskaya
  Geophysical Research Methods Council
    Chairman: Professor M.K. Polshkov
  Geothermal Studies Council
    Chairman: Academician A.N. Tikhonov
  Oil and Gas Formation Council
    Chairman: M.F.Mirchink
  Ore Formation Council
    Chairman: Academician V.T. Smirnov
  Petrography Committee
    Chairman: G.D. Afanasyev
  Sedimentary Rocks Commission
    Chairman: Academician A.V.Sidorenko
  Stratigraphy Committee
    Chairman: Academician D.V.Nalivkin
  Tectonics Committee
    Chairman: M.V.Muratov

Oceanology, Atmospheric Physics and
Geography Department
  Address: Leninsky Prospekt 14, Moscow
    Atmospheric Physics Institute
      Address: Pyzhevsky per 3, Moscow
    Geography Institute
      Address: Staromonetny per.29, Moscow
    Geomorphology Commission
      Chairman: Academician I.P.Gerasimov
    Oceanography Commission
    Oceanology Institute
      Address: Leynaya ul. 1, Lyublino
    Water Problems Institute

Section of Chemistry, Chemical Technology
and Biology
  Address: Leninsky Prospekt 14, Moscow, V71
    Bio-Deterioration Council
      Chairman: Professor B.K.Flerov
    Chemists National Committee
      Chairman: V.N.Kondratyev
    Microbiological Protein Synthesis
      Council
      Chairman: G.K. Skrynabin
    Natural Waters Conservation Commission
      Chairman: Academician N.B.Semenov
    Noginsky Scientific Centre
      Chairman: Academician N.N.Semenov

General and Technical Chemistry Department
  Address: Leninsky Prospekt 14, Moscow V71
    Chemical Kinetics Council
      Chairman: Academician V.N.Kondratyev
    Chemical Physics Institute
      Address: Vorolyevskoe chaussee 2-6
        Moscow
    Chemistry of High Energy Particles Council
      Chairman: V.T. Goldansky

ACADEMY OF SCIENCES OF THE USSR(Continued)
    Chemistry of Mineral Hard Fuel Council
      Chairman: N.M. Karavayev
    Chemistry of Photographic Processes
      Commission
      Chairman: Professor A.P. Terentyev
    Chromatography Council
      Chairman: K.V. Chmutov
    Electrical Chemistry Institute
      Address: Leninsky Prospekt 31, Moscow
    Fuel Elements Council
      Chairman: Academician A.M. Frumkin
    Inorganic Chemistry Council
      Chairman: Academician V.I. Spitsin
    Macro-Molecular Compounds Council
    Macro-Molecular Compounds Institute
      Address: Bolshoy Prospekt 31, Leningrad
    Nomenclature of Chemical Compounds
      Commission
    Oil Chemical Synthesis A.V. Topchiev
      Institute
      Address: Leninsky Prospekt 29, Moscow
    Oil Chemistry Council
      Chairman: Professor N.S. Nametkin
    Oil National Committee
      Chairman: Professor N.S. Nametkin
    Organic and Physical Chemistry A.E.
      Arbuzov Institute
      Address: Kazan, ul.Neftyanikov 8
    Organic Chemistry N.D. Zelinsky
      Institute
      Address: Leninsky Prospekt 47, Moscow
    Organic Elemental Chemistry Council
      Chairman: Academician M.T. Kabachnik
    Organic-Element Compounds Institute
      Address: Ul. Vavilova 14, Moscow
    Physical Chemistry Institute
      Address: Leninsky Prospekt 31, Moscow
    Physics and Chemistry of Semiconductors
      Council
      Chairman: Academician A.M. Frumkin
    Research and Instrumentation Council
    Synthetics and Absorbents Council
      Chairman: Academician M.M. Dubinin

Physical Chemistry and Technology of
Inorganic Materials Department
  Address: Leninsky Prospekt 14, Moscow,V71
    Chemistry of Silicates I.V.
      Grebenshchikov Institute
      Address: Naberezhnaya Makarova 2,
        Leningrad
    General and Inorganic Chemistry N.S.
      Kurnakov Institute
      Address: Leninsky Prospekt 31,Moscow
    High Pure Substances and Physical-
      Chemical Analytical Methods Council
      Chairman: Academician T.P. Alimarin
    Metallurgy A.A. Baikov Institute
      Address: Leninsky Prospekt 47,Moscow
    New Chemical Problems Institute
      Address: Leninsky Prospekt 31,Moscow

ACADEMY OF SCIENCES OF THE USSR (Continued)
- New Heat-Resistant Inorganic Materials Council
  - Chairman: N.N. Semenov
- Physical-Chemical Foundation of Metallurgical Processes Council
- Physical-Chemical Mechanics and Collodial Chemistry Council
  - Chairman: Academician P.A. Rebinder
- Physics and Chemistry of Semiconductors Council
- Structure Materials for New Technology Council
  - Chairman: Academician S.T. Kishkin
- Theoretical Foundation of Chemical Technology Council
  - Chairman: Academician N.M. Zhavoronkov
- Welding National Committee
  - Chairman: Academician N.N. Rykalin

Biochemistry, Biophysics and Chemistry of Physiologically Active Compounds Department
- Agrochemistry and Fertilisers Council
  - Chairman: A.V. Sokolov
- Applied Microelements in Agriculture and Animal Husbandry Council
  - Chairman: Academician Y.V. Peive
- Biochemical All-Union Society
  - Chairman: Academician S.E. Severin
- Biochemistry A.N. Bakh Institute
  - Address: Leninsky Prospekt 35, Moscow
- Biochemistry and Physiology of Micro-organisms Institute
  - Address: Pushchino, Moscow Region
- Biochemists National Committee
  - Chairman: Academician A.I. Oparin
- Biological Physics Institute
  - Address: Akademgorodok, Pushchino/Oka
- Biological Research Centre
  - Director: G.K. Skryabin
- Bio-Organic Chemistry Council
  - Chairman: A.S. Khokhlov
- Biophysics Council
  - Chairman: Academician G.M. Frank
- Chemicals in Plant Growing and Agrarian Chemistry Council
  - Chairman: Y.V. Rakitin
- Chemistry, Biochemistry and Control of Metabolism Joint Council
- Chemistry of Natural Compounds Institute
  - Address: Ul. Vavilova 32, Moscow
- Cytology Council
  - Chairman: Professor A.S. Troshin
- Cytology Institute
  - Address: Prospekt Maklina 2, Leningrad
- Electronic Microscopy Laboratory
  - Director: A.E. Kriss
- Evolution Council
  - Chairman: Academician A.T. Oparin
- Microbiological All-Union Society
  - Chairman: Professor I.L. Rabotnova

ACADEMY OF SCIENCES OF THE USSR (Continued)
- Microbiology Institute
  - Address: Profsoyuznaya 7, Moscow, V-133
- Microphysiology and Biochemistry Council
  - Chairman: Academician A.A. Imshenetsky
- Molecular Biology Council
  - Chairman: Academician V.A. Engelgardt
- Molecular Biology Institute
  - Address: Ul. Vavilova 32, Moscow, V-312
- Photosynthesis Council
  - Chairman: Professor A.A. Nichiporovich
- Photosynthesis Institute
  - Address: Akademgorodok, Pushchino/Oka
- Plant Physiology and Biochemistry Council
  - Chairman: Academician A.L. Kursanov
- Plant Physiology K.A. Timiryazev Institute
  - Address: Leninsky Prospekt 33, Moscow
- Protein Institute
  - Address: Akademgorodok, Pushchino/Oka
- Radiobiology Council
  - Chairman: A.M. Kuzin
- Scientific Foundations of Agrarian Chemistry Joint Council
  - Chairman: Academician S.T. Volfkovich
- Technological Biochemistry Council
  - Chairman: Professor L.V. Metlitsky
- Zoological Biochemistry Council
  - Chairman: Academician A.V. Palladin

Physiology Department
  Address: Leninsky Prospekt 14, Moscow, V-71
- Evolutionary Physiology and Biochemistry I.M. Sechenov Institute
  - Address: Leningrad, K-223, Prospekt M. Toreza, 52
- Higher Nervous Activity and Neurophysiology Institute
  - Address: Pyatnitskaya ul. 48, Moscow
- International Brain Research Organisation Soviet National Committee
  - Chairman: Professor V.S. Rusinov
- Neurophysiology and High Nervous Activity Council
  - Chairman: Professor E.A. Ssratyan
- I.P. Pavlov's Life Works Commission
  - Chairman: Academician P.K. Anokhin
- Physiological All-Union Society - I.P. Pavlov
  - Chairman: Professor L.G. Voronin
- Physiological and Ecological Evolution Council
  - Chairman: Academician E.M. Kreps
- Physiology Joint Council
  - Chairman: Academician V.N. Chernigovsky
- Physiology I.P. Pavlov Institute
  - Address: Naberezhnaya Makarova 6, Leningrad

ACADEMY OF SCIENCES OF THE USSR (Continued)
General Biology Department
　Address: Leninsky Prospekt 14, Moscow
　Biological Foundations of Fauna
　　Council
　　　Chairman: Academician B.E. Bykhovsky
　Biological Foundations of Flora
　　Council
　　　Chairman: Professor G.V. Nikolsky
　Biology of Development Institute
　　Address: Ul. Vavilova 26, Moscow
　　　V-133
　Botanical All-Union Society
　　President: Academician E.M.
　　　Lavrenko
　Botanical Central Garden
　　Address: Botanicheskaya ul.4,
　　　Moscow
　Botanical Institute - V.L. Komarov
　　Address: Ul. Prof. Popova 2, Leningrad
　Entomological All-Union Society
　　President G.Y. Bey-Biyenko
　Evolutionary Morphology and Animal
　　Ecology A.N. Severtsov Institute
　　　Address: Ul. Vavilova 12, Moscow
　Experimental Research Station Gorky
　　Leninskie
　　　Scientific Supervisor: Academician
　　　　T.D. Lysenko
　Forestry Laboratory
　　Address: Zvenigorod
　General Genetics Institute
　　Address: Profsoyuznaya ul.7, Moscow
　Genetics All-Union Society
　　President: Academician B.L. Astaukov
　Helminthologists All-Union Society
　　President: Academician K.I. Skryabin
　Helminthology Laboratory
　　Address: Leninsky Prospekt 35, Moscow,
　　　V-71
　Horticulture Council
　　Chairman: Academician N.T. Tsitsin
　Hydrobiological All-Union Society
　Hydrobiology and Ichthyology Council
　　Chairman: Professor G.V. Nikolsky
　International Biological Programme
　　National Committee
　　　Chairman: Academician B.E. Bykhlovsky
　Ontogeny Council
　　Chairman: Professor M.S. Mitskevich
　Organic Nature and Crop Husbandry Council
　　Chairman: Academician E.M. Lavrenko
　Paleontological Institute
　　Address: Ul. Vavilova 12, Moscow
　Problems of Genetics and Selection Council
　　Chairman: Professor D.K. Beliayev
　Soil Scientists All-Union Society
　　President: Academician I.P. Gerasimov
　Theoretical Foundations of Soil Science
　　Council
　　　Chairman: Professor V.A. Kovda

ACADEMY OF SCIENCES OF THE USSR (Continued)
　Water Conservation Biology Institute
　　Address: Borok
　Zoological Institute
　　Address: Universitetskaya nab 1,
　　　Leningrad
Siberian Division
　Address: Prospekt Nauky 21, Novosibirsk
　　Chairman: Academician M.A. Lavrentyev
　Automation and Electrical Measurements
　　Institute
　　　Address: Akademgorodok, Novosibirsk
　Automation of Research Studies and
　　Instrument-Making Council
　　　Chairman: Professor Y.V. Nesterikmin
　Biological Sciences Joint Council
　　Chairman: Academician A.B. Zhukov
　Biology Institute
　　Address: Ul. Frunze 23-b, Novosibirsk
　Cartography Commission
　　Chairman: V.V. Sochava
　Catalysis Institute
　　Address: Akademgorodok, Novosibirsk
　Central Siberian Botanical Garden
　　Address: Novosibirsk 90
　Chemical Kinetics and Combustion
　　Institute
　　　Address: Akademgorodok, Novosibirsk
　Chemical Sciences Joint Council
　　Chairman: Academician G.K. Boreskov
　Computing Centre
　　Address: Akademgorodok, Novosibirsk
　Cytology and Genetics Institute
　　Address: Akademgorodok, Novosibirsk
　Economics and Management of Industrial
　　Production Institute
　　　Address: Akademgorodok, Novosibirsk
　Economic Sciences Joint Council
　　Chairman: Professor A.G. Aganbegyan
　Endogenetic and Ore Deposits in Siberia
　　and the Far East Council
　　　Chairman: V.A. Kuznetsov
　Explosives in National Economy Council
　　Chairman: Academician M.A. Lavrentyev
　Forestry and Timber V.N. Sukachev
　　Institute
　　　Address: Prospekt Mira 53, Krasnoyarsk
　Forestry Council for Siberia and Far East
　　Chairman: Academician A.B. Zhukov
　Frozen Soils Institute
　　Address: Yakutsk
　Geological, Mineralogical, Geophysical
　　and Geographical Sciences Joint Council
　　　Chairman: Academician A.A. Trofimuk
　Geology and Geophysics Institute
　　Address: Akademgorodok, Novosibirsk
　Geology and Metallogenesis of Pacific
　　Ore Belt Joint Commission
　　　Chairman: Professor E.A. Radkevich
　Geology of Diamond Deposits Joint Council
　　Chairman: I.D. Rozhkov

ACADEMY OF SCIENCES OF THE USSR (Continued)
- Hydrodynamics Institute
  - Address: Akademgorodok, Novosibirsk
- Industrial Catalysis Council
  - Chairman: Academician G.K. Boreskov
- Inorganic Chemistry Institute
  - Address: Akademgorodok, Novosibirsk
- Insects Control Siberian Commission
  - Chairman: A.I. Cherpanov
- Mathematical Council
  - Chairman: Academician S.L. Sobolev
- Mathematics Institute
  - Address: Akademgorodok, Novosibirsk
- Microelements Commission
  - Chairman: O.V. Makeyev
- Mining Institute
  - Address: Krasniy Prospekt 54, Novosibirsk
- Nature Conservation Commission
  - Chairman: V.G. Krylov
- North-Eastern Research Institute
  - Address: Ul. K. Marxa 24, Magadan
- Nuclear Physics Institute
  - Address: Akademgorodok, Novosibirsk
- Organic Chemistry Novosibirsk Institute
  - Address: Akademgorodok, Novosibirsk
- Oriental Studies Commission
  - Chairman: Academician A.P. Okladnikov
- Physical and Chemical Foundations of Mineral Processing Institute
  - Address: Ul. Dzerzhinskogo 18, Novosibirsk
- Physical-Mathematical and Engineering Sciences Joint Council
  - Chairman: Academician M.A. Lavrentyev
- Physics Institute
  - Address: Akademgorodok, Krasnoyarsk
- Physics of Semiconductors Institute
  - Address: Akademgorodok, Novosibirsk
- Physiology Institute
  - Address: Prospekt Nauki 6, Novosibirsk
- Plant Physiology and Biochemistry in Siberia and the Far East Regional Co-ordinating Council
  - Chairman: F.E. Reimers
- Quaternary Era Statistics Commission
  - Chairman: V.N. Saks
- Rock Pressure Council
  - Chairman: T.F. Gorbachev
- Sakhalin Complex Research Institute
  - Address: Novo-Aleksandrovsk, Sakhalin
- Seismic Joint Commission for Siberia and the Far East
  - Chairman: A.A. Treskov
- Siberia and Far East Salt Resources Council
  - Chairman: Academician A.V. Nikolayev
- Siberian and Far Eastern Geology Joint Co-ordinating Commission
  - Chairman: Academician I.A. Kuznetsov
- Siberian Water Resources Protection Commission
  - Chairman: Academician P.Y. Kochina

ACADEMY OF SCIENCES OF THE USSR (Continued)
- Space and Physical Research Council
  - Chairman: V.E. Stepanov
- Spectroscopy Commission
  - Chairman: G.E. Zolotukhin
- Technical Sciences Council
  - Chairman: T.F. Gorbachev
- Tectonics of Siberia and the Far East Council
  - Chairman: Y.A. Kosygin
- Theoretical and Applied Mechanics
  - Address: Akademgorodok, Novosibirsk
- Thermophysics Institute
  - Address: Akademgorodok, Novosibirsk
- Underground Waters in Siberia and the Far East Commission
  - Chairman: E.F. Pinneker
- Volcanology Institute
  - Address: Porgranichaya ul. 3, Petrolpavlovsk/Kamchatka

Branches of the Siberian Department:

East Siberian Branch
  Address: Irkutsk 33
  - Earth's Crust Institute
    - Address: Ul. Favorskogo 1, Irkutsk
  - Energy Institute of Siberia
    - Address: Ul. Lermontova 130, Irkutsk
  - Geochemistry Institute
    - Address: Ul. Favorskogo 1, Irkutsk
  - Geography of Siberia and the Far East Institute
    - Address: Kievskaya ul. 1, Irkutsk
  - Khabarovsk Research Institute
    - Address: Ul. Kim Yu Chena 65, Khabarovsk
  - Limnology Institute
    - Address: Irkutsk, Listvenichnoe Region
  - Organic Chemistry Irkutsk Institute
    - Address: Ul. Favorskogo 1, Irkutsk
  - Plant Physiology and Biochemistry Siberian Institute
    - Address: Irkutsk
  - Terrestrial Magnetism, the Ionosphere and Radio Wave Propagation Institute of Siberia
    - Address: Ul. Lenina 5, Irkutsk

V.L. Komarov Far Eastern Branch
  Address: Ul. Leninskaya 50, Vladivostok
  - Biologically Active Substances Institute
    - Address: Prospekt 100-letiya, Vladivostok 159
  - Biology and Soil Science Institute
    - Address: Prospekt 100-letiya Vladivostok 159
  - Chemistry Department
  - Geology Institute of the Far East
    - Address: Prospekt 100-letiya Vladivostok 159

ACADEMY OF SCIENCES OF THE USSR (Continued)
  Sea Biology Institute
    Address: Prospekt 100-letiya
      Vladivostok 159
  Yakutsk Branch
    Address: Ul.Petrovskogo 36, Yakutsk
      Biology Institute
      Botanical Garden
        Director: Dr. V.N. Dokhunaev
      Energy Department
      Geology Institute
        Address: Yakutsk, Leninsky
          Prospekt 36
      Space Physics Research and
        Aeronomics Institute
        Address: Yakutsk, Leninsky
          Prospekt 61
  Buryat Branch
    Address: Ul.Kirova 35, Ulan-Ude
      Natural Sciences Institute of Buryat
        Director: O. Makeyev

Branches of the Academy of Sciences
of the USSR

  Bashkir Branch
    Address: Ufa, Ul. K. Marksa 6
      Biology Institute
        Director: Professor V.K. Ghifanov
      Mining and Geology Institute
        Director: Professor B.M. Yusupov
      Organic Chemistry Institute
        Director: Academician S.R. Rafikov
  Daghestan Branch
    Address: Ul.Gadzhieva 45, Makhachkala
      Physics Institute
        Director: Academician Kh.I. Amizkhanov
  Karelian Branch
    Address: Pushinskaya Ul.11, Petrozavodsk
      Biology Institute
        Director: Dr.V.I. Ermakov
      Geology Institute
        Director: Dr.V.A. Sokolov
      Water Problems Department
        Director: Dr.I.M. Nesterenko
  S.M. Kirov Kola Branch
    Address: Murmansk Region, P.O. Apatity
      Akademgorodok
      Chemistry and Technology of Rare
        Elements and Minerals Institute
          Director: O.S. Ignatiev
      Geology Institute
        Director: Professor I.V. Belkov
      Marine Biological Institute of
        Murmansk
        Acting Director: Y.I. Galkin
      Mining and Metallurgical Institute
        Director: I.A. Turchaninov
      Polar Alpine Botanical Garden
        Director: T.A. Kozupeyeva

ACADEMY OF SCIENCES OF THE USSR (Continued)
    Polar Geophysics Institute
      Director: S.I. Isaev
  Komi Branch
    Address: Ul.Kommunisticheskaya 24,
      Komi ASSR, Syktyvkar
      Biology Institute
        Director: I.V. Zaboyeva
      Chemistry Department
        Director: V.D. Davydov
      Energy and Water Conservation Department
        Director: L.A. Bratsev
      Geology Institute
        Director: M.V. Fishman
  Ural Branch
    Address: Ul.Pervomaiskaya 91, Sverdlovsk-
      Oblastnoi
      Chemistry Institute
        Director: Professor V.G. Plyusin
      Electrochemistry Institute
        Director: Professor S.V. Karpachev
      Energy and Automation Department
        Director: V.G. Stepanov
      Geology and Geochemistry Institute
        Director: Professor S.N. Ivanov
      Geophysics Institute
        Director: Professor Y.P. Bulashevich
      Metallurgy Institute
        Director: Professor N.A. Vatolin
      Plant and Animal Ecology Institute
        Director: Academician S.S. Swartz

ARMENIAN SSR ACADEMY OF SCIENCES
    Address: Ul.Barekamutyan 24, Erevan
  Physical-Technical and Mathematical
  Sciences Department
    Astro-Physical Observatory of Byurakan
      Address: Ashtarak District, Byurakan
    Computer Centre
      Address: Ul.Gastello 1, Erevan 44
    Mathematics and Mechanics Institute
      Address: Ul.Barekamutyan 24-b,
        Erevan 19
    Optical and Mechanical Laboratory of
      Byurakan
        Address: Ashtarak District, Byurakan
    Physics Institute
      Address: Ul.Barekamutyan 24-a, Erevan 19
    Radio-Physics and Electronics Institute
      Address: Ashtarak
  Chemical Sciences Department
    Biochemistry Institute
      Address: Ul.Gastello 1, Erevan 44
    Chemical Physics Laboratory
      Address: Ul.Moskovyan 22, Erevan 9
    Fine Organic Chemistry Institute
      Address: Azatutyan Prospekt 26,
        Erevan 14

ARMENIAN SSR ACADEMY OF SCIENCES (Continued)

General and Inorganic Chemistry Institute
Address: Ul.Moskovyan 22-a, Erevan 51
Organic Chemistry Institute
Address: Ul.Charentsa 15, Erevan 25

Earth Sciences Department
Geological Sciences Institute
Address: Ul.Barekamutyan 24, Erevan 19
Geophysics and Engineering Seismology Institute
Address: Ul.Leningradyan 5, Leninakan

Biological Sciences Department
Agro-Chemical Problems and Hydroponics Institute
Address: Ul.Noragyukh 108, Erevan 2
Botany Institute
Address: Kanakar, Erevan 63
Experimental Biology Institute
Address: Ul.Nor-Zeitun 128, Erevan 51
Hydrobiological Station of Sevan
Address: Ul.Kirova 192, Sevan
Microbiology Institute
Address: Ul.Charentsa 19, Erevan 25
Physiology L.Orbaeli Institute
Address: Ul.Bratyev Orbeli 3, Erevan 28
Zoology Institute
Address: Ul.Gastello 7, Erevan 44

AZERBAIJAN SSR ACADEMY OF SCIENCES
Address: Ul.Kommunisticheskaya 10, Baku

Physical-Engineering and Mathematical Sciences Department
Astro-Physical Observatory of Shemakha
Address: Shemakha, Pirkulinskoe Post Office, Ul.Sverdlova 75, Baku
Cybernetics Institute
Address: Ul.Ketskhoveli, Kvartal 553, Baku 105
Mathematics and Mechanics Institute
Address: Ul.Ketskhoveli, Kvartal 553, Baku 105
Physics Institute
Address: Prospekt Narimanova 33, Baku 122

Chemical Sciences Department
Chemical Additives Institute
Address: Ul.Telnova 30, Baku 25
Inorganic and Physical Chemistry Institute
Address: Prospekt Narimanova 20, Baku 22

AZERBAIJAN SSR ACADEMY OF SCIENCES (Continued)

Oil Processes Y.G.Mamedaliev Institute
Address: Ul.Telnova 34, Baku 25
Theoretical Problems of Chemical Technology Institute
Address: Prospekt Narimanova 27, Baku 122

Earth Sciences Department
Caspian Sea Commission
Address: Akademgorodok, Baku 122
Deep Oil and Gas Deposits Institute
Address: Ul.Krylova 5, Baku 122
Geographical Society of the Azerbaijan SSR
Chairman: G.A. Aliyev
Geography Institute
Address: Akademgorodok, Baku 122
Geology I.M. Gubkin Institute
Address: Ul. Nizamy 67, Baku 122
Mineralogical Society
Chairman: M.S. Kashkay
Mountain Mud Flows Commission
Address: Akademgorodok, Baku 122
Palaeontological Society
Chairman: K.A. Alizade

Biological Sciences Department
Biochemical Society of Azerbaijan
Chairman: A.A. Gasanov
Botanical Garden
Address: Patamdartskoye Chaussee 40, Baku 23
Botany V.L. Komarov Institute
Address: Lokbatanskoe chaussee 1, Baku
Genetics and Selection Institute
Address: Kommunisticheskaya 5, Baku 1
Genetics and Selection Society of Azerbaijan
Chairman: I.K. Abdullayev
Helminthological Society
Chairman: S.M. Asadov
Nature Preservation Commission
Address: Kommunisticheskaya ul.10, Baku
Physiologists and Pharmacologists Society
Physiology Institute
Address: 73 Akademgorodok, Baku 122
Soil Science and Agrochemistry Institute
Address: Ul.Krylova 5, Baku 122
Soil Scientists Society
Chairman: D.M. Guseinov
Zoology Institute
Address: Ul.Krylova 5, Baku 122

BYELORUSSIAN SSR ACADEMY OF SCIENCES
Address: Leninsky Prospekt 66, Minsk

Physical and Mathematical Sciences Department
Mathematics Institute
Address: Tipografskaya 11, Minsk

BYELORUSSIAN SSR ACADEMY OF SCIENCES (Continued)

Physics Institute
  Address: Leninsky Prospekt 70, Minsk
Physics of Solids and Semiconductors Institute
  Address: Ul.Podlesnaya 17, Minsk

Physical and Engineering Sciences Department
Engineering Cybernetics Institute
  Address: Tipografskaya 11, Minsk
Heat and Mass Exchange Institute
  Address: Ul.Podlesnaya 12, Minsk
Nuclear Energetics Institute
  Address: Sosny Settlement, Minsk
Physical and Engineering Institute
  Address: Ul.Podlesnaya 25, Minsk

Chemical Sciences Department
General and Inorganic Chemistry Institute
  Address: Tipografskaya 9, Minsk
Geochemical Problems Laboratory
  Address: Leninsky Prospekt 66, Minsk
Peat Institute
  Address: Stepianka 81, Minsk
Physical-Organic Chemistry Institute
  Address: Tipografskaya 13, Minsk

Biological Sciences Department
Biophysics and Isotopes Laboratory
  Address: Ul.Akademicheskaya 27, Minsk
Botanical Garden
  Address: Ul.Akademicheskaya 31, Minsk
Experimental Botany Institute
  Address: Ul.Akademicheskaya 27, Minsk
Genetics and Cytology Institute
  Address: Ul.Akademicheskaya 27, Minsk
Gerontology Department
  Address: Ul.Akademicheskaya 26, Minsk
Microbiology Department
  Address: Ul.Akademicheskaya 27, Minsk
Physiology and Systematisation of Primate Plants Department
  Address: Ul. Akademicheskaya 31, Minsk
Physiology Institute
  Address: Ul.Akademicheskaya 26, Minsk
Zoology and Parasitology Department
  Address: Ul.Akademicheskaya 27, Minsk

ESTONIAN SSR ACADEMY OF SCIENCES
Address: Ul.Kokhtu 6, Tallin

Physical-Engineering and Mathematical Sciences Department
Cybernetics Institute
  Address: Lenin Boulevard 10, Tallin
Physics and Astronomy V.Y.Struve Institute
  Address: Tyakhetorn, Tartu
Thermal and Electrical Physics Institute
  Address: Paldiski chaussee 1, Tallin

ESTONIAN SSR ACADEMY OF SCIENCES (Continued)

Chemical Technological and Biological Sciences Department
Botanical Garden of Tallin
  Address: Klustrietsa tee 44, Tallin
Chemistry Institute
  Address: Akademia tee 15, Tallin 26
Experimental Biology Institute
  Address: Kharyu District, Kharku
Geology Institute
  Address: Estoniya Boulevard 7, Tallin
Naturalists Society
  Address: Khirudze 3, Tartu
Nature Preservation Commission
  Address: Ul. Riya 18, Tartu
Zoology and Botany Institute
  Address: Ul. Vanemuyne 21, Tartu

GEORGIAN SSR ACADEMY OF SCIENCES
Address: Ul. Dzerzhinskogo 8, Tbilisi

Mathematics and Physics Department
Astrophysical Observatory of Abastumani
  Address: Adigeni District, Abastumani, Kanobili Mountain
Computing Centre
  Address: Ul. Akurskaya 8, Tbilisi 15
Mathematical Institute Tbilisi A.M. Razmadze
  Address: Ul. Dzerzhiskogo 8, Tbilisi 4
Physics Institute
  Address: Ul. Guramishvili 6 Tbilisi 77

Earth Sciences Department
Clay Studies of Georgia Commission
  Address: Ul. Dzerzhinskogo 8, Tbilisi 4
Geographical Society of Georgia
  Address: Ul.Ketskhoveli 11, Tbilisi 7
Geography Vakhushti Institute
  Address: Ul. Z. Rukhadze 1, Tbilisi 15
Geological Institute
  Address: Ul.Z. Rukhadze 1, Tbilisi 15
Geological Society of Georgia
  Address: Chavchavadze Prospekt Tbilisi 28
Geophysics Institute
  Address: Ul. Z. Rukhadze 1, Tbilisi 15

Applied Mechanics and Control Processes Department
Constructional Mechanics and Seismic Resistance Institute
  Address: Ul. Z. Rukhadze 1, Tbilisi 15

GEORGIAN SSR ACADEMY OF SCIENCES (Continued)

Control Systems Institute
Address: Ul.Pekinskaya 32, Tbilisi 42

Cybernetics Institute
Address: Ul.Chitadze 6, Tbilisi 18

Mechanics of Machines Institute
Address: Ul.Z. Rukhadze 1, Tbilisi 15

Mining Mechanics G.A.Tsulukidze Institute
Address: Ul.Paliashvili 87, Tbilisi 62

Chemistry and Chemical Technology Department

Inorganic Chemistry and Electrical Chemistry Institute
Address: Ul.Z.Rukhadze 1, Tbilisi 15

Metallurgy Institute
Address: Ul.Pavlova 15, Tbilisi 60

Pharmaceutical Chemistry Institute
Address: Ul.Perovskoy 22, Tbilisi

Physical and Organic Chemistry P.G. Melikishvili Institute
Address: Ul.Kamo 14, Tbilisi 2

Biology Department

Biochemical Society of Georgia
Address: Ul. Dzerzhinskogo 8, Tbilisi 4

Botanical Garden of Batumi
Address: Makhinjauri, Batumi

Botanical Garden of Sukhumi
Address: Ul.Chavchavadze 18, Sukhumi

Botanical Society of Georgia
Address: Kodzhorskoye chaussee, Tbilisi 7

Central Botanical Garden
Address: Ul. Botanicheskaya 1, Tbilisi 5

Experimental Morphology A.N.Natishvili Institute
Address: Ul.Kamo 51, Tbilisi 2

Helminthologists of Georgia Society
Address: I.Chavchavadze Prospekt 31, Tbilisi 30

Malignant Tumours Committee
Address: Plekhanov Prospekt 60, Tbilisi 2

Paleobiology Institute
Address: Ul. Potochnaya 4-a, Tbilisi 4

Patho-Anatomists of Georgia Society
Address: Plekhanov Prospekt 60, Tbilisi 2

Physiologists of Georgia Society
Address: Veoenno-Grusinskaya Dorogo 62, Tbilisi 42

Physiology Institute
Address: Voenno-Grusinskaya Doroga 62, Tbilisi 42

GEORGIAN SSR ACADEMY OF SCIENCES (Continued)

Zoological Society of Georgia
Address: Ul.Dzerzhinskogo 8, Tbilisi 4

Zoology Institute
Address: I.Chavchavadze Prospekt 31, Tbilisi 30

KAZAKH SSR ACADEMY OF SCIENCES
Address: Ul.Shevchenko 28, Alma-Ata 2

Physical-Mathematical Sciences Department

Mathematics and Mechanics Institute
Address: Ul.Vinogradova 34, Alma-Ata 2

Nuclear Physics Institute
Address: Nauchny Gorodok, Alma-Ata 82

Earth Sciences and Astrophysics Department

Astrophysical Institute
Address: Alma-Ata 68

Geological Sciences K.I.Satpaev Institute

Hydrogeology and Hydrophysics Institute
Address: Ul. Krasina 94, Alma-Ata 2

Ionosphere Institute
Address: Alma-Ata 68

Mining Institute
Address: Prospekt Lenina 63, Alma-Ata 2

Physical Geography Institute
Address: Ul.Kalinina 65-a, Alma-Ata 2

Chemical-Technological Sciences Department

Chemical-Metallurgical Institute
Address: Sovetskii Prospekt 11-a, Karaganda 61

Chemical Sciences Institute
Address: Ul.Krasina 106, Alma-Ata 2

Metallurgy and Ore-Dressing Institute
Address: Ul. Shevchenko 99, Alma-Ata 2

Biological Sciences Department

Botany Institute
Address: Ul. Kirova 103, Alma-Ata 2

Central Botanical Garden
Address: Alma-Ata 10

Experimental Biology Institute
Address: Prospekt Abaya 38, Alma-Ata 91

Microbiology Institute
Address: Ul. Kirova 103, Alma-Ata 2

Physiology Institute
Address: Ul. Michuzina 80, Alma-Ata 12

KAZAKH SSR ACADEMY OF SCIENCES (Continued)
    Soil Science Institute
        Address: Ul. Kirova 103,
          Alma-Ata 2
    Zoology Institute
        Address: Post Office 57, Alma-Ata

KIRGIZ SSR ACADEMY OF SCIENCES
    Address: Ul. XXII Partsyezda 265-a, Frunze
    Physical-Engineering and Mathematical
    Sciences Department
    Chemical-Technological and Biological
    Sciences Department
        Biochemistry and Physiology Institute
            Address: Ul.XXII Partsyezda 265,
              Frunze 71
        Biology Institute
            Address: Ul.XXII Partsyezda 265,
              Frunze 71
        Botanical Garden
            Address: Ul.Kommunisticheskaya 98
              Frunze
        Inorganic and Physical Chemistry Institute
            Address: Ul.XXII Partsyezda 267,
              Frunze 71
        Organic Chemistry Institute
            Address: Ul.XXII Partsyezda 267,
              Frunze 71

LATVIAN SSR ACADEMY OF SCIENCES
    Address: Ul.Turgeneva 19, Riga
    Physical and Engineering Sciences
    Department
        Electronics and Computing Equipment
          Institute
            Address: Ul.Akademiyas 14, Riga 6
        Energy Institute
            Address: Ul.Ayzkraukles 21, Riga 6
        Mechanics of Polymer Compounds Institute
            Address: Ul.Ayzkraukles 23, Riga 6
        Physical and Engineering Sciences Department
            Address: Salaspils Settlement, Riga
    Chemical and Biological Sciences Department
        Biology Institute
            Address: Ul.Meystaru 10, Riga 47
        Botanical Garden
            Address: Salaspils, Riga District
        Inorganic Chemistry Institute
            Address: Ul. Meystaru 10, Riga 47
        Microbiology Institute
            Address: Ul.Kleystu, Riga 55
        Organic Synthesis Institute
            Address: Ul.Aizkraukles 21, Riga 6
        Wood Chemistry Institute
            Address: Ul.Akademgorodok 27,
              Riga 37

LITHUANIAN SSR ACADEMY OF SCIENCES
    Address: Prospekt Lenina 3, Vilnius
    Physical, Technical and Mathematical
    Sciences Department
        Physical and Engineering Energy
          Institute
            Address: Laysves Alleya 53, Kaunas
        Physics and Mathematics Institute
            Address: Ul.K.Pozhely 52, Vilnius
        Physics and Semiconductors Institute
            Address: Ul.K.Pozhely 54,
              Vilnius
    Chemical and Biological Sciences
    Department
        Biochemistry Institute
            Address: Ul.K.Pozhely 48,Vilnius
        Botanical Garden
            Address: Botanicheskii Prospekt 1,
              Kaunas
        Botany Institute
            Address: Verkyay, Vilnius
        Chemistry and Chemical Technology
          Institute
            Address: Il.K.Pozhely 48, Vilnius
        Zoology and Parasitology Institute
            Address: Ul.K. Pozhely 54, Vilnius

MOLDAVIAN SSR ACADEMY OF SCIENCES
    Address: Prospekt Lenina 1, Kishinev
    Physical Engineering and Mathematical
    Sciences Department
        Applied Physics Institute
            Address: Ul.Akademicheskaya 5,
              Kishinev
        Geography Department
            Address: Leninsky Prospekt 1,
              Kishinev
        Geophysics and Geology Institute
            Address: Kishinev, Ul.
              Akademicheskaya 5
        Mathematics and Computing Centre
          Institute
            Address: Ul.Akademicheskaya 5,
              Kishinev
    Biological and Chemical Sciences
    Department
        Botanical Garden
            Address: Ul.Dunaevskogo 5,
              Kishinev
        Chemistry Institute
            Address: Ul.Akademicheskaya 3,
              Kishinev
        Microbiology Department
            Address: Ul.Akademicheskaya 3,
              Kishinev

MOLDAVIAN SSR ACADEMY OF SCIENCES(Continued)
    Palaeontology and Stratigraphy
      Department
        Address: Ul.Akademicheskaya 5,
          Kishinev
    Plant Genetics Department
      Address: Ul. Akademicheskaya 5,
        Kishinev
    Plant Physiology and Biochemistry
      Institute
      Address: Ul.Akademicheskaya 3,
        Kishinev
    Zoology Institute
      Address: Ul.Akademicheskaya 5,
        Kishinev

TADZHIK SSR ACADEMY OF SCIENCES
    Address: Leninsky Prospekt 33,Dushanbe

  Physical-Engineering and Chemical Sciences
  Department

  Biological Sciences Department
    Botanical Garden
      Address: Ul.Komarova 27,Dushanbe
    Botany Institute
      Address: Ul.Komarova
    Gastro-Enterology Institute
      Address: Parvin Street 12,
        Dushanbe 25
    Plant Physiology and Biophysics
      Institute
      Address: Ul.Khamza-Khakim 17,
        Dushanbe
    Zoology and Parasitology
      Academician E.N.Pavlovsky Institute
      Address: Post Office 70, Dushanbe

TURKMEN SSR ACADEMY OF SCIENCES
    Address: Ul.Gogolya 15, Ashkhabad

  Physical-Engineering and Chemical
  Sciences Department
    Chemistry Institute
      Address: Sad Keshi, Ashkhabad
    Earth and Atmospheric Physics
      Institute
      Address: Ul.Gogolya 16,
        Ashkhabad
    Physical-Engineering Institute
      Address: Ul.Gogolya 15,
        Ashkgabad

  Biological Sciences Department
    Botany Institute
      Address: Prospekt Svobody 81,
        Ashkhabad;
    Central Botanical Garden
      Address: Ul.Timiriazeva,
        Ashkhabad

TURKMEN SSR ACADEMY OF SCIENCES(Continued)
    Deserts Institute
      Address: Sad Keshi, Ashkhabad
    Zoology Institute
      Address: Ul.Engelsa 6, Ashkhabad

UKRAINIAN SSR ACADEMY OF SCIENCES
    Address: Vladimirskaya ul.54, Kiev 1

  Physical-Engineering and Mathematical
  Sciences Division
    Address: Ul.Vladimirskaya 54,
      Kiev 1
  Mathematics, Mechanics and
  Cybernetics Department
    Cybernetics Institute
      Address: Prospekt Nauki 109,
        Kiev 28
    Hydromechanics Institute
      Address: Ul.Zhelyabova 8/4,
        Kiev 57
    Mathematics Institute
      Address: Ul.Repina, Kiev 4
    Mechanics Institute
      Address: Ul.Zhelyabova 2,
        Kiev 57
    Mechanics of Geological Engineering
      Institute
      Address: Ul.K.Marxa 19,
        Dnepropetrovsk
  Physics Department
    Low Temperature Physical-
      Engineering Institute
      Address: Ymovsky tupik 2,
        Kharkov 24
    Physical Engineering Institute
      Address: Bulvar Mira 21, Donetsk
    Physics Institute
      Address: Prospekt Nauki 144,
        Kiev 28
    Physics of Metals Institute
      Address: Vernadsky Prospekt 36,
        Kiev 142
    Radio Physics and Electronics
      Institute
      Address: Ul.Malo-Belgorodskaya 12,
        Kharkov 85
    Semiconductors Institute
      Address: Prospekt Nauki 115,
        Kiev 28
  Physical-Engineering Problems of
  Materials Department
    Electrical Welding E.O.Paton
      Institute
      Address: Ul.Gorkogo 69, Kiev 5
    Foundry Problems Institute
      Address: Svyatoshino,
        Akademgorodok, Kiev 142

UKRAINIAN SSR ACADEMY OF SCIENCES(Continued)
   Physical-Mechanical Institute
      Address: Ul.Nauchnaya 5, Lvov 31
   Problems of Materials Institute
      Address: Svyatoshino, Akademgorodok, Kiev 142

  Earth and Space Problems Department
   Astronomical Observatory
      Address: Goloseevo, Kiev 127
   Geological Sciences Institute
      Address: Ul.Chkalova 55, Kiev 30
   Geology and Geochemistry of Combustible Minerals Institute
      Address: Ul.Kopernika 15, Lvov 6
   Geophysical Joint Committee
      Address: Ul.Novo-Belichanskaya 32, Kiev 68
   Geophysics Institute
      Address: Ul.Novo-Belichanskaya 32, Kiev 68
   Marine Hydrophysical Institute
      Address: Ul.Lenina 28, Sevastopol
   Meteorites Committee
      Address: Ul.Chkalova 55, Kiev

Chemical-Technological and Biological Sciences Division

Chemistry and Chemical Technology Department
   Colloidal Chemistry and Water Institute
      Address: Ul.Leontovicha 9-a, Kiev 30
   Gas Institute
      Address: Ul. Parkhomenko 39, Kiev 113
   General and Inorganic Chemistry Institute
      Address: Akademgorodok, Kiev 68
   Macromolecular Chemistry Institute
      Address: Darnitsa, Ul.Avtomagistralnaya 48, Kiev
   Organic Chemistry Institute
      Address: Murmanskaya 5, Kiev 105
   Physical Chemistry L.V.Pisarzhevsky Institute
      Address: Prospekt Nauki 97, Kiev 28

General Biology Department
   Biology of Southern Seas Institute
      Address: Nakhimov Prospekt 2, Sevastopol
   Botanical Garden
      Address: Ul.Timiryazeva 1, Kiev 14
   Botanical Gardens of Donetsk
      Address: Elevatornaya 19, Donetsk 79
   Botany Institute
      Address: Ul.Repina 4, Kiev 4
   Hydrobiology Institute
      Address: Ul.Vladimirskaya 44, Kiev 3
   Zoology Institute
      Address: Ul.Vladimirskaya 55, Kiev 30

UKRAINIAN SSR ACADEMY OF SCIENCES(Continued)
  Biochemistry, Biophysics and Physiology Department
   Biochemical Society of the Ukraine
      Address: Ul.Leontovicha 9, Kiev 30
   Biochemistry Institute
      Address: Ul. Leontovicha 9, Kiev 30
   Microbiolody and Virology D.K.Zabolotny Institute
      Address: Zabolotny Street 59, Kiev 143
   Physiological Society of the Ukraine
      Address: Ul.Bogomoltsa 4, Kiev 24
   Physiology A.A.Bogomolets Institute
      Address: Ul. Bogomoltsa 4, Kiev 24
   Plant Physiology Institute
      Address: Ul. Vasilkovskaya 49, Kiev 22

UZBEK SSR ACADEMY OF SCIENCES
   Address: Ul. Gogolia 70, Tashkent 47

  Physical-Engineering and Mathematical Sciences Department
   Astronomical Institute
      Address: Ul.Astronomicheskaya 33, Tashkent
   Cybernetics and Computing Centre Institute
      Address: Ul. Volodarskogo 26, Tashkent
   Electronics Institute
      Address: Ul.Observatorskaya 85, Tashkent
   Latitude Station of Ulugbek
      Address: Surkhandaryaoblast, Kitab
   Mathematics V.I.Romanovsky Institute
      Address: Astronomichesky tup. 11, Tashkent
   Mechanics and Seismic Resistance of Constructions Institute
      Address: Akademgorodok, Tashkent 143
   Nuclear Physics Institute
      Address: Tashkent Region, Ulugbek Settlement, Ordjonikidze District
   Physical-Engineering S.V.Starodubtsev Institute
      Address: Ul.Observatorskaya 85, Tashkent

  Earth Sciences Department
   Geology and Geophysics K.M.Abdullaev Institute
      Address: Ul.A.K.Sulaimanovoy 35, Tashkent
   Seismology Institute
      Address: Ul.Observatorskaya 85, Tashkent

  Chemical-Technological and Biological Sciences Department
   Botanical Chemistry Institute
      Address:Akademgorodok, Ul Shastri, Tashkent

UZBEK SSR ACADEMY OF SCIENCES (Continued)
    Botanical Garden
      Address: Ul.D.Abidovoy 272, Tashkent
    Botany Institute
      Address: Akademgorodok, Tashkent 143
    Chemistry Institute
      Address: Akademgorodok, Ul.Shastr
        Tashkent
    Experimental Plant Biology Institute
      Address: Lunacharskoe chaussee,
        Akademgorodok, Tashkent
    Microbiology Department
      Address: Akademgorodok, Tashkent 143
    Zoology and Parasitology Institute
      Address: Sovietskaya 34, Tashkent

ALL UNION V.I.LENIN ACADEMY OF AGRICULTURAL SCIENCES
  Address: Bolshoi Kharitonevsky Per 21, Moscow

  Land Cultivation Division
    Agrochemistry and Fertilisers Department
    Cereals, Leguminous and Groat Plants Department
    General Land Cultivation and Agronomical Soil Science Department
    Meadow Cultivation, Pastoral Farming and Fodder Crops Department
    Plant Protection Department
    Potato Growing Department
    Vegetable Growing Department
    Viticulture Department

  Animal Husbandry Division
    Animal Parasitology Department
    Apiculture and silkworm breeding Department
    Horse Breeding Department
    Infectious Diseases of Animals Department
    Milk and Meat Animal Breeding Department
    Non-Contagious Diseases of Animals Department
    Pig Breeding Department
    Poultry Breeding Department
    Problems of the Chemistry of Animal Husbandry Department
    Sheep Breeding Department

  Mechanisation and Electrification of Agriculture Division

  Hydrotechnics and Land Reclamation Division

  Forestry and Forestry Melioration Division
    Forestry Department
    Forestry Melioration Department

ALL-UNION V.I. LENIN ACADEMY OF AGRICULTURAL SCIENCES (Continued)

  Economics and Management of Agriculture Production Division

RESEARCH INSTITUTES

AGRICULTURAL MICROBIOLOGY RESEARCH INSTITUTE
  Address: Ul. Herzena 42, Leningrad

AGROPHYSICAL RESEARCH INSTITUTE
  Address: Grazhdansky Prospekt 14, Leningrad

ANIMAL HUSBANDRY ALL-UNION RESEARCH INSTITUTE
  Address: Post Office Dubrovitsy, Podolsk District, Moscow Region

ANIMAL PHYSIOLOGY AND BIOCHEMISTRY ALL-UNION RESEARCH INSTITUTE
  Address: Borovsk, Kaluga Region

BOTANICAL GARDEN OF NIKITSKY
  Address: Grymian Region, Yalta

ELECTRIFICATION OF AGRICULTURE ALL UNION RESEARCH INSTITUTE
  Address: 1 Veshnyakovsky proezd, Moscow

EXPERIMENTAL VETERINARY SCIENCE INSTITUTE
  Address: Kuzminki, Moscow

FERTILISERS AND AGRICULTURAL SOIL SCIENCE ALL UNION RESEARCH INSTITUTE
  Address: Ul.Pryanishnikova 31, Moscow A-8

GENETICAL I.V.MICHURIN CENTRAL LABORATORY
  Address: Michurinsk, Tambov Region

HELMINTHOLOGY ALL UNION ACAD. K.I. SKRYABIN INSTITUTE
  Address: Bolshaya Cheremushkinskaya ul. 90, Moscow

MAIZE RESEARCH INSTITUTE
  Address: Ul.Dzerzhinskogo 14, Dnepropetrovsk

MECHANISATION OF AGRICULTURE RESEARCH INSTITUTE
  Address: 2 Institutskaya ul. 20, Moscow

OIL-PRODUCING AND ETHER-BEARING CROPS RESEARCH INSTITUTE
  Address: Ul.Filatova 17, Krasnodar

PLANT GROWING RESEARCH INSTITUTE
Address: Ul.Herzena 44, Leningrad

PLANT PROTECTION RESEARCH INSTITUTE
Address: Ul.Herzena 42, Leningrad

SELECTION AND GENETICS RESEARCH INSTITUTE
Address: Ovidiopolskaya Doroga 2, Odessa

SOIL V.V.DOKUCHAYEV INSTITUTE
Address: Pyzhevsky per 7, Moscow

ACADEMY OF MEDICAL SCIENCES OF THE USSR
Address: Ul.Solyanka 14, Moscow

Hygiene, Microbiology and Epidemiology Department
Epidemiology and Microbiology N.F. Gamalei Institute
Address: Ul.Gamalei 2,Moscow D-182
Experimental Immunobiology Laboratory
Address: Baltiyskaya ul.8, Moscow A-315
General and Municipal Hygiene A.N. Sysin Institute
Address: Pogodinskaya ul.10, Moscow G-117
Industrial Hygiene and Occupational Diseases Institute
Address: Meyerovsky proezd 31, Moscow E-275
New Antibiotics Research Institute
Address: B.Pitogovskays 11, Moscow G-21
Nutrition Institute
Address: Ustynsky proezd 2/14, Moscow ZH-240
Poliomyelitis and Virus Diseases Institute
Address: Kievskoe chaussee, 27 Kilometre, Moskovskaya oblast
Virology D.I.Ivanovsky Institute
Address: I.Shchukinsky proezd 24,Moscow D-98

Clinical Medicine Department
Cardiac and Vascular Surgery Acad.Bakulev. Institute
Address: Leninsky Prospekt 8, Moscow B-49
Cardiology A.L.Myasnikov Institute
Address: Petroverigsku per 10, Moscow B-142
Experimental and Clinical Oncology Institute
Address: Kashirskoe chaussee 6, Moscow M-478
Gerontology Institute
Address: Vishgorodskaya ul.67 Kiev 114
Medical Radiology Institute
Address: Solyanka 14, Kaluga Region, Obninsk and Moscow

ACADEMY OF SCIENCES OF THE USSR(Continued)
Neurology Institute
Address: Volokolamskoe chaussee 80, Moscow D;367
Neorosurgery Acad.N.N.Burdenko Institute
Address: - Tverskoy-Yamskoy per 13/5, Moscow A-47
Obstetrics and Gynaecology Institute
Address: Liniya Mendeleyeva 3, Leningrad B-164
Organ Transplantation Laboratory
Address: Abrikosovsky per 2/6 Moscow G-435
Paediatrics Institute
Address: Lomonosov prospekt 2/40, Moscow V;296
Psychiatry Institute
Address:Zagorodnoe chaussee 2, Moscow M-152
Rheumatism Institute
Address: Petrovka 25, Moscow K-31
Surgery A.V.Vishnevsky Institute
Address: Bolshaya Serpukhovskaya ul. 27, Moscow B-93
Transplantation of Organs and Tissues Institute
Address: Pekhotnaya ul.2/3, Moscow D-436

Medical and Biological Sciences Department
Allergological Laboratory
Address: Leninsky Prospekt, Moscow B-71
Biological and Medical Chemistry Institute
Address: Pogodinskaya ul. 10, Moscow G-117
Gerebrum Institute
Address: Per Obukha 5, Moscow B-120
Experimental Biological Models Laboratory
Address: Moskovskaya oblast, g. Khimki, pos.Svetlie gory
Experimental Endocrinology and Hormone Chemistry Institute
Address: Ul.D.Ulyanova 11,Moscow V-36
Experimental Medicine Institute
Address: Kirovsky Prospekt 69/71, Leningrad P-22
Experimental Pathology and Therapy Institute
Address:Abkhazian ASSR,Sukhumi, Gora Trapetsiya
Experimental Physiology on Re-animation Laboratory
Address: Ul.25 Oktyabrya 9,Moscow K-12
Medical Genetics Institute
Address: Baltiyskaya ul.8,Moscow A-315
Morphology of the Human Being Institute
Address: Ul.Shchepkina 61/2, Moscow 1-110

ACADEMY OF MEDICAL SCIENCES OF THE USSR(Continued)
　　Normal and Pathological Physiology
　　　Institute
　　　　Address: Baltiyskaya ul. 8,
　　　　　Moscow A-315
　　Pharmacology Institute
　　　　Address: Baltiskaya ul. 8,
　　　　　Moscow A-315